Springer Complexity

Springer Complexity is an interdisciplinary program publishing the best research and academic-level teaching on both fundamental and applied aspects of complex systems-cutting across all traditional disciplines of the natural and life sciences, engineering, economics, medicine, neuroscience, social and computer science.

Complex Systems are systems that comprise many interacting parts with the ability to generate a new quality of macroscopic collective behavior the manifestations of which are the spontaneous formation of distinctive temporal, spatial or functional structures. Models of such systems can be successfully mapped onto quite diverse "real-life" situations like the climate, the coherent emission of light from lasers, chemical reaction-diffusion systems, biological cellular networks, the dynamics of stock markets and of the internet, earthquake statistics and prediction, freeway traffic, the human brain, or the formation of opinions in social systems, to name just some of the popular applications.

Although their scope and methodologies overlap somewhat, one can distinguish the following main concepts and tools: self-organization, nonlinear dynamics, synergetics, turbulence, dynamical systems, catastrophes, instabilities, stochastic processes, chaos, graphs and networks, cellular automata, adaptive systems, genetic algorithms and computational intelligence.

The three major book publication platforms of the Springer Complexity program are the monograph series "Understanding Complex Systems" focusing on the various applications of complexity, the "Springer Series in Synergetics", which is devoted to the quantitative theoretical and methodological foundations, and the "Springer Briefs in Complexity" which are concise and topical working reports, case-studies, surveys, essays and lecture notes of relevance to the field. In addition to the books in these two core series, the program also incorporates individual titles ranging from textbooks to major reference works.

Editorial and Programme Advisory Board

Understanding Complex Systems

Founding Editor: S. Kelso

Future scientific and technological developments in many fields will necessarily depend upon coming to grips with complex systems. Such systems are complex in both their composition – typically many different kinds of components interacting simultaneously and nonlinearly with each other and their environments on multiple levels – and in the rich diversity of behavior of which they are capable.

The Springer Series in Understanding Complex Systems series (UCS) promotes new strategies and paradigms for understanding and realizing applications of complex systems research in a wide variety of fields and endeavors. UCS is explicitly transdisciplinary. It has three main goals: First, to elaborate the concepts, methods and tools of complex systems at all levels of description and in all scientific fields, especially newly emerging areas within the life, social, behavioral, economic, neuro- and cognitive sciences (and derivatives thereof); second, to encourage novel applications of these ideas in various fields of engineering and computation such as robotics, nano-technology and informatics; third, to provide a single forum within which commonalities and differences in the workings of complex systems may be discerned, hence leading to deeper insight and understanding.

UCS will publish monographs, lecture notes and selected edited contributions aimed at communicating new findings to a large multidisciplinary audience.

More information about this series at http://www.springer.com/series/5394

Xingjian Jing • Ziqiang Lang

Frequency Domain Analysis and Design of Nonlinear Systems based on Volterra Series Expansion

A Parametric Characteristic Approach

Springer

Xingjian Jing
The Hong Kong Polytechnic University
Hong Kong
PR China

Ziqiang Lang
The University of Sheffield
Sheffield
UK

ISSN 1860-0832 ISSN 1860-0840 (electronic)
ISBN 978-3-319-12390-5 ISBN 978-3-319-12391-2 (eBook)
DOI 10.1007/978-3-319-12391-2
Springer Cham Heidelberg New York Dordrecht London

Library of Congress Control Number: 2015931644

Printed on acid-free paper

Springer is part of Springer Science+Business Media (www.springer.com)

To the authors' families.

Preface

Nonlinearities are ubiquitous and often incur twofold influence, which could be a source of troubles bringing uncertainty, inaccuracy, instability or even disaster in practice, and might also be a superior and beneficial factor for system performance improvement, energy cost reduction, safety maintenance or health monitoring, etc. Therefore, analysis and design of nonlinear systems are important and inevitable issues in both theoretical study and practical applications.

Several methods are available in the literature to this aim including perturbation method, averaging method and harmonic balance method, etc. Nonlinear analysis can also be conducted in the frequency domain based on the Volterra series theory. The latter is a very useful tool with some special and beneficial features to tackle nonlinear problems. It is known that there is a considerably large class of nonlinear systems which allow a Volterra series expansion. Based on the Volterra series, the generalized frequency response function (GFRF) was defined as a multi-variate Fourier transform of the Volterra kernels in the 1950s. This presents a fundamental basis and therefore initiates a totally new theory or area for nonlinear analysis and design in the frequency domain.

The frequency-domain nonlinear analysis theory and methods, based on the Volterra series approach, are observed with a faster development starting from the late 1980s or the early 1990s. Recursive algorithms for computation of the GFRFs for a given parametric nonlinear autoregressive with exogenous input (NARX) model or a given nonlinear differential equation (NDE) model are developed, and output frequency response of nonlinear systems and it properties are investigated accordingly. The area is becoming even more active in recent years. Much more efforts and progress can be seen in the development of application-oriented theory and methods based on the GFRF concept. These include the concepts of nonlinear output spectrum (or output frequency response function) and nonlinear output frequency response function, parametric characteristic analysis, energy transfer properties and various applications in vibration control by exploring nonlinear benefits, fault detection, modelling and identification, data analysis and interpretation, etc.

This book is a systematic summary of some new advances in this area mainly done by the authors in the past years starting from when the first author pursued his Ph.D. degree in the University of Sheffield in 2005. The main results are tried to be formulated uniformly with a parametric characteristic approach, which provides a convenient and novel insight into the nonlinear influence on system output response in terms of characteristic parameters and thus can facilitate nonlinear analysis and design in the frequency domain. The book starts with a brief introduction to the background of nonlinear analysis in the frequency domain, followed by the recursive algorithms for computation of GFRFs for different parametric models, and nonlinear output frequency properties. Thereafter the parametric characteristic analysis method is introduced, which leads to new understanding and formulation of the GFRFs, new concepts about nonlinear output spectrum and new methods for nonlinear analysis and design, etc. Based on the parametric characteristic approach, nonlinear influence in the frequency domain can be investigated with a novel insight, i.e. alternating series, which is followed by some application results in vibration control. Magnitude bounds of frequency response functions of nonlinear systems can also be studied with a parametric characteristic approach, which results in novel parametric convergence criteria for any given parametric nonlinear model whose input–output relationship allows a convergent Volterra series expansion. Although very important and fundamental, these results are summarized at the end of this book.

This book targets those readers (especially Ph.D. students and research staff) who are working in the areas related to nonlinear analysis and design, nonlinear signal processing, nonlinear system identification, nonlinear vibration control and so on. It particularly serves as a good reference for those who are studying frequency-domain methods for nonlinear systems.

Hong Kong, P.R. China
June 2014

Xingjian Jing

Acknowledgements

The first author would like to give his thanks to the Department of Mechanical Engineering, the Hong Kong Polytechnic University, for support in research and to his Ph.D. students for their dedicated efforts in the research group.

The authors are also indebted to all current and/or previous colleagues, most importantly to those in the University of Sheffield (UK) for extensive discussions, invaluable suggestions and generous supports.

The authors wish to acknowledge the support from the Hong Kong GRF project RGC Ref No. 15206514, the China NSFC project No. 61374041 and the UK EPSRC project EP/F017715/1.

Finally, the authors would like to give the greatest gratitude with heart and soul to their families, respectively, for everlasting and abundant understanding and support.

XJ Jing
ZQ Lang

Contents

Chapter 1
Introduction

1.1 Frequency Domain Methods for Nonlinear Systems

Nonlinear analysis takes more and more important roles in system analysis and design in practice from engineering problems to biological systems, and is therefore a very hot topic in the current literature. Several methods are available to this aim including perturbation method, averaging method, harmonic balance, and describing functions etc (Judd 1998; Mees 1981; Gilmore and Steer 1991; Schoukens et al. 2003; Nuij et al. 2006; Pavlov et al. 2007; Jing et al. 2008a, b, c, d, e; Rijlaarsdam et al. 2011; Worden and Tomlinson 2001; Rugh 1981; Doyle et al. 2002).

Nonlinear analysis can also be conducted in the frequency domain. It is known that the analysis and synthesis of linear systems in the frequency domain have been well established. There are many methods and techniques that have been developed to cope with the analysis and design of linear systems in practice such as Bode diagram, root locus, and Nyquist plot (Ogota 1996). Frequency domain methods can often provide more intuitive insights into system linear dynamics or dynamic characteristics and thus have been extensively accepted in engineering practice. For example, the transfer function of a linear system is always a coordinate-free and equivalent description whatever the system model is transformed by any linear transformations; the instability of a linear system is usually associated with at least one right-half-plane pole of the system transfer function; the peak of system output frequency response often happens around the natural resonance frequency of the system, and so on. Therefore, frequency domain analysis and design of engineering systems are often one of the most favourite methodologies in practices and attract extensive studies both in theory and application.

However, frequency domain analysis of nonlinear systems is not straightforward. Nonlinear systems usually have very complicated output frequency

X. Jing, Z. Lang, *Frequency Domain Analysis and Design of Nonlinear Systems based on Volterra Series Expansion*, Understanding Complex Systems,
DOI 10.1007/978-3-319-12391-2_1

characteristics and dynamic response such as harmonics, inter-modulation, chaos and bifurcation, which can transfer system energy between different frequencies to produce outputs at some frequency components that may be quite different from the frequency components of the input. These phenomena complicate the analysis and design of nonlinear systems in the frequency domain, and the frequency domain theory and methods of linear systems cannot directly be extended to nonlinear cases. Existing results in the literature related to analysis and understanding of nonlinear phenomena in the frequency domain are far from full development.

Frequency domain analysis of nonlinear systems has been studied since the fifties of last century. A traditional method was initiated by investigation of global stability of the stationary point within the frames of absolute stability theory, and then frequency domain methods for the analysis of stability of stationary sets and existence of cycles and homo-clinical orbits, as well as the estimation of dimension of attractors etc were developed thereafter (Leonov et al. 1996). Practically, the nonlinear behaviour or characteristics of a specific nonlinear part or nonlinear unit in a system can usually be analyzed by using describing functions or harmonic balance in the frequency domain. The describing function method represents a very powerful mathematical approach for the analysis and design of the behaviour of nonlinear systems with a single nonlinear component (Atherton 1975). It can be effectively applied to the analysis of limit cycle and oscillation for nonlinear systems in which the nonlinearity does not depend on frequency and produces no sub-harmonics etc. Applications to controller designs based on describing function analysis have extensively been reported (Gelb and Vander Velde 1968; Taylor and Strobel 1985). However, limitations of the describing function methods are also noticeable. For example, Engelberg (2002) provides a set of nonlinear systems for which the prediction of limit cycle by using describing functions is erroneous. Simultaneously, some improved methods were developed (Sanders 1993; Elizalde and Imregun 2006; Nuij et al. 2006). Another elegant method for the frequency domain analysis of nonlinear systems is referred to as the harmonic balance (see examples in Solomou et al. 2002; Peyton Jones 2003). This method provides an approximation of the amplitude of the steady state periodic response of a nonlinear system under the assumption that a Fourier series can represent the steady state solution. It can deal with more general problems of nonlinear systems such as the sub-harmonics and jump behaviour etc for both the time domain and frequency domain responses. In addition to these well-established and noticeable methods, there are also some other results for nonlinear system analysis in the frequency domain reported in literature. For example, based on the frequency domain methods for linear systems such as Bode diagrams, singular value decomposition, and the idea of varying eigenvalues or varying natural frequencies, frequency domain methods for the analysis and synthesis of uncertain systems or time-varying systems were studied in Orlowski (2007), Glass and Franchek (1999), Shah and Franchek (1999) and Logemann and Townley (1997); and a frequency response function for convergent systems subject to harmonic inputs was recently proposed in Pavlov et al. (2007) etc.

For a class of nonlinear systems, which have a convergent Volterra series expansion, frequency domain analysis can also be conducted based on the concept of generalized frequency response function (George 1959; Schetzen 1980; Rugh 1981). As studied in Boyd and Chua (1985), nonlinear systems, which are time invariant, causal and have fading memory, can be approximated by a Volterra series of a sufficiently high order. The results in Sandberg (1982, 1983) show that even nonlinear time varying systems have such a locally convergent Volterra series expansion under certain conditions. Therefore, this kind of frequency domain analysis methods can deal with a considerably large class of nonlinear systems which can be driven by any input signals and do not necessarily restrict to consider a specific nonlinear component, and thus is a more general methodology. Although the study on Volterra systems and corresponding frequency domain methods has been done for several decades since the middle of last century, many problems still remain unsolved, related to some theoretical and application issues. The results in this book focus on these problems, and important theory and methods are thus presented, targeting at a systematic and practical method for nonlinear analysis and design in the frequency domain for a wide class of nonlinear systems in engineering practice.

1.2 Frequency Domain Analysis Based on Volterra Series Expansion

The input output relationship of nonlinear systems can be approximated by a Volterra series of a sufficiently high order under certain conditions (Boyd and Chua 1985; Sandberg 1982, 1983), which can be written as

$$y(t) = \sum_{n=1}^{N} \int_{-\infty}^{\infty} \cdots \int_{-\infty}^{\infty} h_n(\tau_1, \cdots, \tau_n) \prod_{i=1}^{n} u(t - \tau_i) d\tau_i \tag{1.1}$$

where N is the maximum order of the series, and $h_n(\tau_1,\cdots,\tau_n)$ is a scalar real valued function of $\tau_1,\cdots,\tau_n$, referred to as the nth order Volterra kernel. Generally, $y(t)$ is a scalar output and $u(t)$ is a scalar bounded input in (1.1). The nth order generalized frequency response function (GFRF) of nonlinear system (1.1) is defined as the multivariate Fourier transformation of $h_n(\tau_1,\cdots,\tau_n)$ (George 1959)

$$H_n(j\omega_1,\cdots,j\omega_n) = \int_{-\infty}^{\infty} \cdots \int_{-\infty}^{\infty} h_n(\tau_1, \cdots, \tau_n) \exp(-j(\omega_1\tau_1 + \cdots + \omega_n\tau_n)) d\tau_1 \cdots d\tau_n \tag{1.2}$$

Equation (1.1) can be regarded as a generalization of the traditional convolution description (e.g., the impulse response) of linear systems. The Volterra series expansion in (1.1) is very useful and convenient in modelling and analysis of a very wide class of nonlinear systems both in deterministic and stochastic (Volterra 1959; Van De Wouw et al. 2002; Rugh 1981). This has been vindicated by a large number of applications of the Volterra series reported in system modelling or identification, control and signal processing for different systems and engineering practices, including electrical systems, biological systems, mechanical systems, communication systems, nonlinear filters, image processing, materials engineering, chemical engineering and so on (Fard et al. 2005; Doyle et al. 2002; French 1976; Boutabba et al. 2003; Friston et al. 2000; Yang and Tan 2006; Raz and Van Veen 1998; Bussgang et al. 1974). Technically, many of these results are related to direct estimation or identification of the kernel $h_n(\tau_1,\cdots,\tau_n)$ or the GFRF $H_n({}_1,\cdots,j\omega_n)$ from input output data (Brilliant 1958; Kim and Powers 1988; Bendat 1990; Nam and Powers 1994; Schetzen 1980; Schoukens et al. 2003; Ljung 1999; Pintelon and Schoukens 2001).

With the existence of Volterra series expansion, the study of nonlinear systems in the frequency domain was initiated by the introduction of the concept of the generalized frequency response functions (GFRFs) as defined in (1.2). This provides a powerful technique for the study of nonlinear systems, which is similar to those frequency domain methods established on the basis of transfer functions of linear systems. Thereafter, a fundamental method, referred to as Probing method (Rugh 1981), greatly promoted the development of this frequency domain theory for nonlinear systems. By using the probing method, the GFRFs for a nonlinear system described by nonlinear differential equations (NDE) or nonlinear autoregressive model with exogenous input (NARX) can directly be obtained from its model parameters. These results were further discussed in Peyton Jones and Billings (1989) and Billings and Peyton-Jones (1990), respectively. With these techniques, many results have been achieved for frequency domain analysis of nonlinear systems. For example, Swain and Billings (2001) extended the computation of GFRFs for SISO models to the case of MIMO nonlinear systems; a derivation of the GFRFs of nonlinear systems with mean level or DC terms was discussed in Zhang et al. (1995); system output spectrum and output frequencies were studied in Lang and Billings (1996, 1997). Moreover, some preliminary results for the bound characteristics of the frequency response functions were given in Zhang and Billings (1996) and Billings and Lang (1996). These bound results were greatly generalized in Jing et al. (2007a, b) where the bound expressions are described into an elegant and compact form which is a polynomial of the first order GFRF with nonlinear model parameters as coefficients. The energy transfer characteristics of nonlinear systems were studied in Billings and Lang (2002) and Lang and Billings (2005) recently, and some diagram based techniques for understanding of higher order GFRFs were discussed in Peyton Jones and Billings (1990) and Yue et al. (2005). Furthermore, the concept of Output Frequency Response Functions of nonlinear systems was proposed in Lang et al. (2006, 2007). These results form a fundamental basis for the development of frequency domain analysis and design methods for nonlinear systems to be presented in this book.

1.3 The Advantages of Volterra Series Based Frequency Domain Methods, and Problems to Be Studied

The frequency domain analysis of nonlinear systems is much more complicated than that for linear systems, because nonlinear systems usually have very complicated nonlinear behaviours such as super-harmonics, sub-harmonics, inter-modulation, and even bifurcation and chaos as mentioned before. These phenomena complicate the study of nonlinear systems in the frequency domain, and the frequency domain theory for linear systems can not directly be extended to nonlinear cases. Although some remarkable results have been developed as discussed above, there is still a great need for further development aiming at a systematic and practical method for the analysis and design of nonlinear systems in the frequency domain.

The study in this book focuses on the frequency domain methods for the class of nonlinear systems which have a convergent Volterra series expansion for its input output relationship in the time domain as described in (1.1) (Sandberg 1982, 1983; Boyd and Chua 1985). By default, the nonlinear systems discussed in what follows belong to this class of nonlinear systems, referred to as Volterra-type nonlinear systems. The computation of the GFRFs and output spectrum is a key step in the frequency domain method based on Volterra series theory. To obtain the GFRFs for Volterra-type nonlinear systems described by NDE or NARX models, the probing method can be used (Rugh 1981). Once the GRFRs are obtained for a practical system, the system output spectrum can then be evaluated (Lang and Billings 1996; Jing 2011). These form a general procedure for this methodology. In practice, the steps in this procedure could be replaced by numerical methods using experimental data, which will be discussed later. The advantages of this method, as mentioned, include the following points:

(a) It is a mathematically elegant method for a considerably large class of nonlinear systems frequently encountered in practices of different fields, not restrict to a specific nonlinear unit or single nonlinear component;
(b) It basically holds for any bounded input signals whatever the input is deterministic or stochastic, not restrict to some specific input signals such as harmonic or triangle or step inputs;
(c) It provides very similar techniques to these for linear systems. For example the GFRFs of nonlinear systems are similar to transfer functions of linear systems, which are familiar to most engineers;
(d) Most importantly, it can directly relate model parameters (or system characteristic parameters including model parameters and input magnitude) to system output frequency response (or nonlinear system output spectrum) since the GFRFs can be recursively computed in terms of model parameters. This can greatly facilitate nonlinear system analysis and design in practice.

(e) The last but not the least, strong nonlinear behaviours such as chaos or bifurcation actually can also be investigated with the Volterra series based methods.

All these points above will be systematically demonstrated and/or discussed in this book. The readers can also refer to other books or publications for harmonic balance and describing function methods for a comparative study.

From previous research results, it can be seen that, high order GFRFs are a sequence of multivariable functions defined in a high dimensional frequency space. The evaluation of the values of the GFRFs higher than the fourth or fifth order can become rather difficult due to the large amount of algebra or symbolic manipulations involved (Yue et al. 2005). The situation may become even worse in the computation of system output spectrum of higher orders, since it involves a series of repetitive computations of the GFRFs from the first to the highest order that are involved. Moreover, the existing recursive algorithms for the computation of the GFRFs and output spectrum can not explicitly and simply reveal the analytical relationship between system time domain model parameters and system frequency response functions in a clear and straightforward manner. These inhibit practical application of the existing theoretical results to such an extent that many problems remain unsolved regarding the nonlinear characteristics of the GFRFs and system output spectrum. For example, how these frequency response functions are influenced by the parameters of the underlying system model, how complex nonlinear behaviours are related to frequency response functions, and so on. From the viewpoint of practical applications, it can be seen that a straightforward analytical expression for the relationship between system time-domain model parameters and system frequency response functions (including the GFRFs and output spectrum) can considerably facilitate the analysis and design of Volterra-type nonlinear systems in the frequency domain.

The following main results are presented in this book to address the problems above:

- Output frequency characteristics of nonlinear systems are investigated, which reveal some novel properties about output frequency generation, energy transferring and cancellation etc. nonlinear effects in the system output frequency response (Chap. 3);
- A parametric characteristic analysis method is proposed, which provides a powerful insight into nonlinear system analysis and design with the framework of the Volterra series based frequency domain method (Chap. 4);
- The parametric characteristics of the GFRFs and nonlinear output spectrum are studied, which clearly demonstrate the relationship between the system time-domain model parameters and frequency response functions (Chap. 5-6);
- A systematic nonlinear characteristic output spectrum method is established, which can greatly facilitate the analysis, design and optimization of nonlinear output spectrum in terms of characteristic parameters (Chap. 7-9);

- Understanding of nonlinear influence in the frequency domain is investigated with a special concept—Alternating series, and applications of the developed theory and methods in vibration control are presented (Chap. 10-12);
- Magnitude bound characteristics of nonlinear system frequency response functions are studied, which lead to new parametric convergent criteria of Volterra-type nonlinear systems in terms of system characteristic parameters (Chap. 13-14).

Moreover, it shall be mentioned that Chap. 2 presents some recursive algorithms for computation of the GFRFs of nonlinear systems. Some special but useful cases for different nonlinear system models are discussed there. A summary and overview section is given thereafter as the conclusion of this book.

Chapter 2
The Generalized Frequency Response Functions and Output Spectrum of Nonlinear Systems

2.1 Volterra Series Expansion and Frequency Response Functions

As discussed in Chap. 1, the input-output relationship for a wide class of nonlinear systems can be approximated by a Volterra series up to a sufficiently high order N as

$$y(t) = \sum_{n=1}^{N} y_n(t) \tag{2.1a}$$

$$y_n(t) = \int_{-\infty}^{\infty} \cdots \int_{-\infty}^{\infty} h_n(\tau_1, \cdots, \tau_n) \prod_{i=1}^{n} u(t - \tau_i) \mathrm{d}\tau_i \tag{2.1b}$$

where $h_n(\tau_1,\cdots,\tau_n)$ is a real valued function of $\tau_1,\cdots,\tau_n$ known as the nth order Volterra kernel. The nth order generalized frequency response function (GFRF) is defined as

$$H_n(j\omega_1,\cdots,j\omega_n) = \int_{-\infty}^{\infty} \cdots \int_{-\infty}^{\infty} h_n(\tau_1, \cdots, \tau_n) \exp\left(-j(\omega_1\tau_1 + \cdots + \omega_n\tau_n)\right) \mathrm{d}\tau_1 \cdots \mathrm{d}\tau_n \tag{2.2}$$

which is the multidimensional Fourier transform of $h_n(\tau_1,\cdots,\tau_n)$. By applying the inverse Fourier transform of the nth order GFRF, (2.1b) can be written as

$$y_n(t) = \frac{1}{(2\pi)^n} \int_{-\infty}^{\infty} \cdots \int_{-\infty}^{\infty} H_n(j\omega_1, \cdots, j\omega_n) \prod_{i=1}^{n} U(j\omega_i) e^{j(\omega_1 + \cdots + \omega_n)t} \mathrm{d}\omega_1 \cdots \mathrm{d}\omega_n$$

X. Jing, Z. Lang, *Frequency Domain Analysis and Design of Nonlinear Systems based on Volterra Series Expansion*, Understanding Complex Systems,
DOI 10.1007/978-3-319-12391-2_2

which can, by denoting $\omega_n = \omega - \omega_1 - \cdots - \omega_{n-1}$, be further written as

$$y_n(t) = \frac{1}{2\pi}\int_{-\infty}^{\infty}\left[\frac{1}{(2\pi)^{n-1}}\underbrace{\int_{-\infty}^{\infty}\cdots\int_{-\infty}^{\infty}}_{n-1} H_n(j\omega_1,\cdots,j\omega_{n-1},j(\omega-\omega_1-\cdots-\omega_{n-1})) \times U_n(j\omega_1,\cdots j\omega_{n-1})\mathrm{d}\omega_1\cdots\mathrm{d}\omega_{n-1}\right] e^{j\omega t}\mathrm{d}\omega$$

where $U_n(j\omega_1,\cdots,j\omega_{n-1}) = U(j\omega_1)\cdots U(j\omega_{n-1})U(j(\omega-\omega_1-\cdots-\omega_{n-1}))$. Therefore the Fourier transform of $y_n(t)$ is obtained as

$$Y_n(j\omega) = \frac{1}{(2\pi)^{n-1}}\underbrace{\int_{-\infty}^{\infty}\cdots\int_{-\infty}^{\infty}}_{n-1} H_n(j\omega_1,\cdots,j\omega_{n-1},j(\omega-\omega_1-\cdots-\omega_{n-1})) \times U_n(j\omega_1,\cdots,j\omega_{n-1})\mathrm{d}\omega_1\cdots\mathrm{d}\omega_{n-1} \tag{2.3}$$

which is referred to as the nth-order output spectrum. The output spectrum of the nonlinear system in (2.1a,b) can then be computed by

$$Y(j\omega) = \sum_{n=1}^{N} Y_n(j\omega) \tag{2.4}$$

Note that in (2.1a,b, 2.3 and 2.4) the input signal u(t) can be any signal with a Fourier transform U(jω). The GFRFs and output spectrum of each order defined above are all referred to as frequency response functions of nonlinear systems in this book. It can be seen that the nonlinear frequency response functions defined above and associated analysis and design methods are important extension and/or natural generalization of existing theory and methods for linear systems to the nonlinear case. A simple comparison is shown in Table 2.1.

2.1.1 The Probing Method

Obviously, the output spectrum of a nonlinear system involves the computation of the GFRFs. Given the parametric model of a nonlinear system, the GFRFs can be derived by using the "harmonic probing" method (Rugh 1981), which can be traced back to Bedrosian and Rice (1971) or earlier. Examples can be seen in Peyton Jones and Billings (1989), and Billings and Peyton-Jones (1990) etc. Consider the excitation of system (2.1a,b) with an input consisting of *n* complex exponentials defined as

Table 2.1 A comparison between the frequency domain theories of linear and nonlinear systems. The nonlinear output spectrum will be discussed in more details in Chaps. 3 and 6–9

Linear systems		Nonlinear systems
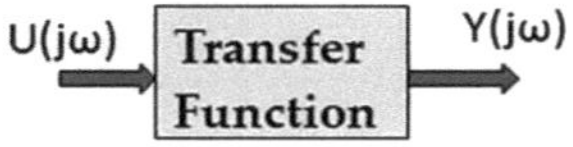		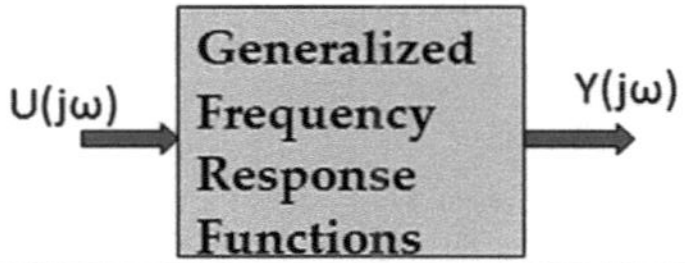
System output: $y(t)=\int_{-\infty}^{\infty}h(\tau)u(t-\tau)d\tau$		System output: $y(t)=\int_{-\infty}^{\infty}h_1(\tau_1)u(t-\tau_1)d\tau_1+\int_{-\infty}^{\infty}\int_{-\infty}^{\infty}h_2(\tau_1,\tau_2)\prod_{i=1}^{2}u(t-\tau_i)d\tau_i$ $+\sum_{n=1}^{3}\int_{-\infty}^{\infty}\cdots\int_{-\infty}^{\infty}h_n(\tau_1,\cdots,\tau_n)\prod_{i=1}^{n}u(t-\tau_i)d\tau_i+\cdots$
Frequency response function or Transfer function $H(j\omega)=\int_{-\infty}^{\infty}h(\tau)\exp(-j(\omega\tau))d\tau$	⇨	Generalized frequency response functions (GFRFs) $H_n(j\omega_1,\cdots,j\omega_n)=\int_{-\infty}^{\infty}\cdots\int_{-\infty}^{\infty}h_n(\tau_1,\cdots,\tau_n)\exp(-j(\omega_1\tau_1+\cdots+\omega_n\tau_n))d\tau_1\cdots d\tau_n$
Output spectrum $Y(j\omega)=H(j\omega)U(j\omega)$		Nonlinear output spectrum $Y(j\omega)=\sum_{n=1}^{N}\frac{1}{\sqrt{n}(2\pi)^{n-1}}\int_{\omega+..+\omega=\omega}H_n(j\omega_1,\cdots,j\omega_n)\prod_{i=1}^{n}U(j\omega_i)d\sigma_\omega$

$$u(t) = \sum_{r=1}^{n} e^{j\omega_r t} \tag{2.5}$$

The output y(t) is given as

$$y(t) = \sum_{n=1}^{N} \sum_{r_1, r_n=1}^{n} H_n(j\omega_{r_1}, \cdots, j\omega_{r_n}) e^{j(\omega_{r_1} + \cdots + \omega_{rn})t} \tag{2.6}$$

Then for the nonlinear system, replacing the input and output by (2.5–2.6), the nth-order GFRF can be obtained by extracting the coefficients of the term $e^{j(\omega_{r_1} + \cdots + \omega_{rn})t}$. For example, consider a static polynomial function,

$$y(t) = f(u(t)) = c_1 u(t) + c_2 u(t)^2 + c_3 u(t)^3 + \ldots + c_n u(t)^n + \ldots \tag{2.7}$$

The *n*th-order GFRF between the input *u*(*t*) and output *y*(*t*) can be derived as

$$H_n(j\omega_1, \cdots, j\omega_n) = c_n \tag{2.8}$$

To show this, the GFRFs for (2.7) can be obtained by directly applying the probing method. Note that (2.6) can be expanded as

$$\begin{aligned} y(t) = {} & n! H_n(j\omega_1, \cdots, j\omega_n) e^{j(\omega_1 + \cdots + \omega_n)t} + (n-1)! H_{n-1}(j\omega_1, \cdots, j\omega_{n-1}) e^{j(\omega_1 + \cdots + \omega_{n-1})t} \\ & + \cdots + H_1(j\omega_1) e^{j\omega_1 t} + \text{(the other terms)} \end{aligned} \tag{2.9}$$

Using (2.5) and (2.9) with n=3 in (2.7), it can be obtained that

$$\begin{aligned} y(t) &= 3! H_3(j\omega_1, \cdots, j\omega_3) e^{j(\omega_1 + \cdots + \omega_3)t} + 2H_2(j\omega_1, j\omega_2) e^{j(\omega_1 + \omega_2)t} + H_1(j\omega_1) e^{j\omega_1 t} + \text{(the other terms)} \\ &= c_1(e^{j\omega_1 t} + e^{j\omega_2 t} + e^{j\omega_3 t}) + c_2(e^{j\omega_1 t} + e^{j\omega_2 t} + e^{j\omega_3 t})^2 + c_3(e^{j\omega_1 t} + e^{j\omega_2 t} + e^{j\omega_3 t})^3 + \ldots \\ &\quad + c_n(e^{j\omega_1 t} + e^{j\omega_2 t} + e^{j\omega_3 t})^n + \ldots \end{aligned}$$

Extracting the coefficients of the term $e^{j\omega_1 t}$ from the equation above, it can be obtained that

$$H_1(j\omega_1) = c_1$$

Extracting the coefficients of $e^{j(\omega_1 + \omega_2)t}$, it can be obtained that

$$H_2(j\omega_1, j\omega_2) = c_2$$

Similarly, from the coefficients of $e^{j(\omega_1 + \cdots + \omega_3)t}$, it can be obtained that

$$H_3(j\omega_1, \cdots, j\omega_3) = c_3$$

Following this method, the nth-order GFRF can be obtained as

$$H_n(j\omega_1,\cdots,j\omega_n) = c_n$$

2.2 The GFRFs for NARX and NDE Models

Nonlinear systems can often be described by different parametric models. To compute the GFRFs, a parametric model of the nonlinear system under study can be given. In this book, two parametric models are focused, i.e., the Nonlinear Auto-Regressive with eXogenous input (NARX) model and Nonlinear Differential Equation (NDE) model.

The NARX model provides a unified and natural representation for a wide class of nonlinear systems, including many nonlinear models as special cases (e.g., Wiener models, Hammerstein models). For this reason, the NARX model has been extensively used in various engineering problems for system identification (Li et al. 2011), signal processing (McWhorter and Scharf 1995; Kay and Nagesha 1994), and control (Sheng and Chon 2003) etc. In practice, most systems are inherently nonlinear and can be identified to obtain a NARX model using several efficient algorithms such as the OLS method (Chen et al. 1989). The NARX model is given by

$$y(t) = \sum_{m=1}^{M} y_m(t)$$

$$y_m(t) = \sum_{p=0}^{m} \sum_{\substack{k_1, k_{p+q} = 1 \\ p+q=m}}^{K} c_{p,q}(k_1,\cdots,k_{p+q}) \prod_{i=1}^{p} y(t-k_i) \prod_{i=p+1}^{p+q} u(t-k_i) \tag{2.10}$$

where $y_m(t)$ is the mth-order output of the NARX model; $\sum_{k_1, k_{p+q}=1}^{K}(\cdot) = \sum_{k_1=1}^{K}(\cdot)\cdots \sum_{k_{p+q}=1}^{K}(\cdot)$; $p+q$ is referred to as the nonlinear degree of parameter $c_{p,q}(\cdot)$, which corresponds to the (p+q)-degree nonlinear terms $\prod_{i=1}^{p} y(t-k_i) \prod_{i=p+1}^{p+q} u(t-k_i)$, *e.g.*, $y(t-1)^p u(t-2)^q$ (p orders in terms of the output and q orders in terms of the input), and k_i is the lag of the ith output when $i \le p$ or the (i-p)th input when $p < i \le m$ with the maximum lag K; $c_{0,1}(.)$ and $c_{1,0}(.)$ of nonlinear degree 1 are referred to as linear parameters, and all the other model parameters are referred to as nonlinear parameters; the model includes all the possible nonlinear combinations in terms of $y(k)$ and $u(k)$ with the maximum order M.

The NDE model can be regarded as a continuous-time version of the NARX model, which is usually obtained by physical modelling and can be given by

$$\sum_{m=1}^{M}\sum_{p=0}^{m}\sum_{k_1,k_{p+q}=0}^{K} c_{p,q}(k_1,\cdots,k_{p+q})\prod_{i=1}^{p}\frac{d^{k_i}y(t)}{dt^{k_i}}\prod_{i=p+1}^{p+q}\frac{d^{k_i}u(t)}{dt^{k_i}}=0 \tag{2.11}$$

where $\left.\frac{d^k x(t)}{dt^k}\right|_{k=0}=x(t)$, and all other notations take similar forms and definitions to those for the NARX model for convenience. But in the NDE model, K is the maximum order of the derivative, and $c_{p,q}(\cdot)$ for $p+q>1$ are referred to as nonlinear parameters corresponding to nonlinear terms in the model of the form $\prod_{i=1}^{p}\frac{d^{k_i}y(t)}{dt^{k_i}}\prod_{i=p+1}^{p+q}\frac{d^{k_i}u(t)}{dt^{k_i}}$, e.g., $y(t)^p u(t)^q$.

2.2.1 Computation of the GFRFs for NARX Models

By using the probing method demonstrated above, a recursive algorithm to compute the nth-order GFRF in terms of model parameters for nonlinear systems described by the NARX model can be developed, which is given as follows (Peyton Jones and Billings 1989; Jing and Lang 2009a):

$$\begin{aligned}&\left(1-\sum_{k_1=1}^{K}c_{1,0}(k_1)\exp(-j(\omega_1+\cdots+\omega_n)k_1)\right)\cdot H_n(j\omega_1,\cdots,j\omega_n)\\&=\sum_{k_1,k_n=1}^{K}c_{0,n}(k_1,\cdots,k_n)\exp(-j(\omega_1k_1+\cdots+\omega_nk_n))\\&+\sum_{q=1}^{n-1}\sum_{p=1}^{n-q}\sum_{k_1,k_n=1}^{K}c_{p,q}(k_1,\cdots,k_{p+q})\exp\left(-j\sum_{i=1}^{q}\omega_{n-q+i}k_{p+i}\right)H_{n-q,p}(j\omega_1,\cdots,j\omega_{n-q})\\&+\sum_{p=2}^{n}\sum_{k_1,k_p=1}^{K}c_{p,0}(k_1,\cdots,k_p)H_{n,p}(j\omega_1,\cdots,j\omega_n)\end{aligned} \tag{2.12}$$

$$H_{n,p}(\cdot)=\sum_{i=1}^{n-p+1}H_i(j\omega_1,\cdots,j\omega_i)H_{n-i,p-1}(j\omega_{i+1},\cdots,j\omega_n)\exp(-j(\omega_1+\cdots+\omega_i)k_p) \tag{2.13}$$

$$H_{n,1}(j\omega_1,\cdots,j\omega_n)=H_n(j\omega_1,\cdots,j\omega_n)\exp(-j(\omega_1+\cdots+\omega_n)k_1) \tag{2.14}$$

Furthermore, define $H_{0,0}(\cdot)=1$, $H_{n,0}(\cdot)=0$ for n>0, $H_{n,p}(\cdot)=0$ for $n<p$, and let

$$\exp\left(\sum_{i=1}^{q}\varepsilon(p)\right)=\begin{cases}1 & q=0,p>1\\0 & q=0,p\le 1\end{cases} \tag{2.15}$$

where $\varepsilon(p)$ is a function of p, and

$$L_n(\omega_1,\cdots,\omega_n) = 1 - \sum_{k_1=1}^{K} c_{1,0}(k_1)\exp(-j(\omega_1+\cdots+\omega_n)k_1) \tag{2.16}$$

Then (2.12) can be written more concisely as

$$H_n(j\omega_1,\cdots,j\omega_n) = \frac{1}{L_n(\omega_1\cdots\omega_n)}\sum_{q=0}^{n}\sum_{p=0}^{n-q}\sum_{k_1,k_{p+q}=1}^{K} c_{p,q}(k_1,\cdots,k_{p+q})e^{-j\sum_{i=1}^{q}(\omega_{n-q+i}k_{p+i})} \times H_{n-q,p}(j\omega_1,\cdots,j\omega_{n-q}) \tag{2.17}$$

Thus the recursive algorithm for the computation of GFRFs is (2.12 or 2.17, 2.13–2.16).

Moreover, $H_{n,p}(j\omega_1,\cdots,j\omega_n)$ in (2.13) can also be written as

$$H_{n,p}(j\omega_1,\cdots,j\omega_n) = \sum_{\substack{r_1\cdots r_p = 1\\ \sum r_i = n}}^{n-p+1} \prod_{i=1}^{p} H_{r_i}(j\omega_{X+1},\cdots,j\omega_{X+r_i}) \exp(-j(\omega_{X+1}+\cdots+\omega_{X+r_i})k_i) \tag{2.18}$$

where $X = \sum_{x=1}^{i-1} r_x$.

2.2.2 Computation of the GFRFs for NDE Models

Similarly, the computation of the GFRFs for the NDE model can be recursively conducted in terms of model parameters as follows (Billings and Peyton-Jones1990; Jing et al. 2008e):

$$\begin{aligned}L_n(j\omega_1+\cdots+j\omega_n)\cdot H_n(j\omega_1,\cdots,j\omega_n) &= \sum_{k_1,k_n=1}^{K} c_{0,n}(k_1,\cdots,k_n)(j\omega_1)^{k_1}\cdots(j\omega_n)^{k_n}\\ &+\sum_{q=1}^{n-1}\sum_{p=1}^{n-q}\sum_{k_1,k_{p+q}=0}^{K} c_{p,q}(k_1,\cdots,k_{p+q})\left(\prod_{i=1}^{q}(j\omega_{n-q+i})^{k_{p+i}}\right)H_{n-q,p}(j\omega_1,\cdots,j\omega_{n-q})\\ &+\sum_{p=2}^{n}\sum_{k_1,k_p=0}^{K} c_{p,0}(k_1,\cdots,k_p)H_{n,p}(j\omega_1,\cdots,j\omega_n)\end{aligned} \tag{2.19}$$

$$H_{n,p}(\cdot) = \sum_{i=1}^{n-p+1} H_i(j\omega_1,\cdots,j\omega_i)H_{n-i,p-1}(j\omega_{i+1},\cdots,j\omega_n)(j\omega_1+\cdots+j\omega_i)^{k_p} \quad (2.20)$$

$$H_{n,1}(j\omega_1,\cdots,j\omega_n) = H_n(j\omega_1,\cdots,j\omega_n)(j\omega_1+\cdots+j\omega_n)^{k_1} \quad (2.21)$$

where

$$L_n(j\omega_1+\cdots+j\omega_n) = -\sum_{k_1=0}^{K} c_{1,0}(k_1)(j\omega_1+\cdots+j\omega_n)^{k_1} \quad (2.22)$$

Moreover, $H_{n,p}(j\omega_1,\cdots,j\omega_n)$ in (2.20) can also be written as

$$H_{n,p}(j\omega_1,\cdots,j\omega_n) = \sum_{\substack{r_1\cdots r_p=1 \\ \sum r_i=n}}^{n-p+1} \prod_{i=1}^{p} H_{r_i}(j\omega_{X+1},\cdots,j\omega_{X+r_i})(j\omega_{X+1}+\cdots+j\omega_{X+r_i})^{k_i} \quad (2.23)$$

where

$$X = \sum_{x=1}^{i-1} r_x \quad (2.24)$$

Similarly, for convenience in discussion, define

$$H_{0,0}(\cdot) = 1, \quad H_{n,0}(\cdot) = 0 \quad \text{for n} > 0, \quad H_{n,p}(\cdot) = 0 \quad \text{for } n < p,$$

$$\text{and} \quad \prod_{i=1}^{q}(\cdot) = \begin{cases} 1 & q=0, p>1 \\ 0 & q=0, p\le 1 \end{cases} \quad (2.25)$$

Then (2.19) can be written in a more concise form as

$$H_n(j\omega_1,\cdots,j\omega_n) = \frac{1}{L_n\left(j\sum_{i=1}^{n}\omega_i\right)} \sum_{q=0}^{n}\sum_{p=0}^{n-q}\sum_{k_1,k_{p+q}=0}^{K} c_{p,q}(k_1,\cdots,k_{p+q}) \times \left(\prod_{i=1}^{q}(j\omega_{n-q+i})^{k_{p+i}}\right) H_{n-q,p}(j\omega_1,\cdots,j\omega_{n-q}) \quad (2.26)$$

Therefore, the recursive algorithm for the computation of the GFRFs is (2.19 or 2.26, 2.25, 2.20–2.23).

Note that the GFRFs above both for the NARX and NDE models are assumed to be asymmetric. Generally, different permutations of the frequency variables

$\omega_1,\cdots,\omega_n$ may lead to different values of $H_n(j\omega_1,\cdots,j\omega_n)$. The symmetric GFRFs can be obtained as

$$H_n^{sym}(j\omega_1,\cdots,j\omega_n) = \frac{1}{n!} \sum_{\substack{\text{all the permutations} \\ \text{of } \{1,2,\ldots,n\}}} H_n(j\omega_1,\cdots,j\omega_n) \tag{2.27}$$

But for computation of nonlinear output spectrum in (2.4), asymmetric GFRFs suffice.

2.3 The GFRFs for a Single Input Double Output Nonlinear System

In many practical cases, nonlinear system models are usually described by a nonlinear state equation with a general nonlinear output function of system states. Sometimes, the output function of interest can also be a nonlinear objective function to optimize. Therefore, the computation of the GFRFs for nonlinear systems in this form would be more relevant in practice. The systems can be classified into several cases: single-input multi-output (SIMO), or multi-input and multi-output (MIMO) etc. The GFRFs for MIMO systems would be more complicated, which can be referred to Swain and Billings (2001). This section addresses a much simpler case, i.e., single-input double-output (SIDO), which is actually frequently encountered in practice. Similar results can be easily extended to the SIMO case (many multiple-degree-of-freedom mechanical systems would belong to this case).

Consider the following SIDO NARX system,

$$x(t) = \sum_{m=1}^{M_1} \sum_{p=0}^{m} \sum_{k_1,k_m=0}^{K} \bar{c}_{p,m-p}(k_1,\cdots,k_m) \prod_{i=1}^{p} x(t-k_i) \prod_{i=p+1}^{m} u(t-k_i) \tag{2.28a}$$

$$y(t) = \sum_{m=1}^{M_2} \sum_{p=0}^{m} \sum_{k_1,k_m=0}^{K} \tilde{c}_{p,m-p}(k_1,\cdots,k_m) \prod_{i=1}^{p} x(t-k_i) \prod_{i=p+1}^{m} u(t-k_i) \tag{2.28b}$$

where M_1, M_2 and K are all positive integers, and $x(t)$, $y(t)$, $u(t) \in \mathbb{R}$. Equation (2.28a) is the system state equation which is still described by a NARX model, and (2.28b) represents the system output which is a nonlinear function of state $x(t)$ and input u(t) in a general polynomial form.

Instead of using the probing method for derivation of the GFRFs for (2.28a,b), an alternative simple method would be adopted here, since the model structure and nonlinear types are known clearly. Note that the expression of the nth-order GFRF in (2.12) for the NARX model (2.10) can be divided into three parts. That is, those arising from pure input nonlinear terms $H_{n_u}(\cdot)$ corresponding to the first part in the right side of (2.12), those from cross product nonlinear terms $H_{n_{uy}}(\cdot)$ corresponding to the second part in the right side of (2.12), and those from pure output nonlinear terms $H_{n_y}(\cdot)$ corresponding to the last part of (2.12). For clarity, (2.12) can also be written as

$$H_n(j\omega_1,\cdots,j\omega_n) = \left(H_{n_u}(\cdot) + H_{n_{uy}}(\cdot) + H_{n_y}(\cdot)\right)/L_n(j(\omega_1 + \cdots + \omega_n)) \quad (2.29)$$

Equation (2.29) shows clearly that different categories of nonlinearities produce different contribution to the system GFRFs. Hence, when deriving the GFRFs of a nonlinear system, what can be done is to combine the different contributions from different nonlinearities without directly using the probing method. This property can be used for the derivation of the GFRFs for (2.28a,b).

To this aim, (2.28a,b) can be regarded as a system of one input u(t) and two outputs x(t) and y(t). Therefore, there are two sets of GFRFs for (2.28a,b) corresponding to the two input-output relationships between the input u(t) and two outputs x(t) and y(t) respectively. Considering the GFRFs from input u(t) to output x(t), there are three categories of nonlinearities as mentioned before. Therefore, the *n*th-order GFRF from input u(t) to output x(t) denoted by $H_n^x(j\omega_1,\cdots,j\omega_n)$ can be directly determined which is the same as (2.12–2.17), i.e.,

$$H_n^x(j\omega_1,\cdots,j\omega_n) = \frac{H_{n_u}^x(j\omega_1,\cdots,j\omega_n) + H_{n_{ux}}^x(j\omega_1,\cdots,j\omega_n) + H_{n_x}^x(j\omega_1,\cdots,j\omega_n)}{L_n(j(\omega_1+\cdots+\omega_n))} \quad (2.30)$$

where, $L_n(j(\omega_1+\cdots+\omega_n)) = 1 - \sum_{k_1=1}^{K} \overline{c}_{1,0}(k_1)\exp(-j(\omega_1+\cdots+\omega_n)k_1)$

$$H_{n_u}^x(j\omega_1,\cdots,j\omega_n) = \sum_{k_1,k_n=0}^{K} \overline{c}_{0,n}(k_1,\cdots,k_n)\exp(-j(\omega_1 k_1 + \cdots + \omega_n k_n)) \quad (2.31a)$$

$$H_{n_{ux}}^x(j\omega_1,\cdots,j\omega_n) = \sum_{q=1}^{n-1}\sum_{p=1}^{n-q}\sum_{k_1,k_{p+q}=0}^{K} \overline{c}_{p,q}(k_1,\cdots,k_{p+q}) \times\exp\left(-j(\omega_{n-q+1}k_{p+1}+\cdots+\omega_n k_{p+q})\right)H_{n-q,p}(j\omega_1,\cdots,j\omega_{n-q}) \quad (2.31b)$$

$$H_{n_x}^x(j\omega_1,\cdots,j\omega_n) = \sum_{p=2}^{n}\sum_{k_1,k_p=0}^{K} \overline{c}_{p,0}(k_1,\cdots,k_p)H_{n,p}(j\omega_1,\cdots,j\omega_n) \quad (2.31c)$$

$$H_{n,p}(j\omega_1,\cdots,j\omega_n) = \sum_{i=1}^{n-p+1} H_i^x(j\omega_1,\cdots,j\omega_i)H_{n-i,p-1}(j\omega_{i+1},\cdots,j\omega_n) \times\exp\left(-j(\omega_1 + \cdots + \omega_i)k_p\right) \quad (2.31d)$$

$$H_{n,1}(j\omega_1,\cdots,j\omega_n) = H_n^x(j\omega_1,\cdots,j\omega_n)\exp(-j(\omega_1+\cdots+\omega_n)k_1) \quad (2.31e)$$

Similarly, consider the GFRFs from input u(t) to output y(t). There are also three categories of nonlinearities in terms of input u(t) and output x(t) (similar to those from input u(t) to output x(t)), and there is one linear output y(t). Note that there are no nonlinearities in terms of y(t), and all the nonlinearities come from input u(t) and output x(t). For this reason, the GFRFs from u(t) to y(t) are dependent on the GFRFs from u(t) to x(t). Therefore, in this case the nth-order GFRF from input u(t) to output y(t) denoted by $H_n^y(j\omega_1,\cdots,j\omega_n)$ is,

$$H_n^y(j\omega_1,\cdots,j\omega_n) = H_{n_u}^y(j\omega_1,\cdots,j\omega_n) + H_{n_{ux}}^y(j\omega_1,\cdots,j\omega_n) + H_{n_x}^y(j\omega_1,\cdots,j\omega_n) \quad (2.32)$$

where the corresponding terms in (2.32) are

$$H_{n_u}^y(j\omega_1,\cdots,j\omega_n) = \sum_{k_1,k_n=0}^{K} \tilde{c}_{0,n}(k_1,\cdots,k_n)\exp(-j(\omega_1 k_1+\cdots+\omega_n k_n)) \quad (2.33a)$$

$$H_{n_{ux}}^y(j\omega_1,\cdots,j\omega_n) = \sum_{q=1}^{n-1}\sum_{p=1}^{n-q}\sum_{k_1,k_{p+q}=0}^{K} \tilde{c}_{p,q}(k_1,\cdots,k_{p+q}) \times\exp(-j(\omega_{n-q+1}k_{p+1}+\cdots+\omega_n k_{p+q}))H_{n-q,p}(j\omega_1,\cdots,j\omega_{n-q}) \quad (2.33b)$$

$$H_{n_x}^y(j\omega_1,\cdots,j\omega_n) = \sum_{p=1}^{n}\sum_{k_1,k_p=0}^{K} \tilde{c}_{p,0}(k_1,\cdots,k_p)H_{n,p}(j\omega_1,\cdots,j\omega_n) \quad (2.33c)$$

Note that p is counted from 1 in (2.33c), different from (2.31c) where p is counted from 2, and $H_{n,p}(j\omega_1,\cdots,j\omega_n)$ in (2.33b, c) is the same as that in (2.31b–d) because the nonlinearities in (2.28b) have no relationship with y(t) but x(t). The results here are developed in a very straightforward manner and provide a concise analytical expression for the GFRFs of the system in (2.28a,b).

Similar results can be obtained for the following SIDO NDE system

$$\sum_{m=1}^{M_1}\sum_{p=0}^{m}\sum_{k_1,k_m=0}^{K} \bar{c}_{p,m-p}(k_1,\cdots,k_m)\prod_{i=1}^{p}\frac{d^{k_i}x(t)}{dt^{k_i}}\prod_{i=p+1}^{m}\frac{d^{k_i}u(t)}{dt^{k_i}} = 0 \quad (2.34a)$$

$$\sum_{m=1}^{M_2}\sum_{p=0}^{m}\sum_{k_1,k_m=0}^{K} \tilde{c}_{p,m-p}(k_1,\cdots,k_m)\prod_{i=1}^{p}\frac{d^{k_i}x(t)}{dt^{k_i}}\prod_{i=p+1}^{m}\frac{d^{k_i}u(t)}{dt^{k_i}} = y(t) \quad (2.34b)$$

where $x(t)$, $y(t)$, $u(t) \in \mathbb{R}$. System (2.34a,b) has similar notations and structure as system (2.28a,b). It can be regarded as an NDE model with two outputs x(t) and y(t), and one input u(t). Hence, following the same idea, the GFRFs for the relationship from u(t) to y(t) are given as

$$H_n^y(j\omega_1,\cdots,j\omega_n) = H_{n_u}^y(j\omega_1,\cdots,j\omega_n) + H_{n_{ux}}^y(j\omega_1,\cdots,j\omega_n) \\ +H_{n_x}^y(j\omega_1,\cdots,j\omega_n) \tag{2.35}$$

where

$$H_{n_u}^y(j\omega_1,\cdots,j\omega_n) = \sum_{k_1,k_n=0}^{K} \widetilde{c}_{0,n}(k_1,\cdots,k_n)(j\omega_1)^{k_1}\cdots(j\omega_n)^{k_n} \tag{2.36a}$$

$$H_{n_{ux}}^y(j\omega_1,\cdots,j\omega_n) = \sum_{q=1}^{n-1}\sum_{p=1}^{n-q}\sum_{k_1,k_{p+q}=0}^{K} \widetilde{c}_{p,q}(k_1,\cdots,k_{p+q}) \\ \times (j\omega_{n-q+1})^{k_{p+1}}\cdots(j\omega_n)^{k_{p+q}} H_{n-q,p}(j\omega_1,\cdots,j\omega_{n-q}) \tag{2.36b}$$

$$H_{n_x}^y(j\omega_1,\cdots,j\omega_n) = \sum_{p=1}^{n}\sum_{k_1,k_p=0}^{K} \widetilde{c}_{p,0}(k_1,\cdots,k_p)H_{n,p}(j\omega_1,\cdots,j\omega_n) \tag{2.36c}$$

$$H_{n,p}(\cdot) = \sum_{i=1}^{n-p+1} H_i^x(j\omega_1,\cdots,j\omega_i)H_{n-i,p-1}(j\omega_{i+1},\cdots,j\omega_n)(j\omega_1+\cdots+j\omega_i)^{k_p} \tag{2.36d}$$

$$H_{n,1}(j\omega_1,\cdots,j\omega_n) = H_n^x(j\omega_1,\cdots,j\omega_n)(j\omega_1+\cdots+j\omega_n)^{k_1} \tag{2.36e}$$

where $H_n^x(j\omega_1,\cdots,j\omega_n)$ is the nth-order GFRF from u(t) to x(t), which is the same as that given in (2.19 or 2.26, 2.25, 2.20–2.23).

Example 2.1 Consider the following nonlinear system,

$$\begin{array}{l} mx(t-2) + a_1x(t-1) + a_2x^2(t-1) + a_3x^3(t-1) + kx(t) = u(t) \\ y(t) = a_1x(t-1) + a_2x^2(t-1) + a_3x^3(t-1) + kx(t) \end{array} \tag{2.37}$$

which can be written into the form of model (2.28a,b) with parameters K=2, $\overline{c}_{1,0}(2) = -m/k, \overline{c}_{1,0}(1) = -a_1/k$, $\overline{c}_{2,0}(11) = -a_2/k, \overline{c}_{3,0}(111) = -a_3/k, \overline{c}_{0,1}(0) = 1/k$ $\widetilde{c}_{1,0}(1) = a_1, \widetilde{c}_{2,0}(11) = a_2, \widetilde{c}_{3,0}(111) = a_3, \widetilde{c}_{1,0}(0) = k$, and all the other parameters are zero. The GFRFs can be computed according to (2.12–2.16). For example,

$$H_{1_u}^x(j\omega_1) = \sum_{k_1=0}^{2} \overline{c}_{0,1}(k_1)\exp(-j\omega_1 k_1) = \overline{c}_{0,1}(0) = 1/k, \quad H_{1_u}^y(j\omega_1) = 0,$$

Because there are no input nonlinearities and cross nonlinearities, thus
$H_{n_u}^x(j\omega_1,\cdots,j\omega_n) = 0$ and $H_{n_u}^y(j\omega_1,\cdots,j\omega_n) = 0$ for n>1
$H_{n_{ux}}^x(j\omega_1,\cdots,j\omega_n) = 0$ and $H_{n_{ux}}^y(j\omega_1,\cdots,j\omega_n) = 0$ for all n

Regarding the output nonlinear terms,

$$H_{1_x}^x(j\omega_1) = 0,$$

$$\begin{aligned}
H_{2_x}^x(j\omega_1, j\omega_2) &= \sum_{p=2}^{2} \sum_{k_1, k_p=1}^{2} \bar{c}_{p,0}(k_1, \cdots, k_p) H_{2,p}(j\omega_1, j\omega_2) \\
&= \sum_{k_1, k_p=1}^{2} \bar{c}_{2,0}(k_1, k_2) H_{2,2}(j\omega_1, j\omega_2) \\
&= \sum_{k_1, k_p=1}^{2} \bar{c}_{2,0}(k_1, k_2) H_1^x(j\omega_1) H_{1,1}(j\omega_2) \exp(-j\omega_1 k_2) \\
&= \sum_{k_1, k_p=1}^{2} \bar{c}_{2,0}(k_1, k_2) H_1^x(j\omega_1) H_1^x(j\omega_2) \exp(-j\omega_2 k_1) \exp(-j\omega_1 k_2) \\
&= -\frac{a_2}{k} H_1^x(j\omega_1) H_1^x(j\omega_2) \exp(-j\omega_2) \exp(-j\omega_1)
\end{aligned}$$

$$\begin{aligned}
H_{1_x}^y(j\omega_1) &= \sum_{k_1}^{2} \tilde{c}_{1,0}(k_1) H_{1,1}(j\omega_1) = \sum_{k_1}^{2} \tilde{c}_{1,0}(k_1) H_1^x(j\omega_1) \exp(-j\omega_1 k_1) \\
&= a_1 H_1^x(j\omega_1) \exp(-j\omega_1) + k H_1^x(j\omega_1)
\end{aligned}$$

$$\begin{aligned}
H_{2_x}^y(j\omega_1, j\omega_2) &= \sum_{p=1}^{2} \sum_{k_1, k_p=0}^{2} \tilde{c}_{p,0}(k_1, \cdots, k_p) H_{2,p}(j\omega_1, j\omega_2) \\
&= \sum_{k_1=0}^{2} \tilde{c}_{1,0}(k_1) H_{2,1}(j\omega_1, j\omega_2) + \sum_{k_1, k_2=0}^{2} \tilde{c}_{2,0}(k_1, k_2) H_{2,2}(j\omega_1, j\omega_2) \\
&= \sum_{k_1=0}^{2} \tilde{c}_{1,0}(k_1) H_2^x(j\omega_1, j\omega_2) \exp(-j(\omega_1 + \omega_2) k_1) \\
&\quad + \sum_{k_1, k_2=0}^{2} \tilde{c}_{2,0}(k_1, k_2) H_1^x(j\omega_1) H_1^x(j\omega_2) \exp(-j\omega_2 k_1) \exp(-j\omega_1 k_2) \\
&= k H_2^x(j\omega_1, j\omega_2) + a_1 H_2^x(j\omega_1, j\omega_2) \exp(-j(\omega_1 + \omega_2) k_1) \\
&\quad + a_2 H_1^x(j\omega_1) H_1^x(j\omega_2) \exp(-j\omega_2) \exp(-j\omega_1)
\end{aligned}$$

Note that

$$\begin{aligned}
L_n(j(\omega_1 + \cdots + \omega_n)) &= 1 - \sum_{k_1=1}^{2} \bar{c}_{1,0}(k_1) \exp(-j(\omega_1 + \cdots + \omega_n) k_1) \\
&= 1 + \frac{a_1}{k} \exp(-j(\omega_1 + \cdots + \omega_n)) + \frac{m}{k} \exp(-j2(\omega_1 + \cdots + \omega_n))
\end{aligned}$$

Hence, by following similar process as above, the GFRFs for x(t) and y(t) can all be computed recursively up to any high orders. For example,

$$H_1^x(j\omega_1) = \frac{H_{1_u}^x(j\omega_1) + H_{1_{ux}}^x(j\omega_1) + H_{1_x}^x(j\omega_1)}{L_1(j\omega_1)} = \frac{1/k}{1 + \frac{a_1}{k}\exp(-j\omega_1) + \frac{m}{k}\exp(-j2\omega_1)}$$

$$H_2^x(j\omega_1,j\omega_2) = \frac{H_{2_u}^x(j\omega_1,j\omega_2) + H_{2_{ux}}^x(j\omega_1,j\omega_2) + H_{2_x}^x(j\omega_1,j\omega_2)}{L_2(j(\omega_1+\omega_2))}$$

$$= \frac{-\frac{a_2}{k}H_1^x(j\omega_1)H_1^x(j\omega_2)\exp(-j\omega_2)\exp(-j\omega_1)}{1 + \frac{a_1}{k}\exp(-j(\omega_1+\omega_2)) + \frac{m}{k}\exp(-j2(\omega_1+\omega_2))}$$

$$H_1^y(j\omega_1) = H_{1_u}^y(j\omega_1) + H_{1_{ux}}^y(j\omega_1) + H_{1_x}^y(j\omega_1) = k + a_1 H_1^x(j\omega_1)\exp(-j\omega_1)$$

$$H_2^y(j\omega_1,j\omega_2) = H_{2_u}^y(j\omega_1,j\omega_2) + H_{2_{ux}}^y(j\omega_1,j\omega_2) + H_{2_x}^y(j\omega_1,j\omega_2)$$

$$= a_1 H_2^x(j\omega_1,j\omega_2)\exp(-j(\omega_1+\omega_2)) + a_2 H_1^x(j\omega_1)$$

$$\times H_1^x(j\omega_2)\exp(-j\omega_2)\exp(-j\omega_1)$$

It can be verified that the first order GFRFs are the frequency response functions in z-space of the linear parts of model (2.37).

Example 2.2 Consider a nonlinear mechanical system shown in Fig. 2.1.

The output property of the spring satisfies $A = kx$, the damper $F = a_1\dot{x} + a_3\dot{x}^3$, and the active unit is described by $F = a_2\dot{x}^2$. u(t) is the external input force. Therefore, the system dynamics is

$$m\ddot{x} = -kx - a_1\dot{x} - a_2\dot{x}^2 - a_3\dot{x}^3 + u(t) \tag{2.38a}$$

with the transmitted force measured on the base as the output

$$y(t) = a_1\dot{x} + a_2\dot{x}^2 + a_3\dot{x}^3 + kx(t) \tag{2.38b}$$

It can be seen that the continuous time model (2.38a,b) is similar in structure to the discrete time model (2.37) in Example 2.1. Therefore, similar results regarding the frequency response functions as demonstrated in Examples 2.1 for the discrete time model (2.37) can be obtained readily for system (2.38a,b).

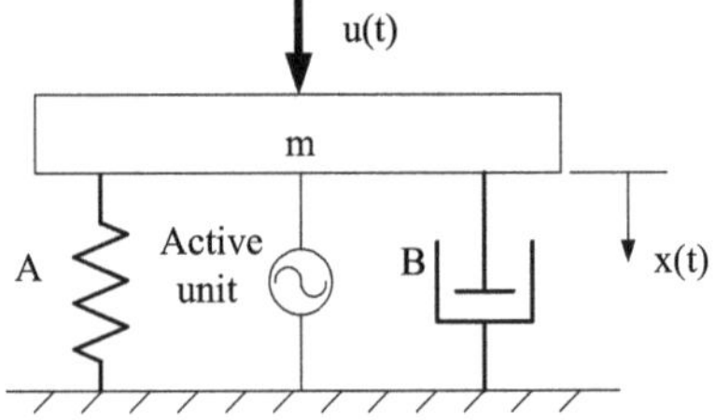

Fig. 2.1 A mechanical system

2.4 The Frequency Response Functions for Block-Oriented Nonlinear Systems

Block-oriented nonlinear systems such as Hammerstein and Wiener models are composed by a cascade combination of a linear dynamic model and a static (memoryless) nonlinear function. Theoretically, any nonlinear systems which have a Volterra expansion can be represented by a finite sum of Wiener models with sufficient accuracy (Korenberg 1982; Boyd and Chua 1985). The Wiener model is shown to be a reasonable model for many chemical and biological processes (Zhu 1999; Kalafatis et al. 1995; Hunter and Korenberg 1986). The magneto-rheological (MR) damping systems can also be well approximated by a Hammerstein model (Huang et al. 1998). Applications of these block-oriented models can be found in many areas such as mechanical systems (Huang et al. 1998), control systems (Bloemen et al. 2001), communication systems (Wang et al. 2010), chemical processes (Kalafatis et al. 1995), and biological systems (Hunter and Korenberg 1986).

Frequently-used block-oriented nonlinear models include Wiener model, Hammerstein model and Wiener-Hammerstein model etc. This section establishes frequency response functions for these nonlinear models under assumption that the nonlinear part allows a polynomial approximation as given in (2.7), which is then extended to a more general case.

2.4.1 *Frequency Response Functions of Wiener Systems*

The GFRFs and nonlinear output spectrum are developed for Wiener systems firstly, and then extended to other models. Consider the Wiener model given by

$$u(t) = g^{\circ}r(t) \text{ and } y(t) = f(u(t)) \tag{2.39a, b}$$

where "°" represents the convolution operator, $g(t)$ is the impulse response of the linear part, and $f(u(t))$ is the static nonlinear part of the system. The linear part is defined as a stable SISO system, which can be described by parametric FIR/IIR models or nonparametric models (See Fig. 2.2).

Note that the GFRFs for (2.39b) are given in (2.8). Equation (2.39a) can be written as

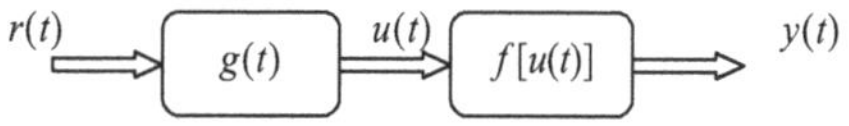

Fig. 2.2 The Wiener model, where $g(t)$ denotes the linear part and $f[\bullet]$ represents the static nonlinear function, both of which could be parametric or nonparametric (Jing 2011)

$$U(\mathrm{j}\omega) = G(\mathrm{j}\omega)R(\mathrm{j}\omega)$$

where $U(\mathrm{j}\omega)$, $G(\mathrm{j}\omega)$ and $R(\mathrm{j}\omega)$ are the corresponding Fourier transforms of $u(t)$, $g(t)$ and $r(t)$ respectively. Using (2.3–2.4), the nth-order output spectrum of (2.39a,b) can be obtained as

$$\begin{aligned}
Y_n(j\omega) &= \frac{1}{(2\pi)^{n-1}}\underbrace{\int_{-\infty}^{\infty}\cdots\int_{-\infty}^{\infty}}_{n-1} c_n U_n(j\omega_1,\cdots,j\omega_{n-1})\mathrm{d}\omega_1\cdots\mathrm{d}\omega_{n-1} \\
&= \frac{1}{(2\pi)^{n-1}}\underbrace{\int_{-\infty}^{\infty}\cdots\int_{-\infty}^{\infty}}_{n-1} c_n\prod_{i=1}^{n}(G(j\omega_i)R(j\omega_i))\mathrm{d}\omega_1\cdots\mathrm{d}\omega_{n-1} \\
&= \frac{1}{(2\pi)^{n-1}}\underbrace{\int_{-\infty}^{\infty}\cdots\int_{-\infty}^{\infty}}_{n-1}\left(c_n\prod_{i=1}^{n}G(j\omega_i)\right)\prod_{i=1}^{n}R(j\omega_i)\mathrm{d}\omega_1\cdots\mathrm{d}\omega_{n-1} \\
&= \frac{c_n}{(2\pi)^{n-1}}\underbrace{\int_{-\infty}^{\infty}\cdots\int_{-\infty}^{\infty}}_{n-1}\left(\prod_{i=1}^{n}G(j\omega_i)\right)\prod_{i=1}^{n}R(j\omega_i)\mathrm{d}\omega_1\cdots\mathrm{d}\omega_{n-1}
\end{aligned} \tag{2.40}$$

where $\omega_n = \omega - \omega_1 - \cdots - \omega_{n-1}$. Comparing the structure of (2.40) with (2.3) gives

$$H_n(j\omega_1,\cdots,j\omega_n) = c_n\prod_{i=1}^{n}G(j\omega_i) \tag{2.41}$$

With the GFRFs given by (2.41), the output spectrum of Wiener system (2.39a,b) can therefore be computed under any input signal with spectrum $R(\mathrm{j}\omega)$ based on (2.3) and (2.41) as,

$$\begin{aligned}
Y_n(j\omega) = \frac{c_n}{(2\pi)^{n-1}}&\underbrace{\int_{-\infty}^{\infty}\cdots\int_{-\infty}^{\infty}}_{n-1} \\
&\left(\prod_{i=1}^{n-1}G(j\omega_i)R(j\omega_i)\right)G(j(\omega-\omega_1-\cdots-\omega_{n-1})) \\
&\times R(j(\omega-\omega_1-\cdots-\omega_{n-1}))\mathrm{d}\omega_1\cdots\mathrm{d}\omega_{n-1}
\end{aligned}$$

Let $\Pi(j\omega_i) = G(j\omega_i)R(j\omega_i)$, then $Y_1(j\omega_i) = c_1\Pi(j\omega_i)$, which represents the linear dynamics of the Wiener system. The equation above can be further written as

$$Y_n(j\omega) = \frac{c_n}{(2\pi)^{n-1}} \underbrace{\int_{-\infty}^{\infty} \cdots \int_{-\infty}^{\infty}}_{n-1} \left(\prod_{i=1}^{n-1} \Pi(j\omega_i) \right)$$

$$\Pi(j(\omega - \omega_1 - \cdots - \omega_{n-1}))d\omega_1 \cdots d\omega_{n-1} \tag{2.42}$$

It can be seen that the nth-order nonlinear output spectrum of Wiener model (2.39a, b) is completely dependent on the frequency response of system linear part. Given the frequency response of the linear part (which can be nonparametric), higher order output spectra can be computed immediately. On the other hand, given higher order output spectra, the linear part of the system could also be estimated consequently. These will be discussed later. The overall output spectrum is a combination of all different order output spectra. Clearly, the nonlinear frequency response functions obtained above can provide an effective insight into the analytical analysis and design of Wiener systems in the frequency domain. Note that the magnitude bound of system output spectrum often provides a useful insight into system dynamics at different frequencies and also into the relationship between frequency response functions and model parameters (Jing et al. 2007a, 2008b, d). With the GFRFs developed above, the bound characteristics of the output spectrum of Wiener system (2.39a,b) can be investigated readily. It is known that output frequencies of nonlinear systems are always more complicated than linear systems including sub- or super-harmonics and inter-modulations (Jing et al. 2010). The GFRFs and output spectrum above could also shed light on the analysis of output frequency characteristics of Wiener-type nonlinear systems.

2.4.2 The GFRFs of Wiener-Hammerstein or Hammerstein Systems

The Wiener-Hammerstein model can be described by

$$u(t) = g^{\circ}r(t);\ x(t) = f(u(t)) \text{ and } y(t) = p^{\circ}x(t) \tag{2.43a, b, c}$$

where p and g denote the linear parts following and preceding nonlinear function $f(\cdot)$ (See Fig. 2.3).

Consider the subsystem from $r(t)$ to $x(t)$, which is the Wiener model in (2.39a,b). According to (2.40–2.41), the nth-order output spectrum of this subsystem is

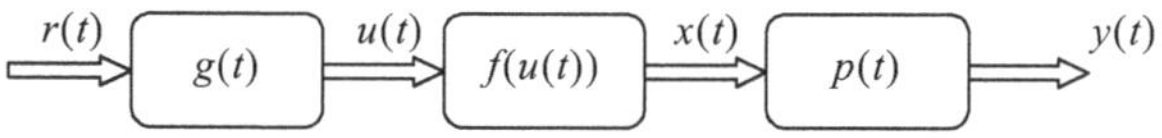

Fig. 2.3 The Wiener-Hammerstein model, where $g(t)$ and $p(t)$ denote the linear parts and $f(\bullet)$ represents the static nonlinear function, g-f is actually a Wiener sub-system and f-p is a Hammerstein sub-system (Jing 2011)

$$X_n(j\omega) = \frac{1}{(2\pi)^{n-1}} \underbrace{\int_{-\infty}^{\infty} \cdots \int_{-\infty}^{\infty}}_{n-1} \left(c_n \prod_{i=1}^{n} G(j\omega_i) \right) \prod_{i=1}^{n} R(j\omega_i) \mathrm{d}\omega_1 \cdots \mathrm{d}\omega_{n-1}$$

where $\omega_n = \omega - \omega_1 - \cdots - \omega_{n-1}$. Then the nth-order output spectrum of (2.43a–c) is

$$\begin{aligned} Y_n(j\omega) &= X_n(j\omega)P(j\omega) \\ &= \frac{c_n P(j\omega)}{(2\pi)^{n-1}} \underbrace{\int_{-\infty}^{\infty} \cdots \int_{-\infty}^{\infty}}_{n-1} \left(\prod_{i=1}^{n} G(j\omega_i) \right) \prod_{i=1}^{n} R(j\omega_i) \mathrm{d}\omega_1 \cdots \mathrm{d}\omega_{n-1} \end{aligned} \tag{2.44}$$

where $P(j\omega)$ is the Fourier transform of $p(t)$. Therefore, the nth-order GFRF for (2.43a–c) is

$$H_n(j\omega_1, \cdots, j\omega_n) = c_n P(j\omega) \prod_{i=1}^{n} G(j\omega_i) \tag{2.45a}$$

Noting that $\omega = \omega_1 + \cdots + \omega_n$ in (2.45a), the equation above can be written as

$$H_n(j\omega_1, \cdots, j\omega_n) = c_n P(j\omega_1 + \cdots + j\omega_n) \prod_{i=1}^{n} G(j\omega_i) \tag{2.45b}$$

Using (2.45b) and noting the Hammerstein model is only a special case ($g(t) = 1$) of the Wiener-Hammerstein model, the GFRFs of Hammerstein systems can be obtained immediately as

$$H_n(j\omega_1, \cdots, j\omega_n) = c_n P(j\omega_1 + \cdots + j\omega_n) \tag{2.46}$$

Note that, the GFRFs and output spectrum of block-oriented nonlinear systems are derived by employing the structure property of the nonlinear output spectrum defined in (2.3) and the structure information of block-oriented models. The resulting frequency response functions are expressed into analytical functions of model parameters, which are not restricted to a specific input but allow any form of input signals. However, many existing frequency-domain results for nonlinear analysis require a specific sinusoidal input signal (Alleyne and Hedrick 1995; Gelb and Velde 1968; Nuij et al. 2006; Schmidt and Tondl 1986; Huang et al. 1998; Baumgartner and Rugh 1975; Krzyzak 1996; Crama and Schoukens 2001; Bai 2003). In the GFRFs, the relationships among the output spectrum, the GFRFs, the system nonlinear parameters, and also the linear dynamics of the system are demonstrated clearly. With these frequency response functions, bound characteristics of the output spectrum and output frequency characteristics etc can all be studied by following the methods in Jing et al. (2006, 2007a, 2008b, d, 2010).

2.4.3 Extension to a More General Polynomial Case

The static nonlinear function of block-oriented models discussed above is only a univariate polynomial function. A more general case is studied in this section. Since both the Wiener and Hammerstein models are special cases, consider the following Wiener-Hammerstein model,

$$u(t) = g^{\circ}r(t);\ x(t) = f(u(t), r(t)) \text{ and } y(t) = p^{\circ}x(t) \tag{2.47a, b, c}$$

where the nonlinear function is defined as a more general multivariate polynomial function as

$$f(u(t),r(t)) = \sum_{m=1}^{M}\sum_{p=0}^{m}\sum_{k_1,k_m=0}^{K} c_{p,m-p}(k_1,\cdots,k_m)\prod_{i=1}^{p}\frac{\mathrm{d}^{k_i}u(t)}{\mathrm{d}t^{k_i}}\prod_{i=p+1}^{m}\frac{\mathrm{d}^{k_i}r(t)}{\mathrm{d}t^{k_i}} \tag{2.47d}$$

where M is the maximum nonlinear degree of the polynomial, K is the maximum order of the derivative and $c_{p,m-p}(k_1,\cdots,k_m)$ is the coefficient of a term $\prod_{i=1}^{p}\frac{\mathrm{d}^{k_i}u(t)}{\mathrm{d}t^{k_i}}\prod_{i=p+1}^{m}\frac{\mathrm{d}^{k_i}r(t)}{\mathrm{d}t^{k_i}}$ in the polynomial.

Obviously, the univariate polynomial function in (2.7) is only a special case of the general form (2.47d). That is, if letting $c_{p,0}(\underbrace{0,\cdots,0}_{p}) = c_p$ for p=1,2,... and the other coefficients in (2.47d) are zero, then (2.47d) will become (2.7). Obviously, (2.47a–d) can represent a wider class of nonlinear systems. For example, if (2.47d) is of sufficiently high degree and includes all possible linear and nonlinear combinations of input $u(t)$ and its derivatives of sufficiently high orders, it will be an equivalent nonlinear IIR model of the Volterra-type nonlinear systems (Kotsios 1997). The following results can be obtained.

Proposition 2.1 The nth-order GFRF of Wiener-Hammerstein system (2.47a–d) is given by

$$\begin{aligned}H_n(j\omega_1,\cdots,j\omega_n) &= P(j\omega_1+\cdots+j\omega_n)(H_{n_r}(j\omega_1,\cdots,j\omega_n)\\ &\quad +H_{n_{ru}}(j\omega_1,\cdots,j\omega_n)+H_{n_u}(j\omega_1,\cdots,j\omega_n))\end{aligned} \tag{2.48}$$

Similarly, for Wiener systems with a general polynomial function (2.47d) it is given by

$$H_n(j\omega_1,\cdots,j\omega_n) = H_{n_r}(j\omega_1,\cdots,j\omega_n)+H_{n_{ru}}(j\omega_1,\cdots,j\omega_n)+H_{n_u}(j\omega_1,\cdots,j\omega_n) \tag{2.49}$$

and for Hammerstein systems with a general polynomial function (2.47d) it is

$$H_n(j\omega_1,\cdots,j\omega_n) = P(j\omega_1 + \cdots + j\omega_n)H_{n_r}(j\omega_1,\cdots,j\omega_n) \tag{2.50}$$

where $H_{n_r}(j\omega_1,\cdots,j\omega_n)$, $H_{n_{ru}}(j\omega_1,\cdots,j\omega_n)$ and $H_{n_u}(j\omega_1,\cdots,j\omega_n)$ are given by

$$H_{n_r}(j\omega_1,\cdots,j\omega_n) = \sum_{k_1,k_n=0}^{K} c_{0,n}(k_1,\cdots,k_n)(j\omega_1)^{k_1}\cdots(j\omega_n)^{k_n} \tag{2.51a}$$

$$H_{n_{ru}}(j\omega_1,\cdots,j\omega_n) = \sum_{q=1}^{n-1}\sum_{p=1}^{n-q}\sum_{k_1,k_{p+q}=0}^{K} c_{p,q}(k_1,\cdots,k_{p+q})\left(j\omega_{n-q+1}\right)^{k_{p+1}}\cdots(j\omega_n)^{k_{p+q}}$$
$$\times H_{n-q,p}\left(j\omega_1,\cdots,j\omega_{n-q}\right) \tag{2.51b}$$

$$H_{n_u}(j\omega_1,\cdots,j\omega_n) = \sum_{p=1}^{n}\sum_{k_1,k_p=0}^{K} c_{p,0}\left(k_1,\cdots,k_p\right)H_{n,p}(j\omega_1,\cdots,j\omega_n) \tag{2.51c}$$

$$H_{n,p}(\cdot) = G(j\omega_1)H_{n-1,p-1}(j\omega_2,\cdots,j\omega_n)(j\omega_1)^{k_p} \tag{2.51d}$$

$$H_{n,1}(j\omega_1,\cdots,j\omega_n) = \begin{cases} G(j\omega_1)(j\omega_1)^{k_1} & \text{n} = 1 \\ 0 & \text{else} \end{cases} \tag{2.51e}$$

Proof See Sect. 2.6.

Since (2.39b) is a special case of (2.47d) ($c_{p,0}(0,\cdots,0)=c_p$ and the others zero in (2.47d)), the nth-order GFRF in (2.41) for Wiener model (2.39a,b) can be shown to be a special case of (2.51a–e). That is, only the parameter $c_{n,0}(\underbrace{0,\cdots,0}_{n}) = c_n$ is not zero and the others are zero in (2.51a–e). Therefore,

$$H_n(j\omega_1,\cdots,j\omega_n)=H_{n_r}(j\omega_1,\cdots,j\omega_n)+H_{n_{ru}}(j\omega_1,\cdots,j\omega_n)+H_{n_u}(j\omega_1,\cdots,j\omega_n) \tag{2.52a}$$

$$H_{n_r}(j\omega_1,\cdots,j\omega_n) = 0, \quad H_{n_{ru}}(j\omega_1,\cdots,j\omega_n) = 0 \tag{2.52b, c}$$

$$H_{n_u}(j\omega_1,\cdots,j\omega_n) = c_{n,0}(0,\cdots,0)H_{n,n}(j\omega_1,\cdots,j\omega_n) \tag{2.52d}$$

$$\begin{aligned} H_{n,n}(\cdot) &= G(j\omega_1)H_{n-1,n-1}(j\omega_2,\cdots,j\omega_n)(j\omega_1)^0 \\ &= G(j\omega_1)G(j\omega_2)\cdots G(j\omega_{n-1})H_{1,1}(j\omega_n) \end{aligned} \tag{2.52e}$$

$$H_{1,1}(j\omega_n) = G(j\omega_n) \tag{2.52f}$$

The nth-order GFRF for (2.39a,b) can now be obtained from (2.52a–f) which is exactly (2.41).

Although the nonlinear frequency response functions above are all developed for continuous time system models, it would be easy to extend them to discrete time systems. In this section, analytical frequency response functions including generalized frequency response functions (GFRFs) and nonlinear output spectrum of block-oriented nonlinear systems are developed, which can demonstrate clearly the relationship between frequency response functions and model parameters, and also the dependence of frequency response functions on the linear part of the model. The nonlinear part of these models can be a more general multivariate polynomial function. These fundamental results provide a significant insight into the analysis and design of block-oriented nonlinear systems. Effective algorithms can also be developed for the estimation of nonlinear output spectrum and for parametric or nonparametric identification of nonlinear systems, which can be referred to Jing (2011).

2.5 Conclusions

The computation of the GFRFs and/or output spectrum for a given nonlinear system described by NARX, NDE or Block-oriented models is a fundamental task for nonlinear analysis in the frequency domain. This chapter summarizes the results for the computation of the GFRFs and output spectrum for several frequently-used parametric models, and shows that the GFRFs are the explicit functions of model parameters (of different nonlinear degrees), which can be regarded as an important extension of the transfer function concept of linear systems to the nonlinear case.

2.6 Proof of Proposition 2.1

As the Wiener system is a sub-system of the Wiener-Hammerstein system in (2.47a–d), the GFRFs for the sub-Wiener system can be derived firstly and then it will be easily extended to the other block-oriented models as demonstrated in Sect. 2.41–2.42. To derive the GFRFs for Wiener systems with the general polynomial function (2.47d), i.e.,

$$u(t) = g^{\circ} r(t) \quad \text{and} \quad y(t) = f(u(t), r(t)) \qquad \text{(A1, A2)}$$

$$f(u(t), r(t)) = \sum_{m=1}^{M} \sum_{p=0}^{m} \sum_{k_1, k_m=0}^{K} c_{p,m-p}(k_1, \cdots, k_m) \prod_{i=1}^{p} \frac{d^{k_i} u(t)}{dt^{k_i}} \prod_{i=p+1}^{m} \frac{d^{k_i} r(t)}{dt^{k_i}} \qquad \text{(A3)}$$

the model can be regarded as a nonlinear differential equation model with two outputs $u(t)$ and $y(t)$, and one input $r(t)$. Note that the frequency response function from the input $r(t)$ to the intermediate output $u(t)$ is the Fourier transform of the impulse response function g(t), i.e., $G(j\omega)$, which is a linear dynamics; while the frequency responses from the input $r(t)$ to the output $y(t)$ involve nonlinear dynamics. The latter are the GFRFs to be derived. The terms in the polynomial function $f(u(t),r(t))$ can be categorized into three groups, i.e., pure input terms $c_{0,m}(k_1, \cdots, k_m) \prod_{i=1}^{m} \frac{d^{k_i} r(t)}{dt^{k_i}}$, pure output terms $c_{p,0}(k_1, \cdots, k_p) \prod_{i=1}^{p} \frac{d^{k_i} u(t)}{dt^{k_i}}$, and input-output cross terms $c_{p,m-p}(k_1,\cdots,k_m)\prod_{i=1}^{p}\frac{\mathrm{d}^{k_i}u(t)}{\mathrm{d}t^{k_i}}\prod_{i=p+1}^{m}\frac{\mathrm{d}^{k_i}r(t)}{\mathrm{d}t^{k_i}}$ $(0<p<m)$.

Therefore, by applying the probing method, each group of terms corresponds to a specific form of contributions to the nth-order GFRF of Wiener system (A1–A3), which can be written as

$$H_n(j\omega_1,\cdots,j\omega_n) = H_{n_r}(j\omega_1,\cdots,j\omega_n) + H_{n_{ru}}(j\omega_1,\cdots,j\omega_n) + H_{n_u}(j\omega_1,\cdots,j\omega_n) \quad \text{(A4)}$$

where $H_{n_r}(j\omega_1,\cdots,j\omega_n)$ represents the contribution from the pure input terms and similar notations are used for the other terms. This is a special case of the system studied in Sect. 2.3 or Jing et al. (2008c). Therefore following the method there, the equations in (2.51a–e) can be obtained.

Similarly, the corresponding GFRFs for the Hammerstein model and Wiener-Hammerstein model with the general polynomial function defined in (2.47d) can be derived respectively. Note that only input nonlinearity is involved in the Hammerstein model. The extended polynomial function for the Hammerstein model can be written as

$$x(t) = f(r(t)) = \sum_{m=1}^{M}\sum_{k_1,k_m=0}^{K} c_{0,m}(k_1,\cdots,k_m)\prod_{i=1}^{m}\frac{d^{k_i}r(t)}{dt^{k_i}} \quad \text{(A5)}$$

Following the same line, the nth-order GFRF for the extended Hammerstein model is given by

$$H_n(j\omega_1,\cdots,j\omega_n) = P(j\omega_1+\cdots+j\omega_n)\sum_{k_1,k_n=0}^{K} c_{0,n}(k_1,\cdots,k_n)(j\omega_1)^{k_1}\cdots(j\omega_n)^{k_n} \quad \text{(A6)}$$

The nth-order GFRF for the Wiener-Hammerstein model (2.47a–d) can be obtained as

$$H_n(j\omega_1,\cdots,j\omega_n) = P(j\omega_1+\cdots+j\omega_n)(H_{n_r}(j\omega_1,\cdots,j\omega_n) + H_{n_{ru}}(j\omega_1,\cdots,j\omega_n) + H_{n_u}(j\omega_1,\cdots,j\omega_n)) \quad \text{(A7)}$$

where $H_{n_r}(j\omega_1,\cdots,j\omega_n)$, $H_{n_{ru}}(j\omega_1,\cdots,j\omega_n)$ and $H_{n_u}(j\omega_1,\cdots,j\omega_n)$ are given by (2.51a–e). This completes the proof. ∎

Chapter 3
Output Frequency Characteristics of Nonlinear Systems

3.1 Introduction

An important phenomenon for nonlinear systems in the frequency domain is that there are always very complicated output frequencies, appearing as super-harmonics, sub-harmonics, inter-modulation, and so on. This usually makes it rather difficult to analyze and design output frequency response of nonlinear systems, compared with linear systems. Output frequencies of nonlinear systems have been studied by several authors (Raz and Van Veen 1998; Lang and Billings 1997, 2000; Bedrosian and Rice 1971; Wu et al. 2007; Wei et al. 2007; Bussgang et al. 1974; Frank 1996). These results provide different viewpoints for computation and prediction of output frequencies of nonlinear systems. It is shown that Volterra-type nonlinear systems can effectively be used to account for super-harmonics and inter-modulation in nonlinear output spectrum.

In this chapter, some interesting properties of output frequencies of Volterra-type nonlinear systems are particularly investigated. These results provide a very novel and useful insight into the super-harmonic and inter-modulation phenomena in output frequency response of nonlinear systems, with consideration of the effects incurred by different nonlinear components in the system. The new properties theoretically demonstrate several fundamental output frequency characteristics and unveil clearly the mechanism of the interaction (or coupling effects) between different harmonic behaviors in system output frequency response incurred by different nonlinear components. These results have significance in the analysis and design of nonlinear systems and nonlinear filters in order to achieve a specific output spectrum in a desired frequency band by taking advantage of nonlinearities. They can provide an important guidance to modeling, identification, control and signal processing by using the Volterra series theory in practice.

X. Jing, Z. Lang, *Frequency Domain Analysis and Design of Nonlinear Systems based on Volterra Series Expansion*, Understanding Complex Systems,
DOI 10.1007/978-3-319-12391-2_3

3.2 Output Frequencies of Nonlinear Systems

As discussed in Chap. 2, the output spectrum of nonlinear system (2.1) subject to a general input can be computed by (2.3–2.4). For convenience, the output spectrum of system (2.1) in (2.3-2.4) can also be written as (Lang and Billings 1996)

$$\begin{aligned} Y(j\omega) &= \sum_{n=1}^{N} Y_n(j\omega) \\ Y_n(j\omega) &= \frac{1}{\sqrt{n}(2\pi)^{n-1}} \int_{\omega_1+\cdots+\omega_n=\omega} H_n(j\omega_1,\cdots,j\omega_n)\prod_{i=1}^{n} U(j\omega_i)d\sigma_\omega \end{aligned} \tag{3.1}$$

where $\int_{\omega_1+\cdots+\omega_n=\omega} (\cdot)d\sigma_\omega$ represents the integration on the super plane $\omega_1+\cdots+\omega_n=\omega$. $Y_n(j\omega)$ is referred to as the nth-order output spectrum. For a specific case, when the system is subject to a multi-tone input

$$u(t) = \sum_{i=1}^{\overline{K}} |F_i| \cos(\omega_i t + \angle F_i) \tag{3.2}$$

the system output spectrum is

$$\begin{aligned} Y(j\omega) &= \sum_{n=1}^{N} Y_n(j\omega) \\ Y_n(j\omega) &= \frac{1}{2^n} \sum_{\omega_{k_1}+\cdots+\omega_{k_n}=\omega} H_n(j\omega_{k_1},\cdots,j\omega_{k_n})F(\omega_{k_1})\cdots F(\omega_{k_n}) \end{aligned} \tag{3.3}$$

where $\overline{K}(>0) \in \mathbb{Z}$, $F_i \in \mathbb{C}$, $F(\omega_{k_i})$ can be written explicitly as

$$F(\omega_{k_i}) = |F_{|k_i|}|e^{j\angle F_{|k_i|}\cdot \mathrm{sgn1}(k_i)} \quad \text{for } k_i \in \{\pm 1,\cdots,\pm K\}, \text{ and}$$

$$\mathrm{sgn}\ 1(a) = \begin{cases} 1 & a > 0 \\ 0 & a = 0 \\ -1 & a < 0 \end{cases} \quad \text{for } a \in \mathbb{R}.$$

Nonlinear systems usually have complicated output frequencies, which are quite different from linear systems having output frequencies completely identical to the input frequencies. From (3.1) and (3.3), it can be seen that the output frequencies in the nth-order output spectrum, denoted by W_n and simply referred to as the nth-order output frequencies, are completely determined by

$$\omega = \omega_1 + \omega_2 + \cdots + \omega_n \text{ or } \omega = \omega_{k_1} + \omega_{k_2} + \cdots + \omega_{k_n}$$

which produce super-harmonics and inter-modulation in system output frequencies.

In this chapter, the input $u(t)$ in (3.1) is considered to be any continuous and bounded input function in $t \geq 0$ with Fourier transform $U(j\omega)$ whose domain (input frequency range) is denoted by V, *i.e.*, $\omega \in V$, and V can be regarded as any closed set in the real. The multi-tone function (3.2) is only a special case of this. Therefore, for the input $U(j\omega)$ defined in V, the nth-order output frequencies are

$$W_n = \left\{\omega = \omega_1 + \omega_2 + \cdots + \omega_n \middle| \omega_i \in \overline{V}, i = 1, 2, \ldots, n\right\} \tag{3.4a}$$

or for the multi-tone input (3.2),

$$W_n = \left\{\omega = \omega_{k_1} + \omega_{k_2} + \cdots + \omega_{k_n} \middle| \omega_{k_i} \in \overline{V}, i = 1, 2, \ldots, n\right\} \tag{3.4b}$$

where $\overline{V} = -V \cup V$. This is an analytical expression for the super-harmonics and inter-modulations in the nth-order output frequencies of nonlinear systems. All the output frequencies up to order N, denoted by W, can be written as

$$W = W_1 \cup W_2 \cup \cdots \cup W_N \tag{3.4c}$$

In (3.4a–c), $\overline{V}$ represents the theoretical input frequency range including both positive and negative frequencies contributing to high order (larger than 1) output spectra (involving only positive frequencies), V is the physical input frequency range contributing to every order output spectrum and W_1 represents the output frequencies incurred by the linear part of the system. For example, V may be a real set $[a, \text{b}] \cup [\text{c}, \text{d}]$, thus $\overline{V} = [-d, -c] \cup [-\text{b}, -a] \cup [a, \text{b}] \cup [c, d]$, where $d \geq c \geq b \geq a > 0$. When the system is subject to the multi-tone input (3.2), then the input frequency range is $\overline{V} = \{\pm\omega_1, \pm\omega_2, \cdots, \pm\omega_{\overline{K}}\}$, which is obviously a special case of the former one.

3.3 Fundamental Properties of Nonlinear Output Frequencies

In this section, fundamental properties of the output frequencies of system (2.1) with assumption that V is any closed set of frequency points in the real are developed. Importantly, the periodicity of output frequencies is revealed. Although some results about the computation of system output frequencies for the case with V=[a,b] has been studied in Raz and Van Veen (1998) and Lang and Billings (1997), for the multi-tone case in Lang and Billings (2000), Wei et al. (2007) and Yuan and Opal (2001) and for the multiple narrow-band signals in Bussgang et al. (1974), and some of the properties discussed in this section can be partly concluded from these previous results for the case V=[a,b] and multi-tone case

$V = \{\omega_1, \omega_2, \cdots, \omega_{\overline{K}}\}$, all the properties of this section are established in a uniform manner based on the analytical expressions (3.4a–c) for the input domain V which is any closed set in the real.

The following Property is straightforward from (3.4a, b).

Property 3.1 Consider the nth-order output frequency W_{n},

(a) Expansion, i.e., $W_{n-2} \subseteq W_n$;
(b) Symmetry, i.e., $\forall \Omega \in W_n$, then $-\Omega \in W_n$;
(c) n-multiple, i.e., $\max(W_n) = n \cdot \max(V)$ and $\min(W_n) = -n \cdot \max(V)$. □

Property 3.1 shows that the output frequency range will expand larger and larger with the increase of the nonlinear order (Property 3.1a), the expansion is symmetric around zero point (Property 3.1b) and its rate is n-multiple of the input frequency range (Property 3.1c). These are some fundamental properties which may be known in literature for some cases and clearly stated here for Volterra-type nonlinear systems subject to the mentioned class of input signals. Property 3.1a shows that, the (n−2)th order output frequencies $W_{\mathrm{n}-2}$ are completely included in the nth-order output frequencies W_{n}. This property can be used to facilitate the computation of output frequencies for nonlinear systems. That is, only the highest order of W_{n} in odd number and the highest order in even number, of which the corresponding GFRFs are not zero, are needed to be considered in (3.4c) (e.g., $W = W_{N-1} \cup W_N$). For example, suppose the system maximum order N=10, then only W_{10} and W_9 are needed to be computed if $H_{10}(.)$ and $H_9(.)$ are not zero, and the system output frequencies are $W = W_9 \cup W_{10}$ (in case that $H_9(.)$ is zero, W_9 should be replaced by the output frequencies corresponding to the highest odd order of nonzero GFRFs). For the case that V=[a,b], Property 3.1a can also be concluded from the results in Lang and Billings (1997). Here the conclusion is shown to hold for any V.

The following proposition theoretically demonstrates another fundamental and very useful property for the output frequencies of nonlinear systems, and provides a novel and interesting insight into system output frequency characteristics.

Proposition 3.1 (Periodicity Property) The frequencies in W_{n} can be generated periodically as follows

$$W_n = \bigcup_{i=1}^{n+1} \Pi_i(n) \tag{3.5a}$$

$$\Pi_i(n) = \left\{ \omega = \omega_1 + \omega_2 + \cdots + \omega_n \middle| \begin{array}{l} \omega_j \in \overline{V}, \\ \omega_j < 0 \text{ for } 1 \le j \le i-1, \\ \omega_j > 0 \text{ for } j \ge i \end{array} \right\} \tag{3.5b}$$

or

$$\Pi_i(n) = \left\{ \omega = \omega_{k_1} + \omega_{k_2} + \cdots + \omega_{k_n} \middle| \begin{array}{l} \omega_{k_j} \in \overline{V}, \\ \omega_{k_j} < 0 \text{ for } 1 \le j \le i-1, \\ \omega_{k_j} > 0 \text{ for } j \ge i \end{array} \right\} \tag{3.5c}$$

The process above has the following properties

$$\max(\Pi_i(n)) = -(i-1)\min(V) + (n-i+1)\max(V) \tag{3.6a}$$

and

$$\min(\Pi_i(n)) = -(i-1)\max(V) + (n-i+1)\min(V) \tag{3.6b}$$

$$\begin{aligned}\max(\Pi_{i-1}(n)) - \max(\Pi_i(n)) &= \min(\Pi_{i-1}(n)) - \min(\Pi_i(n)) \\ &= \min(V) + \max(V)\end{aligned} \tag{3.4c}$$

$$\Delta(n) = \max(\Pi_i(n)) - \min(\Pi_i(n)) = n \cdot (\max(V) - \min(V)) \tag{3.6d}$$

Moreover, when the system is subject to the class of input signals $U(j\omega)$ defined in [a,b] or specially subject to the multi-tone input (3.2) with $\omega_{i+1} - \omega_i = const > 0$ for $i = 1, \ldots, \overline{K} - 1\left(\overline{K} > 1\right)$, then

$$\Pi_i(n) = \Pi_{i-1}(n) - T \quad \text{for } i = 2, \ldots, n+1 \tag{3.6e}$$

where $T = \min(V) + \max(V)$ is the length of the frequency generation period, $\Pi_i(n) - T$ is a set whose elements are the elements in $\Pi_i(n)$ minus T, and $\Delta(n)$ is referred to as the frequency span in each period.

Proof See Sect. 3.6. □

Note that (3.6e) is a very useful property which can be used to simplify the computation of the output frequencies in applicable cases, because only one period of frequencies are needed to be computed and the other frequencies can be simply obtained by subtracting the length T. However, this property cannot hold for any input cases. The following corollary is straightforward.

Corollary 3.1 When the system is subject to the class of input $U(j\omega)$ which is specially defined in

$$\bigcup_{i=1}^{Z} [a + (i-1)\varepsilon, b + (i-1)\varepsilon]$$

where $b \geq a, \varepsilon \geq (b-a)$ and Z is a positive integer, then (3.6e) holds. □

Property 3.2 Consider $\Pi_i(n)$ in W_n, which corresponds to the frequencies in the ith frequency generation period,

(a) If the system input is the multi-tone function (3.2), then for any two frequencies Ω and Ω' in $\Pi_i(n)$ and any two frequencies ω and ω' in V, $\min(\Omega - \Omega') = \min(\omega - \omega')$.

(b) If $\Delta(n) > T$, then $\max(\Pi_{i+1}(n)) > \min(\Pi_i(n))$ for $i = 1, \ldots, n$. That is, there is overlap between the successive periods of frequencies in W_n.

Proof (a) is obvious from the proof for Proposition 3.1. Note that $\max(\Pi_{i+1}(n)) = \max(\Pi_i(n)) - T$, thus it can be derived that $\max(\Pi_{i+1}(n)) - \min(\Pi_i(n)) = \max(\Pi_i(n)) - \min(\Pi_i(n)) - T = \Delta(n) - T > 0$. Thus (b) is proved. □

The results above theoretically demonstrate some interesting properties of nonlinear system output frequencies. The periodicity of such output frequencies can be used to simplify the computation of the output frequencies for some special cases as mentioned above (where only one period of output frequencies need to be computed), and facilitate the computation of the output frequencies in the general case. Importantly, it theoretically reveals a novel insight into the output frequencies of nonlinear systems and helps understanding of the nonlinear behaviors in output frequency response of Volterra-type nonlinear systems. Interesting results based on this property will be further discussed later.

Example 3.1 Consider a simple nonlinear system as follows

$$y = -0.01\dot{y} + au^2 + bu^3 + cu - dy^3$$

The input is a multi-tone function u(t)=sin(6t)+sin(7t)+sin(8t). In order to demonstrate the properties above clearly, only several simple cases of the system are considered here. Firstly, consider $c=d=0$. That is, there are only nonlinear components (of nonlinear degrees 2 and 3) related only to the input (in short, input nonlinearities). Therefore, there will be only finite output frequencies because only the first, second and third order GFRFs are not zero and all the other orders are zero. The output spectra are given in Figs. 3.1 and 3.2 under different cases of input nonlinearities. As mentioned, because there are only input nonlinearities with order 2 and 3 in the system, the system output frequencies of the system are totally the same as the second and third order output frequencies. That is, $W=W_2$ for $a=1$ and $b=c=d=0$; $W=W_3$ for $a=c=d=0$ and $b=1$; and $W=W_2 \cup W_3$ for $a=1$, $b=1$, and $c=d=0$. Specifically, for the case $a=c=d=0$ and $b=1$ (Fig. 3.1), noting that $V=\{6,7,8\}$, and according to Proposition 1, there should be four periods in the output frequencies, two of which are positive, *i.e.*, $\Pi_1(3)=\{18,19,20,21,22,23,24\}$ and $\Pi_2(3)=\{4,5,6,7,8,9,10\}$; the period is T=6+8=14; the frequency span in each period is $\Delta(3)=\max(\Pi_i(3))-\min(\Pi_i(3))=3\cdot(\max(V)-\min(V))=6$. Figures 3.1 and 3.2 demonstrate the results in Properties 3.1c–3.2a and Proposition 3.1. It is also shown that the system output frequencies are simply the accumulation of all the output frequencies in each order output spectrum when the involved nonlinearities have no coupling effect and no overlap as stated in Property 3.2b. When and how there are coupling effects between different nonlinearities will be discussed in the next section. Note that the output response spectrum shown in the figures is 2|Y| not |Y|, because 2|Y| represents the physical magnitude of the system output.

When considering more complicated cases that there are different nonlinearities existing in the input and output of the system, the periodicity of the output frequencies can still be observed, but the interaction (i.e., the coupling effects) at some frequencies among different output harmonic responses incurred by different nonlinearities usually produce very complicated output spectrum. However, proper design of these different nonlinear terms can result in very special desirable output behaviour.

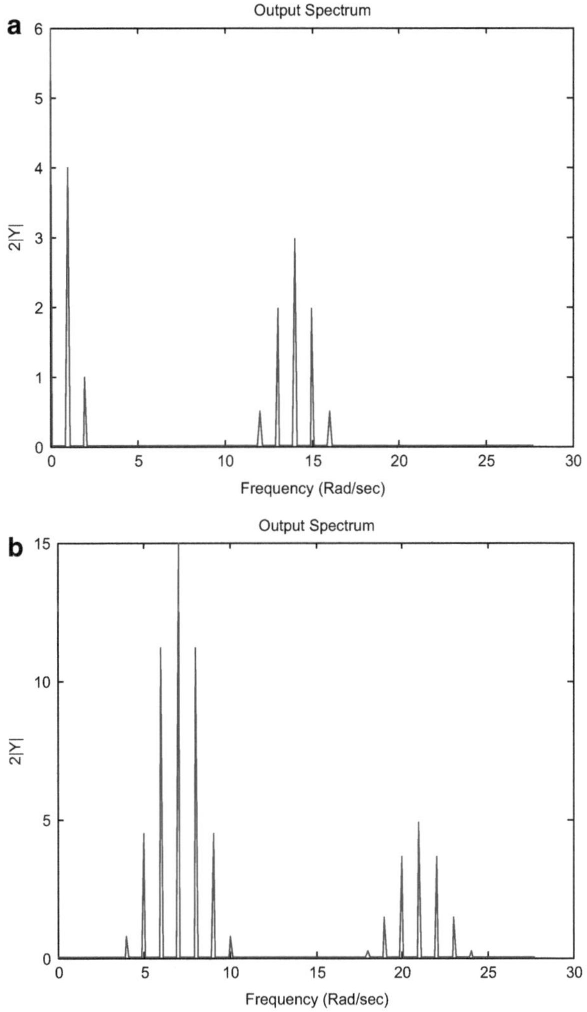

Fig. 3.1 Output frequencies when (**a**) a = 1, b = c = d = 0 and (**b**) a = c = d = 0, b = 1 (Jing et al. 2010)

See Fig. 3.3a–c. The output harmonics incurred by different single cubic nonlinear term in input or output both demonstrate the periodicity. The output harmonics incurred by single input nonlinearity are finite (see Fig. 3.3a) while the ones incurred by output nonlinearity are infinite (Fig. 3.3b). When both the two cubic nonlinearities work together, the output harmonics are still periodic but demonstrate coupling effect such that the output magnitude at the frequencies 6, 7, 8 are coincidently 1.

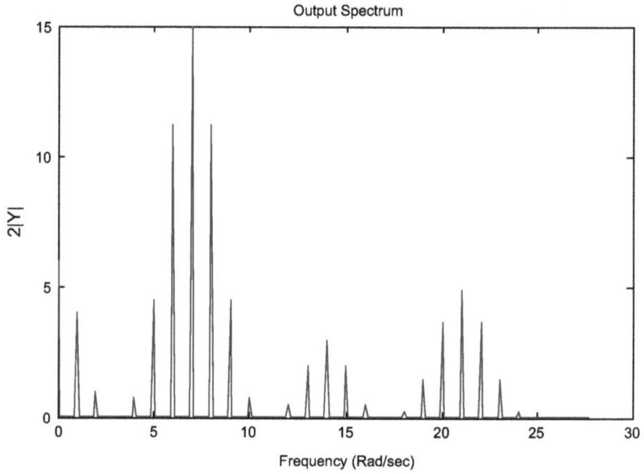

Fig. 3.2 Output frequencies when a = 1, b = 1, c = d = 0 (Jing et al. 2010)

3.4 Nonlinear Effect in Each Frequency Generation Period

The periodicity of output frequencies is revealed and demonstrated in the last section. In this section, the nonlinear harmonic or inter-modulated effect on system output spectrum in each frequency generation period, are studied. Firstly, a general property for each period of output frequencies of nonlinear systems is given. Then the interaction between different output harmonics incurred by different input nonlinearities is investigated as a case study. It is well known in literature and also as demonstrated in Example 3.1 that different nonlinearities can usually interact with each other such that the output harmonics incurred by different nonlinearities have coupling effects and become complicated. However, the mechanism about what the coupling effect is and how the different output harmonics interact with each other is seldom reported. In this section, after a general property is discussed, the interaction mechanism between different input nonlinearities is studied in detail based on the periodicity and under an assumption that the nonlinearities only exist in system input for simplicity in discussion. This does not only demonstrate the usefulness of the novel perspective revealed by the periodicity property, but also provide some useful results for the analysis and design of nonlinear FIR filters, which can be referred to the topic discussed in Billings and Lang (2002). Moreover, although it is convenient to analyze the input nonlinearities because the input nonlinearities only bring finite order output spectrum and less coupling effect when there are no system nonlinearities related to the output. More complicated results in this topic for the other kind of nonlinearities can be developed by following the similar method.

Firstly, from (3.1) and (3.3), it can be seen that the operators $\int\limits_{\omega_1+\cdots+\omega_n=\omega}(\cdot)d\sigma_\omega$ and $\sum\limits_{\omega_{k_1}+\cdots+\omega_{k_n}=\omega}(\cdot)$ have an important role in the frequency characteristics of the nth-order

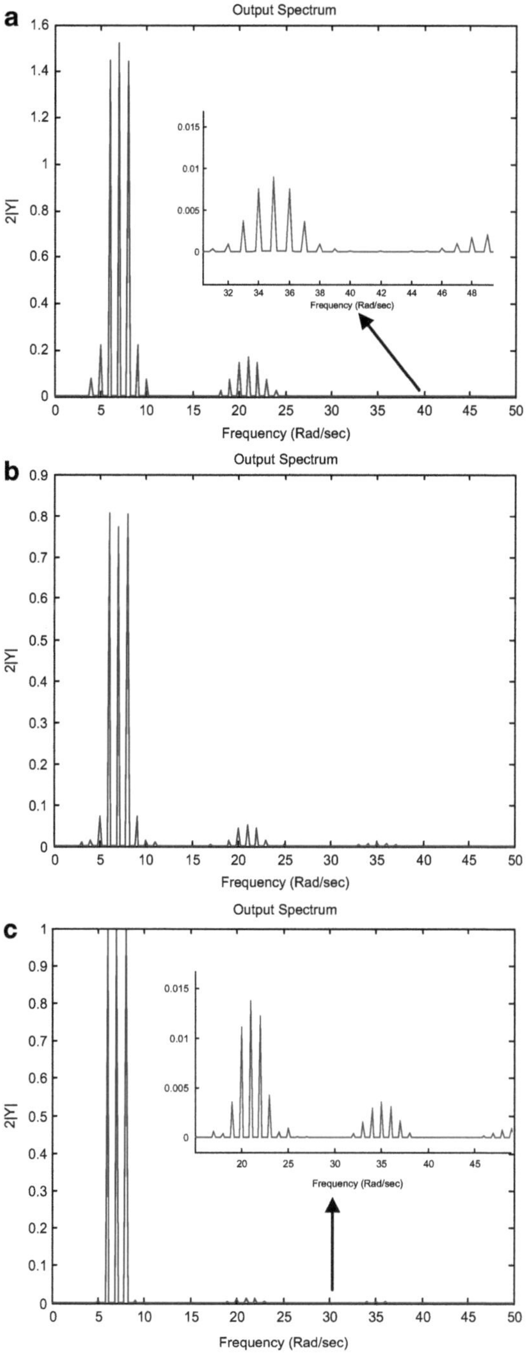

Fig. 3.3 Output frequencies when (**a**) a = 0, c = 1, b = 0.1, d = 0; (**b**) a = 0, c = 1, b = 0, d = 0.1; (**c**) a = 0, c = 1, b = 0.1, d = 0.1 (Jing et al. 2010)

output spectrum in each frequency generation period. The following property can be obtained.

Property 3.3 For $\omega \in \Pi_i(n)$ $(1 \le i \le \lfloor (n+1)/2 \rfloor)$, $\sum\limits_{\omega_{k_1}+\cdots+\omega_{k_n}=\omega} 1$, reaches its maximum at the central frequency $(\max(\Pi_i(n)) + \min(\Pi_i(n)))/2$ or around the central frequency if the central frequency is not in $\Pi_i(n)$, and has its minimum value at frequencies $\max(\Pi_i(n))$ and $\min(\Pi_i(n))$, i.e.,

$$\min_{\omega \in \Pi_i(n)} \left(\sum_{\omega_{k_1}+\cdots+\omega_{k_n}=\omega} 1 \right) = \sum_{\omega_{k_1}+\cdots+\omega_{k_n}=\max(\Pi_i(n))} 1 = \sum_{\omega_{k_1}+\cdots+\omega_{k_n}=\min(\Pi_i(n))} 1 = C_n^{i-1}$$

where $C_n^k = \frac{n(n-1)\cdots(n-k+1)}{k!}$ $(0 \le k \le n)$ $\left(C_n^0 = 1\right)$. Moreover, for $\omega \in \Pi_i(n)$ $(2 \le i \le \lfloor (n+1)/2 \rfloor)$,

$$\sum_{\omega_{k_1}+\cdots+\omega_{k_n}=\omega} 1 > \sum_{\omega_{k_1}+\cdots+\omega_{k_n}=<\omega+T>} 1$$

Especially, for the multi-tone input case with $\omega_{i+1} - \omega_i = const > 0$ for $i = 1, \ldots, \overline{K} - 1$,

$$\sum_{\omega_{k_1}+\cdots+\omega_{k_n}=\max(\Pi_i(n))-k'\cdot const} 1 = \sum_{\omega_{k_1}+\cdots+\omega_{k_n}=\min(\Pi_i(n))+k'\cdot const} 1$$

for $0 \le k' \le \Delta(n)/const$. Where, $\lfloor (n+1)/2 \rfloor$ is the largest integer which is not more than $(n+1)/2$, $<\omega+T>$ is the frequency in $\Pi_{i-1}(n)$ which is the nearest to $\omega+T$. The similar results also hold for the input defined in Corollary 3.1 by replacing $\sum\limits_{\omega_{k_1}+\cdots+\omega_{k_n}=\omega} 1$ with $\int\limits_{\omega_1+\cdots+\omega_n=\omega} 1 d\sigma_\omega$.

Proof Note that $\sum\limits_{\omega_{k_1}+\cdots+\omega_{k_n}=\omega} 1$ is equal to the number of all the combinations satisfying $\omega_{k_1} + \cdots + \omega_{k_n} = \omega$ and with the n frequency variables satisfying the conditions in $\Pi_i(n)$, thus the conclusions in this property can be obtained by using the combinatorics, which are straightforward. When the values of ω_1 and $\omega_{\overline{K}}$ are fixed and $\overline{K}$ is approaching infinity such that *const* approaches zero, the multi-tone frequencies will become a continuous closed set $\left[\omega_1, \omega_{\overline{K}}\right]$. The input frequencies defined in Corollary 3.1 are further extended from these two cases. Hence, the conclusions holding for the multi-tone case can be easily extended to the input case defined in Corollary 3.1. This completes the proof. □

Property 3.3 shows that in each frequency generation period, the effect of the operator $\int\limits_{\omega_1+\cdots+\omega_n=\omega} (\cdot) d\sigma_\omega$ and $\sum\limits_{\omega_{k_1}+\cdots+\omega_{k_n}=\omega} (\cdot)$ on system output spectrum tends naturally to be more complicated at the central frequency. That is, there is only one combination for the frequency variables in the operator $\sum_{\omega_{k_1}+\cdots+\omega_{k_n}=\omega}(\cdot)$ at the two boundary frequency of each period, it reaches the maximum at the central frequency of the same period and tends to be decreasing in different period with the frequency increasing. These can be regarded as the natural characteristics of the output frequencies that cannot be designed (This can be seen in the Figures of Examples 3.1–3.2).

3.4.1 Nonlinear Effect of Different Input Nonlinearities

As mentioned, different nonlinearities may have quite different effect on system output spectrum and there will be many coupling effects at the same frequency from different nonlinearities. This will make the output spectrum at the frequencies of interest to be enhanced or suppressed. For example, different nonlinearities (*e.g.*, $u(t)^3$ and $\dot{u}(t)u(t)^2$ are both input nonlinearity with nonlinear degree 3) may bring the same output frequencies according to Jing et al. (2006). However, the effect from different nonlinearities at the same frequency generation period may counteract with each other such that the output spectrum may be suppressed in some periods and others enhanced. This property is of great significance in the design of nonlinear systems for suppressing output vibration (Zhou and Misawa 2005; Jing et al. 2008a). As discussed, the periodicity property reveals a useful perspective for understanding of interaction mechanism between different system nonlinearities. In order to demonstrate this, the nonlinear effect between the harmonics incurred by different input nonlinearities is studied in this subsection.

Consider Volterra-type nonlinear systems described by the NDE model in (2.11). For convenience, it is given here as

$$\sum_{m=1}^{M}\sum_{p=0}^{m}\sum_{l_1,l_{p+q}=0}^{K} c_{p,q}(l_1,\cdots,l_{p+q})\prod_{i=1}^{p}\frac{d^{l_i}y(t)}{dt^{l_i}}\prod_{i=p+1}^{p+q}\frac{d^{l_i}u(t)}{dt^{l_i}}=0 \tag{3.7}$$

There are three kinds of nonlinearities in (3.7): input nonlinearity with coefficient $c_{0,q}(.)$ $(q>1)$, output nonlinearity with coefficient $c_{p,0}(.)$ $(p>1)$, and input output cross nonlinearity with coefficient $c_{p,q}(.)$ $(p+q>1$ and $p>0)$. Here, consider that there are only input nonlinearities in the NDE model above, i.e., $c_{p,q}(.)=0$ for all $p+q>1$ and $p>0$. In this case, the GFRFs can be written as

$$H_n(j\omega_1,\cdots,j\omega_n)=\frac{1}{L_n(j\omega_1+\cdots+j\omega_n)}\sum_{l_1,l_n=0}^{K} c_{0,n}(l_1,\cdots,l_n)(j\omega_1)^{l_1}\cdots(j\omega_n)^{l_n} \tag{3.8}$$

where

$$L_n(j\omega_1 + \cdots + j\omega_n) = -\sum_{l_1=0}^{K} c_{1,0}(l_1)(j\omega_1 + \cdots + j\omega_n)^{l_1} \tag{3.9}$$

From (3.8, 3.9) and (3.3), the nth-order output spectrum under the multi-tone input (3.2) can be obtained as

$$Y_n(j\omega) = \frac{1}{2^n} \sum_{\omega_{k_1}+\cdots+\omega_{k_n}=\omega} \left(\frac{F(\omega_{k_1})\cdots F(\omega_{k_n})}{L_n(j\omega_{k_1} + \cdots + j\omega_{k_n})} \times \sum_{l_1, l_n=1}^{K} c_{0,n}(l_1, \cdots, l_n)(j\omega_{k_1})^{l_1}\cdots(j\omega_{k_n})^{l_n} \right)$$

$$= \frac{1}{2^n L_n(j\omega)} \sum_{\omega_{k_1}+\cdots+\omega_{k_n}=\omega} (F(\omega_{k_1})\cdots F(\omega_{k_n}) \times \sum_{l_1, l_n=1}^{K} c_{0,n}(l_1, \cdots, l_n)(j\omega_{k_1})^{l_1}\cdots(j\omega_{k_n})^{l_n}) \tag{3.10}$$

To study the nonlinear effect from input nonlinearity in each frequency generation period, the following results can be obtained.

Definition 3.1 (Opposite Property) Considering two input nonlinear terms of the same degree with coefficients $c_{0,n}(l_1, \cdots, l_n)$ and $c_{0,n}(l_1', \cdots, l_n')$, if there exist two nonzero real number c_1 and c_2 satisfying $c_{0,n}(l_1, \cdots, l_n) = c_1$ and $c_{0,n}(l_1', \cdots, l_n') = c_2$, such that at a given frequency Ω?>?0,

$$\sum_{\omega_{k_1}+\cdots+\omega_{k_n}=\Omega} \left(F(\omega_{k_1})\cdots F(\omega_{k_n}) \cdot \left(c_1 (j\omega_{k_1})^{l_1}\cdots(j\omega_{k_n})^{l_n} + c_2 (j\omega_{k_1})^{l_1'}\cdots(j\omega_{k_n})^{l_n'} \right) \right) = 0$$

with respect to a multi-tone input (3.2), then the two terms are referred to as being opposite at frequency Ω under $c_{0,n}(l_1, \cdots, l_n) = c_1$ and $c_{0,n}(l_1', \cdots, l_n') = c_2$, whose effects in frequency domain counteract with each other at Ω.

The following definition will be used in what follows:

$$\text{sgn}(a + bj) = [\text{sgn1}(a) \quad \text{sgn1}(b)] \quad \text{for } a, b \in \mathbb{R}.$$

Proposition 3.2 summarized the cancellation property of input nonlinearities.

Proposition 3.2 (Cancellation Effect of Input Nonlinearity) Consider nonlinear systems with only input nonlinearities subject to multi-tone input, in which there are two nonlinear terms with coefficients $c_{0,n}(l_1,\cdots,l_n)$ and $c_{0,n}(l_1',\cdots,l_n')$. If there exists a non-negative integer $m \le \lfloor (n+1)/2 \rfloor - 1$ such that $\mathrm{sgn}(F(\omega_{k_1})\cdots F(\omega_{k_n}))$ is constant with respect to all the combinations of $\omega_{k_1},\cdots,\omega_{k_n} \in \{\pm\omega_1,\cdots,\pm\omega_{\overline{K}}\}$ satisfying $\omega_{k_1}+\cdots+\omega_{k_n} \in \Pi_{m+1}(n)$, then the two nonlinear terms can be designed to be opposite at any frequency in the (m+1)th frequency generation period $\Pi_{m+1}(n)$ with proper parametric values of the two coefficients, if and only if $\sum_{i=1}^{n} l_i$ and $\sum_{i=1}^{n} l_i'$ are both odd integers or even integers simultaneously (The proof is given in Sect. 3.6). □

Note that if two nonlinear terms satisfying the conditions in Proposition 3.2 are opposite in $\Pi_{m+1}(n)$, this does imply that the effects from these two nonlinear terms on system output spectrum can be counteracted with each other completely at any given frequency in $\Pi_{m+1}(n)$ but not implies that they can be counteracted completely at all the other frequencies in $\Pi_{m+1}(n)$ at the same time. Examples for that $\mathrm{sgn}(F(\omega_{k_1})\cdots F(\omega_{k_n}))$ is constant with respect to all the combinations of $\omega_{k_1},\cdots,\omega_{k_n} \in \{\pm\omega_1,\cdots,\pm\omega_{\overline{K}}\}$ satisfying $\omega_{k_1}+\cdots+\omega_{k_n} \in \Pi_{m+1}(n)$, are that $\overline{K}=1$ or F_i is a real number in (3.2). Proposition 3.2 shows that what input nonlinear terms of the same nonlinear degree can be opposite and thus provides guidance about how to choose from input nonlinear terms to achieve a proper output spectrum.

Moreover, from (3.10), it can be seen that the magnitude of $Y_n(j\omega)$ is dependent on three terms: $L_n(j\omega)$ and $F(\omega_{k_1})\cdots F(\omega_{k_n}) \sum_{l_1,l_n=1}^{K} c_{0,n}(l_1,\cdots,l_n)(j\omega_{k_1})^{l_1}\cdots(j\omega_{k_n})^{l_n}$, and the function operator $\sum_{\omega_{k_1}+\cdots+\omega_{k_n}=\omega}(\cdot)$. $\sum_{\omega_{k_1}+\cdots+\omega_{k_n}=\omega}(\cdot)$ represents a natural characteristic of the system which cannot be designed as mentioned. The first term $L_n(j\omega)$ represents the influence from the linear part of the system and the second term represents the nonlinear influence from input nonlinearities. These two terms can be designed purposely in practice. Therefore, the results in Proposition 3.2 provide a useful insight into the design of input nonlinearities to achieve a specific output spectrum in practice. The following corollaries are straightforward from Proposition 3.2.

Corollary 3.2 Suppose the conditions in Proposition 3.2 are satisfied for two nonlinear terms but they are not opposite at a frequency when $c_{0,n}(l_1,\cdots,l_n)=c_1$ and $c_{0,n}(l_1',\cdots,l_n')=c_2$, then they must be opposite at this frequency when $c_{0,n}(l_1,\cdots,l_n)=c_1$ and $c_{0,n}(l_1',\cdots,l_n')=-\mathrm{sgn}1(c_2)c_3$ for a proper value of c_3. □

Corollary 3.3 Suppose the conditions in Proposition 3.2 are satisfied for two nonlinear terms with nonzero coefficients $c_{0,n}(l_1,\cdots,l_n)$ and $c_{0,n}(l_1',\cdots,l_n')$. For a proper value of $c_{0,n}(l_1, \cdots, l_n)/c_{0,n}(l_1', \cdots, l_n') > 0$, they are opposite in $\Pi_{m+1}(n)$ if for a real $\Omega > 0$,

$$\operatorname{sgn1}\left(\sum_{\substack{\omega_{k_1} + \cdots + \omega_{k_n} = (n-2m)\cdot\Omega \\ \omega_{k_1}, \cdots, \omega_{k_n} \in \{+\Omega, -\Omega\}}} (\omega_{k_1})^{l_1} \cdots (\omega_{k_n})^{l_n} \right) (-1)^{\left\lfloor \frac{\left| l_1 - l_1' + \cdots + l_n - l_n' \right|}{2} \right\rfloor}$$

$$= -\operatorname{sgn1}\left(\sum_{\substack{\omega_{k_1} + \cdots + \omega_{k_n} = (n-2m)\cdot\Omega \\ \omega_{k_1}, \cdots, \omega_{k_n} \in \{+\Omega, -\Omega\}}} (\omega_{k_1})^{l_1'} \cdots (\omega_{k_n})^{l_n'} \right) \tag{3.11}$$

Proof See the proof in Sect. 3.6. □

Note that similar results can be extended for the other kinds of nonlinearities.

Example 3.2 Consider a simple nonlinear system as follows

$$y = -0.01\dot{y} + au^5 + bu^3\dot{u}^2$$

The input is a multi-tone signal u(t)=0.8sin(7t)+0.8sin(8t)+ sin(9t), which can be written as u(t)=0.8cos(7t−90°)+ 0.8cos(8t−90°)+cos(9t−90°). Therefore, F$(\omega_{\pm 1}) = \mp 0.8j, F(\omega_{\pm ?2}) = \mp 0.8$j and F$(\omega_{\pm ?3}) = \mp$j. Firstly, it can be verified that, $\operatorname{sgn}(F(\omega_{k_1})\cdots F(\omega_{k_5})))$ is constant in each period $\Pi_i(5)$ for i=1,..,6. Secondly, the involved coefficients are $c_{0,5}(0,0,0,0,0)=a$ and $c_{0,5}(0,0,0,1,1)=b$ which satisfy that $\sum_{i=1}^{n} l_i (= 0)$ and $\sum_{i=1}^{n} l_i' (= 2)$ are both even integers. Therefore, the two nonlinear terms au^5 and $bu^3\dot{u}^2$ can be opposite in each frequency generation period by properly designing a and b. These can be verified by simulations (See Figs. 3.4, 3.5, 3.6, and 3.7). It can be seen that, by choosing carefully the values of a and b, the effects on output frequency response from the two nonlinear terms can be counteracted with each other in each frequency period. It shall be noted that when the output spectrum is suppressed in one period, it may be enhanced in the other.

For a specific frequency period and under specific values of a and b, Corollary 3.2 and Corollary 3.3 can be used to check whether it is suppressed or not. The frequencies in the second frequency generation period $\Pi_2(5) = \{19, 20, \ldots, 29\}$ is taken as an example to illustrate this for the case a=1.3, b=0.1. For the nonlinear term au^5, (3.11) can be checked as (for n=5)

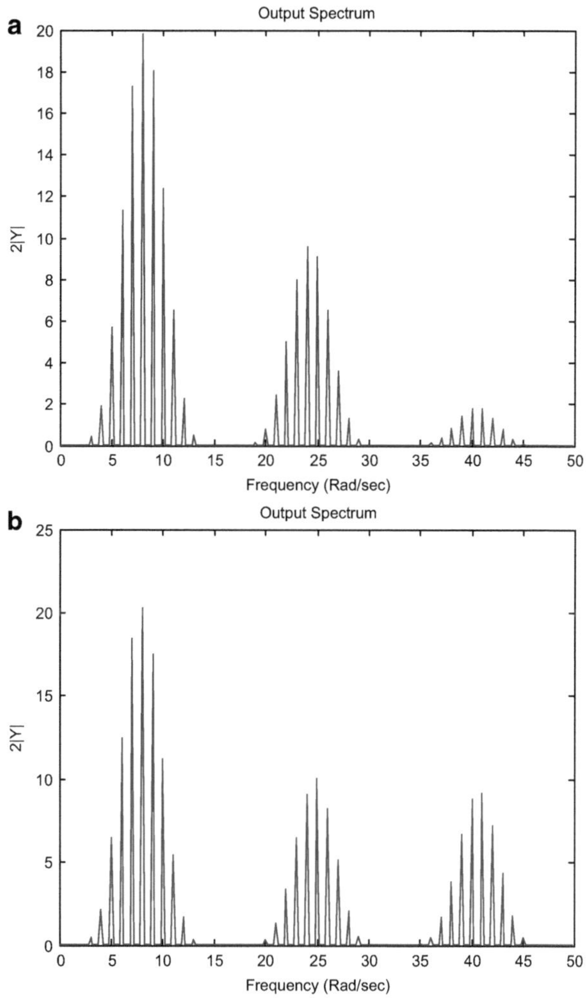

Fig. 3.4 Output spectrum when (**a**) a = 1.3, b = 0 and (**b**) a = 0, b = 0.1 (Jing et al. 2010)

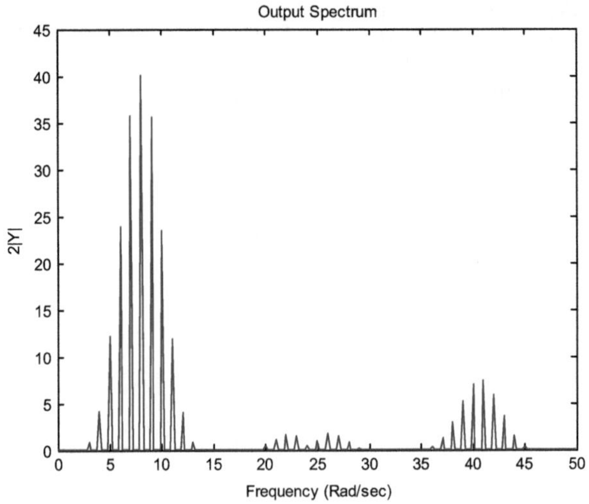

Fig. 3.5 Output spectrum when a = 1.3, b = 0.1(Jing et al. 2010)

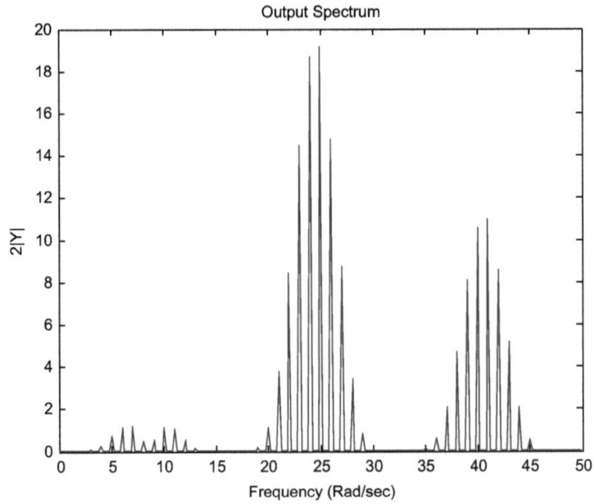

Fig. 3.6 Output spectrum when a = −1.3, b = 0.1 (Jing et al. 2010)

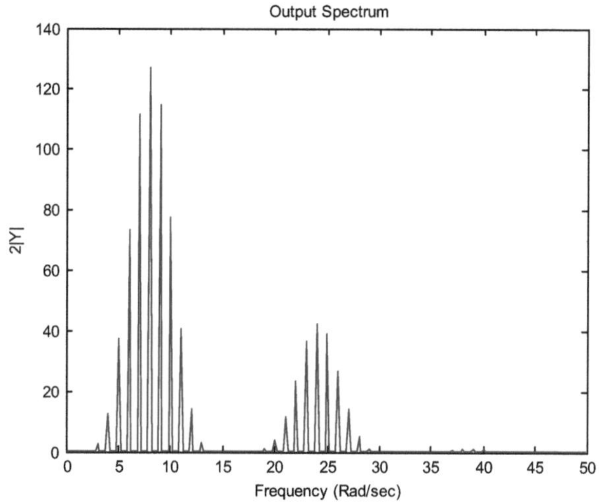

Fig. 3.7 Output spectrum when a = 7, b = 0.1 (Jing et al. 2010)

$$\mathrm{sgn1}\left(\sum_{\substack{\omega_{k_1}+\cdots+\omega_{k_n}=(n-2m)\cdot\Omega\\ \omega_{k_1},\cdots,\omega_{k_n}\in\{+\Omega,-\Omega\}}}(\omega_{k_1})^{l_1}\cdots(\omega_{k_n})^{l_n}\right)(-1)^{\left\lfloor\frac{\left|l_1-l_1'+\cdots+l_n-l_n'\right|}{2}\right\rfloor}$$

$$=\mathrm{sgn1}\left(\sum_{\substack{\omega_{k_1}+\cdots+\omega_{k_5}=(5-2)\cdot\Omega\\ \omega_{k_1},\cdots,\omega_{k_5}\in\{+\Omega,-\Omega\}}}(\omega_{k_1})^{0}\cdots(\omega_{k_5})^{0}\right)(-1)^{1}$$

$$=\mathrm{sgn1}\left(\sum_{\substack{\omega_{k_1}+\cdots+\omega_{k_5}=3\cdot\Omega\\ \omega_{k_1},\cdots,\omega_{k_5}\in\{+\Omega,-\Omega\}}}1\right)(-1)^{1}=-1$$

For the nonlinear term $bu^3\dot{u}^2$,

$$-\mathrm{sgn1}\left(\sum_{\substack{\omega_{k_1}+\cdots+\omega_{k_n}=(n-2m)\cdot\Omega\\ \omega_{k_1},\cdots,\omega_{k_n}\in\{+\Omega,\,-\Omega\}}}(\omega_{k_1})^{l_1}\cdots(\omega_{k_n})^{l_n}\right)$$

$$=-\mathrm{sgn1}\left(\sum_{\substack{\omega_{k_1}+\cdots+\omega_{k_5}=(5-2)\cdot\Omega\\ \omega_{k_1},\cdots,\omega_{k_5}\in\{+\Omega,\,-\Omega\}}}(\omega_{k_4})^{1}(\omega_{k_5})^{1}\right)$$

$$=-\mathrm{sgn1}\left(\sum_{\substack{\omega_{k_1}+\cdots+\omega_{k_5}=3\Omega\\ \omega_{k_1},\cdots,\omega_{k_5}\in\{+\Omega,\,-\Omega\}}}(\omega_{k_1}\omega_{k_2})\right)$$

Note that there are five combinations for $\omega_{k_1}+\cdots+\omega_{k_5}=3\Omega$, $\omega_{k_1},\cdots,\omega_{k_5}\in\{+\Omega,\,-\Omega\}$, i.e., $-\Omega,\Omega,\Omega,\Omega,\Omega$; $\Omega,-\Omega,\Omega,\Omega,\Omega$; $\Omega,\Omega,-\Omega,\Omega,\Omega$; $\Omega,\Omega,\Omega,-\Omega,\Omega$; $\Omega,\Omega,\Omega,\Omega,-\Omega$; Therefore

$$-\mathrm{sgn1}\left(\sum_{\substack{\omega_{k_1}+\cdots+\omega_{k_5}=3\Omega\\ \omega_{k_1},\cdots,\omega_{k_5}\in\{+\Omega,\,-\Omega\}}}(\omega_{k_1}\omega_{k_2})\right)=-\mathrm{sgn1}\left(\Omega^2\right)=-1$$

Equation (3.11) is satisfied.

From Fig. 3.5 it can be seen that, the counteraction between the effects from the two input nonlinear terms when a=1.3, b=0.1 results in the suppression of the output spectrum in the second period, but the enhancement for the first period and little suppression for the third period, compared with the output spectrum under single nonlinear term au^5. Similar results can be seen in Figs. 3.6 and 3.7 under different parameter values.

Moreover, it is obvious that given system model and input, the system output spectrum can be analytically computed from (3.1–3.3). On the other hand, given system model in the multi-tone input case, the input function can be obtained from the output spectrum at a specific frequency generation period for example $\Pi_1(n)$. Because each output frequency in $\Pi_i(n)$ can be explicitly determined, thus a series of equations can be obtained in terms of $F(\omega_{k_1})\cdots F(\omega_{k_n})$, and then $F(\omega_1),\cdots F(\omega_n)$ can be solved. That is, the original input signal can be recovered from the received signal in a specific frequency generation period. This is another interesting property based on the periodicity and is worth further investigating.

3.5 Conclusion

The super-harmonics and inter-modulations in output frequencies of nonlinear systems are theoretically studied and demonstrated, and some interesting properties of system output frequencies are revealed explicitly in a general and analytical form. These properties provide a useful insight into the nonlinear behavior in the output spectrum of Volterra-type nonlinear systems such as the periodicity and opposite property in the frequency domain. Especially, the interaction mechanism among different output harmonics in system output spectrum incurred by different input nonlinearities is demonstrated using the new perspective revealed by the periodicity property. There are few results having been reported in this topic. These results can be used for the design of nonlinear systems or nonlinear FIR filters to achieve a special output spectrum by taking advantage of nonlinearities, and thus provide an important and significant guidance to the analysis and design of nonlinear systems in the frequency domain. Further study will focus on these theoretical and practical issues. For example, given a specific frequency interval, how to achieve a suppressed output spectrum by using the opposite property; how to extend the current results for only input nonlinearities to more general complicated cases, and so on.

Moreover, output frequency characteristics can also be studied with a parametric characteristic analysis, which can indicate how different nonlinear parameters affect output frequencies to certain extent. This will be studied in the following chapters.

3.6 Proofs

A. Proof of Proposition 3.1

Consider multi-tone input case only. Then the same results can be extended to the general input case readily. From (3.4b), it can be seen that the frequencies in W_{n} are determined by $\omega = \omega_{k_1} + \omega_{k_2} + \cdots + \omega_{k_n}$. When all the frequency variable $\omega_{k_i} \in \overline{V}$ (for $\mathrm{i} = 1, \ldots, n$) are positive, *i.e.*, $\omega_{k_i} > 0$ for $i = 1, \ldots, n$, the computed frequencies are obviously those in $\Pi_1(n)$. Then $\Pi_2(n)$ can be computed by setting that there is only one frequency variable (for example ω_{k_1}) to be negative and all the other frequency variables to be positive, i.e.,

$$\Pi_2(n) = \left\{ \omega = \omega_{k_1} + \omega_{k_2} + \cdots + \omega_{k_n} \middle| \begin{array}{l} \omega_{k_i} \in \overline{V}, \omega_{k_1} < 0, \omega_{k_i} > 0, \\ i = 2, 3, \ldots, n \end{array} \right\}$$

Similarly, $\Pi_3(n)$ can be computed by setting that there is only two frequency variables (for example ω_{k_1} and ω_{k_2}) are negative and all the other frequency variables are positive, i.e.,

$$\Pi_3(n) = \left\{ \omega = \omega_{k_1} + \omega_{k_2} + \cdots + \omega_{k_n} \middle| \begin{array}{l} \omega_{k_i} \in \overline{V}, \omega_{k_1} < 0, \omega_{k_2} < 0, \\ \omega_{k_i} > 0, i = 2, 3, \ldots, n \end{array} \right\}$$

Proceed with this process until that all the frequency variables are negative. There are totally n negative frequencies (or frequency variables) in $\overline{V}$, thus it is obvious that the repetitive number of the computation process above is n.

From (3.5c), it can be obtained that

$$\begin{aligned} \max(\Pi_i(n)) &= -(i-1)\min(V) + (n-i+1)\max(V) \text{ and } \min(\Pi_i(n)) \\ &= -(i-1)\max(V) + (n-i+1)\min(V) \end{aligned}$$

Therefore,

$$\begin{aligned} \max(\Pi_{i-1}(n)) - \max(\Pi_i(n)) &= -(i-2)\min(V) + (n-i+2)\max(V) \\ &\quad + (i-1)\min(V) - (n-i+1)\max(V) \\ &= \min(V) + \max(V) = T \end{aligned}$$

and

$$\begin{aligned} \min(\Pi_{i-1}(n)) - \min(\Pi_i(n)) &= -(i-2)\max(V) + (n-i+2)\min(V) \\ &\quad + (i-1)\max(V) - (n-i+1)\min(V) \\ &= \max(V) + \min(V) = T \end{aligned}$$

Moreover, the specific width that the frequencies span in $\Pi_i(n)$ is

$$\begin{aligned} \Delta(n) &= \max(\Pi_i(n)) - \min(\Pi_i(n)) \\ &= -(i-1)\min(V) + (n-i+1)\max(V) \\ &\quad + (i-1)\max(V) - (n-i+1)\min(V) \\ &= n \cdot (\max(V) - \min(V)) \end{aligned}$$

which is a constant.

Now consider the case that the input is the multi-tone (3.2) with $\omega_{i+1} - \omega_i = const > 0$ for i $= 1, \ldots, \overline{K} - 1$. In this case, it can be shown that the difference between any two successive frequencies in $\Pi_i(n)$ is *const*. For example, for any $\Omega \in \Pi_i(n)$, let $\Omega = \omega_{k_1} + \omega_{k_2} + \cdots + \omega_{k_n}$. Without specialty, suppose $\min(V) \leq \omega_{k_1} < \max(V)$, then the smallest frequency that is larger than Ω must be Ω' which can be computed as $\omega'_{k_1} + \omega_{k_2} + \cdots + \omega_{k_n}$ where $\omega'_{k_1} = \omega_{k_1} + const$. Hence, there exists an integer number $0 \leq \alpha \leq \Delta(n)/const$ such that $\Omega = \min(\Pi_i(n)) + \alpha \cdot const$ for $\forall \Omega \in \Pi_i(n)$. Considering $\forall \Omega \in \Pi_i(n)$ with $\Omega = \min(\Pi_i(n)) + \alpha\Delta(n)$, it can be derived that

$$\begin{aligned}\Omega + T &= \min(\Pi_i(n)) + \alpha\Delta(n) + T = -(i-1)\max(V) + (n-i+1)\min(V)\\ &\quad + \alpha\Delta(n) + \max(V) + \min(V)\\ &= -(i-2)\cdot\max(V) + (n-i+2)\min(V) + \alpha\Delta(n)\\ &= \min(\Pi_{i-1}(n)) + \alpha\Delta(n) \in \Pi_{i-1}(n)\end{aligned}$$

Therefore, for $\forall\Omega\in\Pi_i(n)$ there exists a frequency $\Omega'\in\Pi_{i-1}(n)$ such that $\Omega'=\Omega+T$ and vice versa. This gives (3.6e). When $\omega_1=a$, $\omega_{\overline{K}}=b$ and $\overline{K}\to\infty$ such that $\omega_{i+1}-\omega_i = const \to 0$ for i = 1, …, $\overline{K}-1$, it will become the case of a general input $U(j\omega)$ defined in [a,b]. The proposition is proved. □

B. Proof of Proposition 3.2

When the multi-tone input satisfies that $\mathrm{sgn}(F(\omega_{k_1})\cdots F(\omega_{k_n}))$ is constant with respect to all the combinations of $\omega_{k_1},\cdots,\omega_{k_n}\in\{\pm\omega_1,\cdots,\pm\omega_{\overline{K}}\}$ satisfying $\omega_{k_1}+\cdots+\omega_{k_n}\in\Pi_{m+1}(n)$ (for example $\overline{K}=1$ or F_i is a real number in (3.2)), then the opposite condition according to Definition 3.1 is that, there exist two nonzero real number c_1 and c_2 such that at a given frequency $\Omega'\in\Pi_{m+1}(n)$,

$$\sum_{\omega_{k_1}+\cdots+\omega_{k_n}=\Omega'}\left(c_1(j\omega_{k_1})^{l_1}\cdots(j\omega_{k_n})^{l_n} + c_2(j\omega_{k_1})^{l'_1}\cdots(j\omega_{k_n})^{l'_n}\right)=0 \tag{B0}$$

(B0) can also be written as

$$\sum_{\omega_{k_1}+\cdots+\omega_{k_n}=\Omega'}\left(\frac{c_1}{c_2}(j)^{\sum_{i=1}^{n}\left(l_i-l'_i\right)}(\omega_{k_1})^{l_1}\cdots(\omega_{k_n})^{l_n}\right) = -\sum_{\omega_{k_1}+\cdots+\omega_{k_n}=\Omega'}\left((\omega_{k_1})^{l'_1}\cdots(\omega_{k_n})^{l'_n}\right) \tag{B1}$$

Note that given two specific nonlinear parameters $c_{0,n}(l_1,\cdots,l_n)$ and $c_{0,n}(l'_1,\cdots,l'_n)$, it can be seen that $(\omega_{k_1})^{l_1}\cdots(\omega_{k_n})^{l_n}$ and $(\omega_{k_1})^{l'_1}\cdots(\omega_{k_n})^{l'_n}$ are both nonzero for $\omega_{k_1},\cdots,\omega_{k_n}\in\{\pm\omega_1,\cdots,\pm\omega_{\overline{K}}\}$ satisfying $\omega_{k_1}+\cdots+\omega_{k_n}\in\Pi_{m+1}(n)$, and the right side of (B1) is real, therefore

$$(j)^{\sum_{i=1}^{n}\left(l_i-l'_i\right)}\ \text{must be nonzero real} \tag{B2}$$

On the other hand, if (B2) holds, whatever the value of $-\sum_{\omega_{k_1}+\cdots+\omega_{k_n}=\Omega'}\left((\omega_{k_1})^{l'_1}\cdots(\omega_{k_n})^{l'_n}\right)$ is, there always exist two real number c_1 and c_2

such that (B1) holds. Hence, the opposite condition above now is equivalent to be that (B2) holds. That (B2) holds is equivalent to be that $\sum_{i=1}^{n}\left(l_i - l_i'\right)$ is an even integer. This is further equivalent to be that $\sum_{i=1}^{n} l_i$ and $\sum_{i=1}^{n} l_i'$ are both odd integers or even integers simultaneously. □

C. Proof of Corollary 3.3

Noting that $\sum_{i=1}^{n}\left(l_i - l_i'\right)$ is an even integer, then from (3.11), it can be derived that

$$\operatorname{sgn}\left(\sum_{\omega_{k_1}+\cdots+\omega_{k_n}=\Omega'}\left(c_1 (j\omega_{k_1})^{l_1}\cdots(j\omega_{k_n})^{l_n}\right)\right) = -\operatorname{sgn}\left(\sum_{\omega_{k_1}+\cdots+\omega_{k_n}=\Omega'}\left(c_2 (j\omega_{k_1})^{l_1'}\cdots(j\omega_{k_n})^{l_n'}\right)\right) \tag{B3}$$

where $\omega_{k_1},\cdots,\omega_{k_n} \in \{+\Omega, -\Omega\}$ and $\Omega' = (n-2m)\Omega$ for any $\Omega > 0$. (B3) implies that there exist two nonzero real number c_1 and c_2 satisfying $c_1/c_2 > 0$ such that at a given frequency $\Omega' \in \Pi_{m+1}(n) = \{(n-2m)\Omega\}$, (B0) holds. Note that $\Pi_{m+1}(n) = \{(n-2m)\Omega\}$ is the case that the input is a single tone function *i.e.*, $\overline{K} = 1$. Hence, (3.11) implies that (B0) holds for $\overline{K} = 1$. To finish the proof, it needs to prove that, if (3.11) holds, then (B0) holds for all $\Omega' \in \Pi_{m+1}(n)_{\overline{K}>1}$ (note that when $\overline{K} > 1$ there are more than one elements in $\Pi_{m+1}(n)_{\overline{K}>1}$). By using the mathematical induction and combination theory, it can be proved that

$$\operatorname{sgn}\left(\sum_{\substack{\omega_{k_1}+\cdots+\omega_{k_n}=\Omega' \\ \Omega' \in \Pi_{m+1}(n)_{\dot{K}=1}}}\left(c_1 (j\omega_{k_1})^{l_1}\cdots(j\omega_{k_n})^{l_n}\right)\right) = \operatorname{sgn}\left(\sum_{\substack{\omega_{k_1}+\cdots+\omega_{k_n}=\Omega' \\ \Omega' \in \Pi_{m+1}(n)_{\dot{K}>1}}}\left(c_2 (j\omega_{k_1})^{l_1}\cdots(j\omega_{k_n})^{l_n}\right)\right)$$

For paper limitation, this is omitted. Therefore, if (3.11) holds, (B0) holds for all $\Omega' \in \Pi_{m+1}(n)_{\overline{K}>1}$. □

Chapter 4
Parametric Characteristic Analysis

4.1 Separable Functions

Definition 4.1 A function $h(s;\, x)$ is said to be separable with respect to parameter x if it can be written as $h(s; x) = g(x) \cdot f_1(s) + f_0(s)$, where $f_i(.)$ for i=0,1 are functions of variable s but independent of the parameter x. □

A function $h(s;\, x)$ satisfying Definition 4.1 is referred to as x-separable function or simply separable function, where x is referred to as the parameter of interest which may be a parameter to be designed for a system, and s represents other parameters or variables, which may be a reference variable (or independent variable) of a system such as time or frequency.

Remark 4.1 In the definition of an x-separable function $h(s;\, x)$, x may be a vector including all the separable parameters of interest, and s denotes not only the independent variables of $h(.)$, but also may include all the other un-separable and uninterested parameters in $h(.)$. The parameter x and s are real or complex valued, but the detailed properties of the function $h(.)$ and its parameters are not necessarily considered here. Note also that in Definition 4.1, $f_0(s)$ and $f_1(s)$ are invariant with respect to x and $g(x)$. Thus $h(s;\, x)$ can be regarded as a pure function of x for any specific s. In this case, if $g(x)$ is known, and additionally the values of $h(s;\, x)$ and $g(x)$ under some different values of x, for example x_1 and x_2, can be obtained by certain methods (simulations or experimental tests), then the values of $f_0(s)$ and $f_1(s)$ can be achieved by the Least Square method, i.e.,

X. Jing, Z. Lang, *Frequency Domain Analysis and Design of Nonlinear Systems based on Volterra Series Expansion*, Understanding Complex Systems,
DOI 10.1007/978-3-319-12391-2_4

$$\begin{cases} h(s;x_1) = g(x_1)\cdot f_1(s) + f_0(s) \\ h(s;x_2) = g(x_2)\cdot f_1(s) + f_0(s) \end{cases} \Rightarrow \begin{bmatrix} f_0(s) \\ f_1(s) \end{bmatrix} = \begin{bmatrix} 1 & g(x_1) \\ 1 & g(x_2) \end{bmatrix}^{-1} \begin{bmatrix} h(s;x_1) \\ h(s;x_2) \end{bmatrix} \tag{4.1}$$

Thus the function $h(s; x)$ at a given s can be obtained which is an analytical function of the parameter x. This provides a numerical method to determine the relationship between the parameters of interest and the corresponding function. □

An x-separable function $h(s; x)$ at a given point s is denoted as $h(x)|_s$, or simply as $h(x)_s$.

Consider a parameterized function series

$$H(s;x) = g_1(x)f_1(s) + g_2(x)f_2(s) + \cdots + g_n(x)f_n(s) = G\cdot F^T \tag{4.2}$$

where $n>1$, $f_i(s)$ and $g_i(x)$ for $i=1,\ldots,n$ are all scalar functions, let $F=[f_1(s), f_2(s),\cdots,f_n(s)]$ and $G=[g_1(x), g_2(x),\cdots, g_n(x)]$, x and s are both parameterized vectors including the interested parameters and other parameters, respectively. The series is obviously x-separable, thus $H(x)_s$ is completely determined by the parameters in x or the values of $g_1(x), g_2(x),\cdots, g_n(x)$. Note that at a given point s, the characteristics of the series $H(s; x)$ is completely determined by G, and how the parameters in x are included in $H(s; x)$ is completely demonstrated in G, too. Therefore, the parametric characteristics of the series $H(s; x)$ can be totally revealed by the function vector G. The vector G is referred to as the parametric characteristic vector of the series. If the characteristic vector G is determined, then following the method mentioned in Remark 4.1, the function $H(x)_s$ which shows the analytical relationship between the concerned parameter x and the series is achieved, and consequently the effects on the series from each parameter in x can be studied. The function $H(x)_s$ is referred to as parametric characteristic function of the series $H(s; x)$. Based on the discussions above, the following result can be concluded.

Lemma 4.1 If $H(s; x)$ is a separable function with respect to the parameter x, then there must exist a parametric characteristic vector G and an appropriate function vector F, such that $H(s; x)= G\cdot F^T$, where the elements of G are functions of x and independent of s, and the elements of F are functions of s but independent of x. □

According to the definition and discussion above, it will be seen that the nth-order GFRF of the NDE model in (2.11) and NARX model in (2.10) is separable with respect to any nonlinear parameters of the corresponding models. As mentioned, in order to study the relationship between an interested function $H(s; x)$ and its separable parameters x, the parametric characteristic vector G should be obtained. For a simple parameterized function, it may be easy to obtain parameterized vector G. But for a complicated function series with recursive computations, this is not straightforward. To this aim, and more importantly for the purpose of the parametric characteristic analysis for the nth-order GFRF and output spectrum of Volterra-type nonlinear systems described by (2.10) or (2.11), a novel operator is introduced in the following section for the extraction of any parameters of interest involved in a separable parameterized polynomial function series.

4.2 Coefficient Extractor

Let C_s be a set of parameters which takes values in $\mathbb{C}$, let P_c be a monomial function set defined in C_s, *i.e.*, $P_c = \{c_1^{r_1} c_2^{r_2} \cdots c_I^{r_I} | c_i \in C_s, r_i \in \mathbb{Z}_0, I = |C_s|\}$, where $|C_s|$ is the number of the parameters in C_s, $\mathbb{Z}_+$ denotes all the positive integers. Let W_s be another parameter set similar to C_s but $W_s \cap C_s = \phi$, and let P_f be a function set defined in W_s, *i.e.*, $P_f = \{f(w_1, \cdots, w_I) | w_i \in W_s, I = |W_s|\}$. Let Ξ denote all the finite order function series with coefficients in P_c timing some functions in P_f. A series in Ξ can be written as

$$H_{CF} = s_1 f_1 + s_2 f_2 + \cdots + s_\sigma f_\sigma \in \Xi \tag{4.3}$$

where $s_i \in P_c, f_i \in P_f$ for $i=1, \ldots, \sigma \in \mathbb{Z}_+$, $C=[s_1, s_2, \ldots, s_\sigma]$, and $F=[f_1, f_2, \ldots, f_\sigma]^{\mathrm{T}}$. Obviously, this series is separable with respect to the parameters in C_s and W_s. Define a **Coefficient Extraction** operator $CE : \Xi \rightarrow P_c{}^\sigma$, such that

$$CE(H_{CF}) = [s_1, s_2, \cdots, s_\sigma] = C \in P_c{}^\sigma \tag{4.4}$$

where $P_c^\sigma = \{[s_1, s_2, \cdots, s_\sigma] | s_1, \cdots, s_\sigma \in P_c\}$. This operator has the following properties:

(1) Reduced vectorized sum "$\oplus$".

$$CE(H_{C_1F_1} + H_{C_2F_2}) = CE(H_{C_1F_1}) \oplus CE(H_{C_2F_2}) = C_1 \oplus C_2 = \left[C_1, C_2'\right]$$

and $C_2' = VEC(\overline{C}_2 - \overline{C}_1 \cap \overline{C}_2)$, where $\overline{C}_1 = \{C_1(i) | 1 \leq i \leq |C_1|\}$, $\overline{C}2 = \{C_2(i) | 1 \leq i \leq |C_2|\}$, $VEC(.)$ is a vector consisting of all the elements in set (.). C_2' is a vector including all the elements in C_2 except the same elements as those in C_1.

(2) Reduced Kronecker product "$\otimes$".

$$\begin{aligned} &CE(H_{C_1F_1} \cdot H_{C_2F_2}) = CE(H_{C_1F_1}) \otimes CE(H_{C_2F_2}) \\ &= C_1 \otimes C_2 \triangleq VEC\left\{ c \left| \begin{array}{l} C_3 = [C_1(1)C_2, \cdots, C_1(|C_1|)C_2] \\ c = C_3(i), 1 \leq i \leq |C_3| \end{array} \right. \right\} \end{aligned}$$

which implies that there are no repetitive elements in $C_1 \otimes C_2$.

(3) Invariant.

(i) $CE(\alpha \cdot H_{CF}) = CE(H_{CF}), \forall \alpha \notin C_s$; (ii) $CE(H_{CF_1} + H_{CF_2}) = CE\left(H_{C(F_1+F_2)}\right) = C$.

(4) Unitary. (i) If $\frac{\partial H_{CF}}{\partial c} = 0$ for $\forall c \in C_s$, then $CE(H_{CF}) = 1$; (ii) if $H_{CF}=0$ for $\forall c \in C_s$, then $CE(H_{CF})=0$. When there is a unitary 1 in $CE(H_{CF})$, there is a nonzero constant term in the corresponding series H_{CF} which has no relation with the parameters in C_s.

(5) Inverse. $CE^{-1}(C)=H_{CF}$. This implies any a vector C consisting of the elements in P_c should correspond to at least one series in Ξ.
(6) $CE(H_{C_1F_1}) \approx CE(H_{C_2F_2})$ if the elements of C_1 are the same as those of C_2, where "$\approx$" means equivalence. That is, both series are in fact the same result considering the order of s_if_i in the series has no effect on the value of a series H_{CF}. This further implies that the CE operator is also commutative and associative, for instance, $CE(H_{C_1F_1}+H_{C_2F_2}) = C_1 \oplus C_2 \approx CE(H_{C_2F_2}+H_{C_1F_1}) = C_2 \oplus C_1$. Hence, the results by the CE operator may be different but all may correspond to the same function series and are thus equivalent.
(7) Separable with respect to parameters of interest only. A parameter in a series can only be extracted if the parameter is of interest and the series is separable with respect to this parameter. Thus the operation result is different for different purposes.

Note that from the definition of the CE operator above, all the operations are in terms of the parameters in C_s, and the CE operator sets up a mapping from Ξ to P_c^{σ}. For convenience, let $\underset{(*)}{\otimes}(\cdot)$ and $\underset{(*)}{\oplus}(\cdot)$ denote the multiplication and addition by the reduced Kronecker product "$\otimes$" and vectorized sum "$\oplus$" of the terms in (.) satisfying (*), respectively; and $\overset{k}{\underset{i=1}{\otimes}} C_{p,q} = C_{p,q} \otimes \cdots \otimes C_{p,q}$ can be simply written as $C_{p,q}^k$. For model (2.11), define the ($p+q$)th degree nonlinear parameter vector as

$$C_{p,q} = [c_{p,q}(0,\cdots,0), c_{p,q}(0,\cdots,1),\cdots, c_{p,q}(\underbrace{K,\cdots,K}_{p+q=m})] \tag{4.5}$$

which includes all the nonlinear parameters of the form $c_{p,q}(.)$ in model (2.11). A similar definition for model (2.10) as

$$C_{p,q} = [c_{p,q}(1,\cdots,1), c_{p,q}(1,\cdots,2),\cdots, c_{p,q}(\underbrace{K,\cdots,K}_{p+q=m})] \tag{4.6}$$

Note that $C_{p,q}$ can also be regarded as a set of the ($p+q$)th degree nonlinear parameters of the form $c_{p,q}(.)$. Moreover, if all the elements of CE(H_{CF}) are zero, *i.e.*, CE(H_{CF})=0, then CE(H_{CF}) is also regarded as empty.

The CE operator provides a useful tool for the analysis of the parametric characteristics of separable functions. It can be shown that the nonlinear parametric characteristics of the GFRFs for (2.10) or (2.11) can be obtained by directly substituting the operations "+" and "." by "$\oplus$" and "$\otimes$" in the corresponding recursive algorithms, respectively, and neglecting the corresponding multiplied frequency functions. This is demonstrated by the following example.

Example 4.1 Computation of the parametric characteristics of the second order GFRF of model (2.11). The second order GFRF from (2.19–2.24) is

$$L(n)\cdot H_n(j\omega_1,\cdots,j\omega_n)=\sum_{k_1,k_n=1}^{K} c_{0,n}(k_1,\cdots,k_n)(j\omega_1)^{k_1}\cdots(j\omega_n)^{k_n}$$
$$+\sum_{q=1}^{n-1}\sum_{p=1}^{n-q}\sum_{k_1,k_{p+q}=0}^{K} c_{p,q}\left(k_1,\cdots,k_{p+q}\right)\left(j\omega_{n-q+1}\right)^{k_{p+1}}\cdots(j\omega_n)^{k_{p+q}}H_{n-q,p}\left(j\omega_1,\cdots,j\omega_{n-q}\right)$$
$$+\sum_{p=2}^{n}\sum_{k_1,k_p=0}^{K} c_{p,0}\left(k_1,\cdots,k_p\right)H_{n,p}(j\omega_1,\cdots,j\omega_n) \tag{4.7}$$

for n=2, where $L(2)=-\sum_{k_1=0}^{K} c_{1,0}(k_1)(j\omega_1+j\omega_2)^{k_1}$, $H_{1,1}(j\omega_1)=H_1(j\omega_1)(j\omega_1)^{k_1}$, $H_{2,2}(\cdot)=H_1(j\omega_1)H_{1,1}(j\omega_2)(j\omega_1)^{k_2}$.

Applying the CE operator to (4.7) for nonlinear parameters and using the notation in (4.5), it can be obtained that

$$\begin{aligned}CE(H_2(\cdot))&=CE(L(2)\cdot H_2(\cdot))\\&=C_{0,2}\oplus\left(\mathop{\oplus}_{q=1}^{2-1}\mathop{\oplus}_{p=1}^{2-q}\left(C_{p,q}\otimes CE\left(H_{2-q,p}(\cdot)\right)\right)\right)\oplus\left(\mathop{\oplus}_{p=2}^{2}C_{p,0}\otimes CE\left(H_{2,p}(\cdot)\right)\right)\\&=C_{0,2}\oplus\left(C_{1,1}\otimes CE(H_{1,1}(\cdot))\right)\oplus\left(C_{2,0}\otimes CE(H_{2,2}(\cdot))\right)\end{aligned}$$

Note that $H_1(.)$ has no relationship with nonlinear parameters, from the definition of CE operator, it can be obtained that $CE(H_1(.))=1$. Similarly, it can be obtained that $CE(H_{2,2}(.))=1$. Therefore, the parametric characteristic vector of the second order GFRF is

$$CE(H_2(\cdot)) = C_{0,2}\oplus C_{1,1}\oplus C_{2,0} \tag{4.8}$$

Equation (4.8) shows clearly that nonlinear parameters in $C_{0,2}$, $C_{1,1}$ and $C_{2,0}$ have independent effects on the second order GFRF without interference, and no any other nonlinear parameters have any influence on the second order GFRF. This provides an explicit insight into the relationship between the second order GFRF and nonlinear parameters. For example, if $H_2(.)$ is required to have a special amplitude or phase, only the parameters in $C_{0,2}$, $C_{1,1}$ and $C_{2,0}$ may need to be designed purposely. □

Example 4.1 shows that the CE operator is very effective for the derivation of the parametric characteristic vector of a separable function series about the parameters of interest. It provides a fundamental technique for the study of parametric effects on the involved parameter-separable function series for any systems. In the present study, in most cases, the CE operator will be applied for all the nonlinear parameters in model (2.10) or model (2.11). When the CE operator is applied for a specific nonlinear parameter c, the parametric characteristic of the nth-order GFRF will be denoted by $CE(H_n(.))_c$.

4.3 Case Study: Parametric Characteristics of Output Frequencies

There are three categories of nonlinearities in model (2.10) or (2.11): input nonlinearity with coefficient $c_{0,q}(.)$ ($q>1$), output nonlinearity with coefficient $c0_{p,0}(.)$ ($p>1$), and input output cross nonlinearity with coefficient $c_{p,q}(.)$ ($p+q>1$ and $p>0$) (where p and q are integers). Different category and degree of nonlinearity in a system can bring different output frequencies to the system. How a nonlinear term affects system output frequencies and what the effect is, are very interesting and important topics. However, few results have been reported for this. As an example for application of the parametric characteristic analysis established in this chapter, this section provides some useful results for this topic based on the output frequency properties developed in Chap. 3.

Consider the NDE system in (2.11). What model parameters contribute to a specific order GFRF and how model parameters affect the GFRFs can be revealed by using the parametric characteristic analysis. From (3.1)–(3.3), it can be seen that the nth-order output frequencies W_n are also determined by the nth order GFRF. If the nth order GFRF is zero, then $W_n=[]$. It is known from Chap. 2 that the nth order GFRF is dependent on its parametric characteristics, thus the nth-order output frequencies are also determined by the parametric characteristics of the nth-order GFRF. Therefore, (3.4a,b) can be written as

$$W_n=\left\{\omega=(\omega_1+\omega_2+\cdots+\omega_n)\cdot\left(1-\bar{\delta}(CE(H_n(\omega_1,\cdots,\omega_n)))\right)\middle|\omega_i\in\bar{V},i=1,2,\ldots,n\right\} \tag{4.9a}$$

and

$$W_n=\left\{\omega=(\omega_{k_1}+\omega_{k_2}+\cdots+\omega_{k_n})\cdot\left(1-\bar{\delta}(CE(H_n(\omega_{k_1},\cdots,\omega_{k_n})))\right)\middle|\omega_{k_i}\in\bar{V},i=1,2,\ldots,n\right\} \tag{4.9b}$$

where $\delta(x)=\begin{cases}1 & x=0 \text{ or } 1\\ 0 & \text{else}\end{cases}$. In (4.9a,b), suppose W_n is empty when $\bar{\delta}(CE(H_n(.)))=1$.

Equations (4.9a,b) demonstrate the parametric characteristics of the output frequencies for Volterra-type nonlinear systems described by (2.10) and (2.11), by which the effect on the system output frequencies from different nonlinearities can be studied. Since negative output frequencies are symmetrical with positive output frequencies with respect to zero (Property 3.1b), thus for convenience only non-negative output frequencies are considered in what follows.

Property 4.1 Regarding nonlinearities of odd and even degrees,

(a) when there are no nonlinearities of even degrees, the output frequencies generated by the nonlinearities with odd degrees happen at central frequencies $(2l+1)T/2$ for $l=0,1,2,\ldots$ with certain frequency span, where T is the frequency generation period (Chap. 3);
(b) when there are only input nonlinearities of even degrees, the output frequencies happen at central frequencies $l \cdot T$ for $l=0,1,2,\ldots$ with certain frequency span;
(c) in other cases, the output frequencies happen at central frequencies $l{\cdot}T/2$ for $l=0,1,2,\ldots$ with certain frequency span.

The frequency span is $\Delta(n)$ corresponding to the nth order output frequencies if applicable.

Proof See Sect. 4.5 for the proof. □

Property 4.1 shows that odd degrees of nonlinearities bring quite different output frequencies to the system from those brought by even degrees of nonlinearities.

Property 4.2 Regarding different categories of nonlinearities,

(a) when there are only input nonlinearities of largest nonlinear degree n, the non-negative output frequencies are in the closed set $[0, n{\cdot}\max(V)]$;
(b) in other cases, the output frequencies span to infinity.

Proof (a) From the GFRFs in Chap. 2 (and the corresponding parametric characteristics to be further discussed in Chap. 5), only the GFRFs of orders equal to the nonlinear degrees of the non-zero input nonlinearities are not zero since there are no other kinds of nonlinearities in the system. Thus the largest order of non-zero GFRFs is n. The conclusion is therefore straightforward from Property 3.1c. (b) If there are other kinds of nonlinearities, the largest order of nonzero GFRFs will be infinite, because for any parameter $c_{p,q}(.)$ with $p>0$ and $p+q>1$, it can form a monomial with any high nonlinear degree ($c_{p,q}(.)^{n}$) and thus contribute to any high order GFRF (this will be more clear from the parametric characteristics of the GFRFs in Chap. 5). Thus the output frequencies can span to infinity. This completes the proof. □

The input nonlinearities of a finite degree can independently produce output frequencies in a finite frequency band.

Property 4.3 Regarding different categories and degrees of nonlinearities,

(a) when there are only input nonlinearities, a nonlinear term of degree n can only produce output frequencies W_n, and there are no crossing effect on output frequencies between different degrees of input nonlinearities;
(b) in other cases, a nonlinear term of degree n contributes to not only output frequencies W_n but also some high order output frequencies W_m for $m > n$ due to crossing effect with other nonlinearities.

Proof (a) Considering a nonlinear coefficient $c_{0,n}(.)$, it can be seen from the GFRFs in Chap. 2 that, only $CE(H_n(.))$ is not empty, if all the other degree and type of nonlinear parameters are zero. That is, $c_{0,n}(.)$ only contributes to $H_n(.)$ in this case. If there are other input nonlinearities, it can be known from Proposition 3.1 in Chap. 3 that only nonlinear parameters from input nonlinearities cannot form an effective monomial which is an element of any order GFRF. That is there are no crossing effects between different degrees of input nonlinearities. (b) When there are output or input-output cross nonlinearities, it can be seen from the GFRFs in Chap. 2 (and the corresponding parametric characteristics to be further discussed in Chap. 5) that there are crossing effects between different nonlinearities, and the nonlinear degree of any effective monomial (e.g. $c_{1,q}(.)c_{0,q}(.)^n$ ($q>1$)) formed by the coefficients from the crossing nonlinearities can be infinity. Thus a nonlinear parameter of degree n, for example $c_{0,n}(.)$, has contribution not only to $H_n(.)$, but also to some higher order GFRFs, for example $c_{1,n}(.)c_{0,n}(.)^z$ is an element of CE ($H_m(.)$) where $m=z\cdot n+n+1-z$. This completes the proof. □

From Property 4.3, the crossing effect usually happens easily between the output nonlinearities and the input-output cross nonlinearities.

Properties 4.1–4.3 provide some novel and interesting results about the output frequencies for nonlinear systems when the effects from different nonlinearities are considered, based on the GFRFs in Chap. 2 (and the corresponding parametric characteristics to be further discussed in Chap. 5). Property 4.1 shows that odd degrees of nonlinearities have quite different effect on system output frequencies from even degrees of nonlinearities. Especially, it is shown from the properties above that input nonlinearities have special effect on system output frequencies compared with the other categories of nonlinearities. That is, input nonlinearities can move the input frequencies to higher frequency bands without interference between different frequency generation periods. These properties may have significance in design of nonlinear systems for some special purposes in practices. For example, some proper input nonlinearities can be used to design a nonlinear filter such that input frequencies are moved to a place of higher frequency or lower frequency as discussed in Billings and Lang (2002). The results in this section have also significance in modelling and identification of nonlinear systems. For example, if a nonlinear system has only output frequencies which are odd multiples of the input frequency when subjected to a sinusoidal input, the system may have only nonlinearities of odd degree according to Property 4.1. Obviously, the results in this section provide a useful guidance to the structure determination and parameter selection for the design of novel nonlinear filters and also for system modelling or identification.

Example 4.2 Consider a simple nonlinear system as follows

$$y = -0.01\dot{y} + au^5 - by^3 - cy^2$$

The input is a multi-tone function $u(t)=\sin(6t)+\sin(7t)+\sin(8t)$. The output spectra under different parameter values are given in Figs. 4.1, 4.2, and 4.3, which demonstrate the results in Properties 4.1–4.3. For the input nonlinearity, the readers can also refer to Figs. 3.1–3.7.

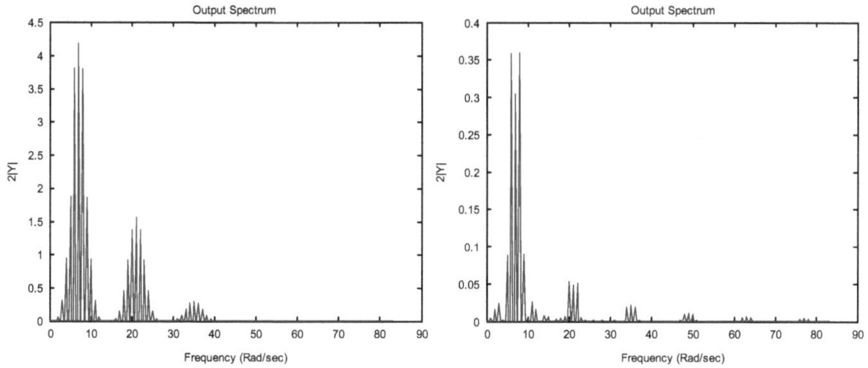

Fig. 4.1 Output frequencies when $a = 0.1$, $b = 0$, $c = 0$ (*left*) and $a = 0$, $b = 5$, $c = 0$ (*right*)

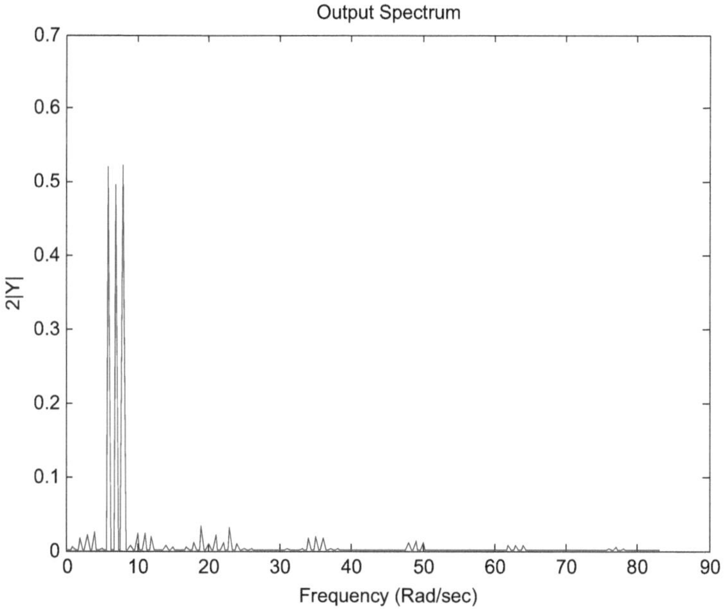

Fig. 4.2 Output frequencies when $a = 0.1$, $b = 5$, $c = 0$

When there are only odd nonlinearities, the output frequencies happen at around central frequencies 7*(2k+1). When there are even nonlinearities, the output frequencies appear at around central frequencies 7*k. The input nonlinearities only produce independently the output frequencies within a finite frequency band. The periodicity of the output frequencies can also be seen clearly from these figures.

Especially, it is worthy pointing out from Figs. 3.1, 3.2 and 4.1 that there can be no coupling effects between proper chosen input nonlinearities as mentioned before, which cannot be realized by the other categories of nonlinearities.

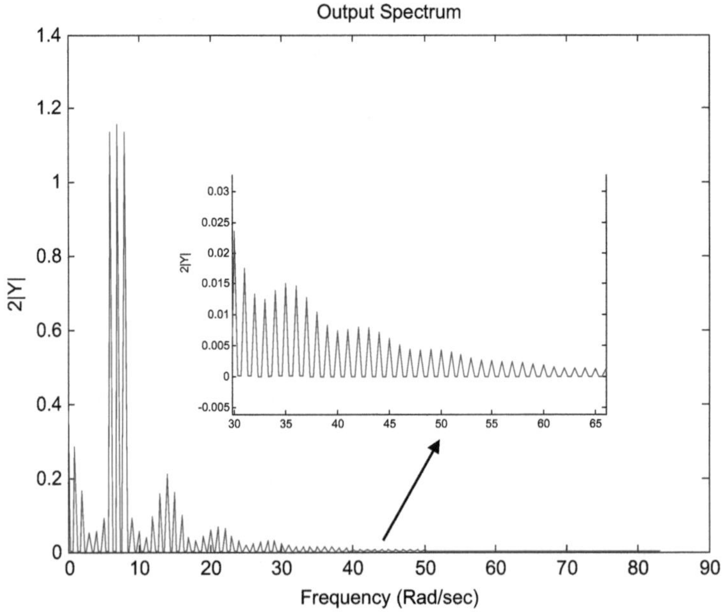

Fig. 4.3 Output frequencies when $a = 0$, $b = 0$, $c = 0.09$

Thus the input frequencies can be moved to higher frequency periodically without interference between different periods and then decoded by using some methods. This property may have significance when a system is designed to achieve a special output spectrum at a desired frequency band in practices by using nonlinearities.

4.4 Conclusions

The parametric characteristic analysis given in this chapter is to reveal how the parameters of interest in a separable parameterized function series or polynomial affect the function series or polynomial and what the possible effects are. This can provide a novel and convenient approach to investigate nonlinear effects incurred by different type and degree of nonlinearities in the frequency response functions of nonlinear systems. Using this method, the GFRF and nonlinear output spectrum can all be studied in a parametric way and eventually formulated into a more practical form for nonlinear system analysis, design and optimization in the frequency domain.

Importantly, the CE operator provides an important and fundamental technique for this parametric characteristic analysis method. As shown in Sect. 4.3, the parametric characteristic analysis based on the CE operator can allow a convenient way to analyze the output frequency characteristics in terms of different nonlinear terms. More interesting and useful results will be developed and demonstrated in the following chapters for the parametric characteristics of the GFRFs using this novel CE operator.

4.5 Proof of Property 4.1

The proof needs the parametric characteristic result in Chap. 5, which are directly cited here.

(a) According to the GFRFs in Chap. 2 (and the corresponding parametric characteristics to be further discussed in Chap. 5), the elements of $CE(H_n(.))$ must be monomial functions of the coefficients of the nonlinear terms, i.e., $c_{p_1,q_1}(\cdot)\cdots c_{p_L,q_L}(\cdot)$ for some $L \geq 1$. Note that there are only nonlinearities of odd degrees, i.e., 2k+1 (k=0,1,2,...), thus the nonlinear degree of any monomial in this case is (Proposition 5.1 in Chap. 5) $n = \sum_{i=1}^{L}(p_i + q_i) - L + 1 = \sum_{i=1}^{L}(2k_i + 1) - L + 1 = 2\sum_{i=1}^{L}k_i + 1$. Clearly, n is still an odd number. That is the nonlinearities in the system of this case can only contribute to odd order GFRFs. Thus all the even order GFRFs are zero, i.e., $CE(H_n(.))=0$ for n is even. Therefore, W_n may not be empty only when n is odd, otherwise it is empty.

Suppose n is an odd integer and $CE(H_n(.))\neq 0$ and 1. That is, there are nonzero elements in $CE(H_n(.))$ and all the elements in $CE(H_n(.))$ consist of the coefficients of some nonlinear terms of the studied case. According to Proposition 3.1, the first period in W_n must be $\Pi_1(n) \subseteq [n \cdot \min(V), n \cdot \max(V)]$, whose central point is obviously $n \cdot T/2$ and of which the frequency span is $\Delta(n)$. Also from Proposition 7.1, the kth period in W_n must be $\Pi_k(n) \subseteq [n \cdot \min(V) - (k-1)T, n \cdot \max(V) - (k-1)T]$, whose central point is obviously $n \cdot T/2-(k-1)T=(n-2(k-1))T/2$ and of which the frequency span is still $\Delta(n)$. Note that n−2(k−1) is an odd integer for k=1,2,.... The first point of the property is proved.

(b) Consider the case that there are only input nonlinearities of even degrees. In this case, it can be verified from the parametric characteristics in Chap. 5 that only the GFRFs of orders equal to the nonlinear degrees of the non-zero input nonlinearities are not zero. That is, only some GFRFs of even orders are not zero. Suppose n is an even integer and $CE(H_n(.))\neq 0$ and 1. According to Proposition 3.1, the kth period in W_n must be $\Pi_k(n) \subseteq [n \cdot \min(V) - (k-1)T, n \cdot \max(V) - (k-1)T]$, whose central point is obviously $n \cdot T/2-(k-1)T=(n-2(k-1))T/2$ and of which the frequency span is $\Delta(n)$. Note that n−2(k−1) is an even integer for k=1,2,.... This second point of the property is proved.

(c) The conclusion is straightforward since there are non-zero GFRFs of even and odd orders. This completes the proof. □

Chapter 5
The Parametric Characteristics of the GFRFs and the Parametric Characteristics Based Analysis

5.1 The GFRFs and Notations

The concept of the GFRFs provides a basis for the study of nonlinear systems in the frequency domain. For a specific parametric model of nonlinear systems such as NARX, NDE, Block-oriented models, the GFRFs can be derived with the probing methods as discussed in Chap. 2. For convenience of discussions, the computation of the nth-order GFRF for the NDE model (2.11) is given here:

$$
\begin{aligned}
L_n(j\omega_1+\cdots+j\omega_n)\cdot H_n(j\omega_1,\cdots,j\omega_n) &= \sum_{k_1,k_n=1}^{K} c_{0,n}(k_1,\cdots,k_n)(j\omega_1)^{k_1}\cdots(j\omega_n)^{k_n} \\
&+\sum_{q=1}^{n-1}\sum_{p=1}^{n-q}\sum_{k_1,k_{p+q}=0}^{K} c_{p,q}(k_1,\cdots,k_{p+q})\left(\prod_{i=1}^{q}(j\omega_{n-q+i})^{k_{p+i}}\right)H_{n-q,p}(j\omega_1,\cdots,j\omega_{n-q}) \\
&+\sum_{p=2}^{n}\sum_{k_1,k_p=0}^{K} c_{p,0}(k_1,\cdots,k_p)H_{n,p}(j\omega_1,\cdots,j\omega_n)
\end{aligned}
\tag{5.1}
$$

$$
H_{n,p}(\cdot)=\sum_{i=1}^{n-p+1} H_i(j\omega_1,\cdots,j\omega_i)H_{n-i,p-1}(j\omega_{i+1},\cdots,j\omega_n)(j\omega_1+\cdots+j\omega_i)^{k_p} \tag{5.2}
$$

X. Jing, Z. Lang, *Frequency Domain Analysis and Design of Nonlinear Systems based on Volterra Series Expansion*, Understanding Complex Systems,
DOI 10.1007/978-3-319-12391-2_5

$$H_{n,1}(j\omega_1,\cdots,j\omega_n) = H_n(j\omega_1,\cdots,j\omega_n)(j\omega_1+\cdots+j\omega_n)^{k_1} \tag{5.3}$$

where

$$L_n(j\omega_1+\cdots+j\omega_n) = -\sum_{k_1=0}^{K} c_{1,0}(k_1)(j\omega_1+\cdots+j\omega_n)^{k_1} \tag{5.4}$$

Moreover, $H_{n,p}(j\omega_1,\cdots,j\omega_n)$ in (3.2) can also be written as

$$H_{n,p}(j\omega_1,\cdots,j\omega_n) = \sum_{\substack{r_1\cdots r_p=1 \\ \sum r_i = n}}^{n-p+1} \prod_{i=1}^{p} H_{r_i}(j\omega_{X+1},\cdots,j\omega_{X+r_i})(j\omega_{X+1}+\cdots+j\omega_{X+r_i})^{k_i} \tag{5.5}$$

where

$$X = \sum_{x=1}^{i-1} r_x \tag{5.6}$$

Furthermore, if defining the following notations,

$$H_{0,0}(\cdot) = 1, \tag{5.7}$$

$$H_{n,0}(\cdot) = 0 \quad \text{for n} > 0, \tag{5.8}$$

$$H_{n,p}(\cdot) = 0 \quad \text{for } n < p, \tag{5.9}$$

and

$$\prod_{i=1}^{q}(\cdot) = \begin{cases} 1 & q=0, p>1 \\ 0 & q=0, p\le 1 \end{cases} \tag{5.10}$$

then (5.1) can be written in a more concise form as

$$H_n(j\omega_1,\cdots,j\omega_n) = \frac{1}{L_n\left(j\sum_{i=1}^{n}\omega_i\right)} \sum_{q=0}^{n}\sum_{p=0}^{n-q}\sum_{k_1,k_{p+q}=0}^{K} c_{p,q}(k_1,\cdots,k_{p+q}) \left(\prod_{i=1}^{q}(j\omega_{n-q+i})^{k_{p+i}}\right) H_{n-q,p}(j\omega_1,\cdots,j\omega_{n-q}) \tag{5.11}$$

Therefore, the recursive algorithm for the computation of the GFRFs is (5.8 or 5.11, 5.10, 5.2–5.5).

Importantly, comparing the nth-order GFRF in (5.11) for the NDE model and that for the NARX model in (2.17), i.e.,

$$H_n(j\omega_1,\cdots,j\omega_n)=\frac{1}{L_n(\omega_1\cdots\omega_n)}\sum_{q=0}^{n}\sum_{p=0}^{n-q}\sum_{k_1,k_{p+q}=1}^{K}c_{p,q}(k_1,\cdots,k_{p+q})e^{-j\sum_{i=1}^{q}(\omega_{n-q+i}k_{p+i})}H_{n-q,p}(j\omega_1,\cdots,j\omega_{n-q}) \tag{5.12}$$

both (5.11) and (5.12) have the same structure and notations. Therefore, the parametric characteristics of the GFRFs for the NDE model are the same as for the NARX model.

From the recursive algorithm for the computation of the GFRFs in (5.8 or 5.11, 5.10, 5.2–5.5), it can be seen that the nth-order GFRF is a parameter-separable polynomial function with respect to the nonlinear parameters in model (2.10 or 2.11). For convenience, let

$$C(n,K)=\left(c_{p,q}(k_1,\cdots,k_{p+q})\left|\begin{array}{c}p=0\cdots m,\ p+q=m,\\ 2\le m\le n\\ k_i=0\cdots K,\ i=1\cdots p+q\end{array}\right.\right) \tag{5.13}$$

which includes all the nonlinear parameters from nonlinear degree 2 to n. Obviously, $C(M,K)$ includes all the nonlinear parameters involved in model (2.10 or 2.11). In the following sections, the CE operator will be applied to all the nonlinear parameters in $C(n,K)$. Note also the notations defined in (4.5) and (4.6), which will be used frequently throughout this book without further explanation.

5.2 Parametric Characteristics of the GFRFs

A fundamental result can be obtained firstly for the parametric characteristic of the nth-order GFRF in (5.11) or (5.12), which provides an important basis for the parametric characteristic analysis of the frequency response functions in the following studies.

Proposition 5.1 Consider the GFRFs in (5.1). There exists a complex valued function vector with appropriate dimension $f_n(j\omega_1,\cdots,j\omega_n)$ which is a function of $j\omega_1,\cdots,j\omega_n$ and the linear parameters of the NDE model (2.11), such that

$$H_n(j\omega_1,\cdots,j\omega_n)=CE(H_n(j\omega_1,\cdots,j\omega_n))\cdot f_n(j\omega_1,\cdots,j\omega_n) \tag{5.14}$$

where $CE(H_n(j\omega_1,\cdots,j\omega_n))$ is the parametric characteristic vector of the nth-order GFRF, and its elements include and only include all the nonlinear parameters in

$C_{0,n}$ and all the parameter monomials in $C_{p,q} \otimes C_{p_1,q_1} \otimes C_{p_2,q_2} \otimes \cdots \otimes C_{p_k,q_k}$ for $0 \le k \le n-2$, whose subscripts satisfy

$$p + q + \sum_{i=1}^{k} (p_i + q_i) = n + k,\ \ 2 \le p_i + q_i \le n - k,\ \ 2 \le p + q \\ \le n - k \text{ and } 1 \le p \le n - k \tag{5.15}$$

Proof Equation (5.14) is directly followed from Lemma 4.1 and the corresponding discussions in Chap. 4. It can be derived by applying the CE operator to Eqs. (5.1)–(5.4) that

$$CE(H_n(j\omega_1, \cdots, j\omega_n)) = C_{0,n} \oplus \left(\mathop{\oplus}_{q=1}^{n-1} \mathop{\oplus}_{p=1}^{n-q} C_{p,q} \otimes CE\big(H_{n-q,p}(\cdot)\big) \right) \\ \oplus \left(\mathop{\oplus}_{p=2}^{n} C_{p,0} \otimes CE\big(H_{n,p}(\cdot)\big) \right) \tag{5.16a}$$

$$CE\big(H_{n,p}(\cdot)\big) = \mathop{\oplus}_{i=1}^{n-p+1} CE(H_i(\cdot)) \otimes CE\big(H_{n-i,p-1}(\cdot)\big) \text{ or } CE\big(H_{n,p}(\cdot)\big) \\ = \mathop{\oplus}_{\substack{r_1 \cdots r_p = 1 \\ \sum r_i = n}}^{n-p+1} \mathop{\otimes}_{i=1}^{p} CE(H_{r_i}(\cdot)) \tag{5.16b}$$

$$CE(H_{n,1}(\cdot)) = CE(H_n(\cdot)) \tag{5.16c}$$

Obviously, $C_{0,n}$ is the first term in Eq. (5.16a). For clarity, consider a simpler case that there is only output nonlinearities in (5.16a), then (5.16a) is reduced to the last term of Eq. (5.16a), *i.e.*, $\mathop{\oplus}_{p=2}^{n} C_{p,0} \otimes CE\big(H_{n,p}(\cdot)\big) = \mathop{\oplus}_{p=2}^{n} C_{p,0} \otimes \mathop{\oplus}_{\substack{r_1 \cdots r_p = 1 \\ \sum r_i = n}}^{n-p+1} \mathop{\otimes}_{i=1}^{p} CE(H_{r_i}(\cdot))$.

Note that $\mathop{\oplus}_{\substack{r_1 \cdots r_p = 1 \\ \sum r_i = n}}^{n-p+1} \mathop{\otimes}_{i=1}^{p} CE(H_{r_i}(\cdot))$ includes all the combinations of $(r_1, r_2, \ldots, r_p)$ satisfying $\sum_{i=1}^{p} r_i = n$, $1 \le r_i \le n-p+1$, and $2 \le p \le n$. Moreover, $CE(H_1(\cdot)) = 1$ since there are no nonlinear parameters in it, and any repetitive combinations have no contribution. Hence, $\mathop{\oplus}_{\substack{r_1 \cdots r_p = 1 \\ \sum r_i = n}}^{n-p+1} \mathop{\otimes}_{i=1}^{p} CE(H_{r_i}(\cdot))$ must include all the possible non-repetitive combinations of $(r_1, r_2, \ldots, r_k)$ satisfying $\sum_{i=1}^{k} r_i = n - p + k$, $2 \le r_i \le n - p + 1$ and $1 \le k \le p$. So does $CE(H_n(j\omega_1, \cdots, j\omega_n))$.

Each of the subscript combinations corresponds to a monomial of the involved nonlinear parameters. Thus, by including the term $C_{p,0}$ and considering the range of each variable (*i.e.*, r_i, p, and k), CE($H_n(j\omega_1, \cdots, j\omega_n)$) must include all the possible non-repetitive monomial functions of the nonlinear parameters of the form $C_{p0} \otimes C_{r_1 0} \otimes C_{r_2 0} \otimes \cdots \otimes C_{r_k 0}$ satisfying $p + \sum_{i=1}^{k} r_i = n + k$, $2 \le r_i \le n - k$, $0 \le k \le n-2$ and $2 \le p \le n-k$.

When the other types of nonlinearities are considered, by extending the results above to a more general case such that the nonlinear parameters appear in the form $C_{pq} \otimes C_{p_1 q_1} \otimes C_{p_2 q_2} \otimes \cdots \otimes C_{p_k q_k}$ and the subscripts satisfy $p + q + \sum_{i=1}^{k} (p_i + q_i) = n + k$, $2 \le p_i + q_i \le n - k$, $0 \le k \le n-2$, $2 \le p + q \le n - k$ and $1 \le p \le n - k$, the same conclusion can be reached. Hence, the proposition is proved. □

Remark 5.1 In Proposition 5.1, $f_n(j\omega_1, \cdots, j\omega_n)$ is not a function of $CE(H_n(j\omega_1, \cdots, j\omega_n))$ and is invariant at a specific point $(\omega_1, \cdots, \omega_n)$ if the linear parameters of model (1.5) are fixed. Proposition 5.1 provides for the first time an explicit analytical expression for the *n*th-order GFRF which reveals a straightforward relationship between the nonlinear parameters of model (1.5) and the system GFRFs, and is an explicit function of the nonlinear parameters at any specific frequency point $(\omega_1, \cdots, \omega_n)$. Equation (5.14) is referred to as the parametric characteristic function of the *n*th-order GFRF, which is denoted by $H_n(C(n,K))_{(\omega_1, \cdots, \omega_n)}$. □

Remark 5.2 As mentioned in Chap. 4, the CE operator sets up a mapping from Ξ to $P_c{}^{\sigma}$ (see the definitions in Sect. 4.2). When applying the CE operator to the GFRFs of the NDE model (2.11),

$$
\begin{aligned}
C_s &= C(M,K), \\
W_s &= \{\omega_1, \cdots, \omega_N\} \cup \{c_{1,0}(k_1), c_{0,1}(k_1) | 0 \le k_1 \le K\}, \\
P_c &= \left\{ c_1^{r_1} c_2^{r_2} \cdots c_I^{r_I} | c_i \in C(M,K), r_i \in {}_0, I = |C(M,K)| \right\}
\end{aligned}
$$

and

$$
\Xi = \{H_n(\cdot) | 1 \le n \le N\}.
$$

The condition described by (5.15) in Proposition 5.1 provides a sufficient and necessary condition on what nonlinear parameters of model (2.11) can appear in the *n*th-order GFRF, and also how the GFRF is determined by these parameters. □

For a better understanding of the parametric characteristic $CE(H_n(j\omega_1, \cdots, j\omega_n))$, the following properties of $CE(H_n(j\omega_1, \cdots, j\omega_n))$ for the NDE model (2.11) can be obtained, based on Proposition 5.1.

Definition 5.1 If a nonlinear parameter monomial $\prod_{i=1}^{k} c_{p_i q_i}^{j_i}(\cdot)$ (k>0, $j_i \geq 0$) is an element of $CE(H_n(j\omega_1,\cdots,j\omega_n))$, then it has an independent contribution to $H_n(j\omega_1,\cdots,j\omega_n)$, and is referred to as a complete monomial of order n (simply as n-order complete); otherwise, if it is part of an n-order complete monomial, then it is referred to as n-order incomplete.

Obviously, all the elements in $CE(H_n(j\omega_1,\cdots,j\omega_n))$ are n-order complete.

Property 5.1 The largest nonlinear degree of the nonlinear parameters appearing in $CE(H_n(j\omega_1,\cdots,j\omega_n))$ is n corresponding to nonlinear parameters $c_{p,q}(.)$ with $p+q=n$, and the n-degree nonlinear parameters of form $c_{p,q}(.)$ $(p+q=n)$ are all n-order complete.

Proof In (5.15) when $p+q=n$, then $p+q+\sum_{i=1}^{k}(p_i+q_i)=n+\sum_{i=1}^{k}(p_i+q_i)=n+k$, which further yields $\sum_{i=1}^{k}(p_i+q_i)=k$. Note that $2\leq p_i+q_i\leq n-k$ and $0\leq k\leq n-2$, thus $k=p_i=q_i=0$. Therefore, the property is proved. □

Property 5.2 $c_{p,q}(.)$ is j-order incomplete for $j>p+q$. That is, for a nonlinear parameter $c_{p,q}(.)$, it will appear in all the GFRFs of order larger than $p+q$.

Proof This property can be seen from the recursive Eqs. (5.16a–c) and can also be proved from Proposition 5.1. Suppose $c_{p,q}(.)$ does not appear in $H_n(j\omega_1,\cdots,j\omega_n)$, where n>p+q. Consider a monomial $c_{p,q}(.)c_{2,0}^{k}(.)$ with $k=n\text{-}p\text{-}q$. It can be verified from Proposition 5.1 that $c_{p,q}(.)c_{2,0}^{n-p-q}(.)$ is n-order complete. This results in a contradiction. □

Properties 5.1–5.2 show that only the nonlinear parameters of degree from 2 to n have contribution to $CE(H_n(j\omega_1,\cdots,j\omega_n))$, and the n-degree nonlinear parameters contribute to all the GFRFs of order $\geq n$.

Property 5.3 If $2\leq p_i+q_i$, $1\leq k$ and there is at least one p_i satisfying $1\leq p_i$ except for $k=1$, then $c_{p_1q_1}(\cdot)c_{p_2q_2}(\cdot)\cdots c_{p_kq_k}(\cdot)$ is Z-order complete, where $Z=\sum_{i=1}^{k}(p_i+q_i)-k+1$. Moreover, $\prod_{i=1}^{k} c_{p_iq_i}(\cdot)$ are j-order incomplete for $j>Z$, and have no effect on the GFRFs of order less than Z. □

The proof of Property 5.3 is given in Sect. 5.5. Given any monomial $c_{p_1q_1}(\cdot)c_{p_2q_2}(\cdot)\cdots c_{p_kq_k}(\cdot)$, it can be easily determined from Property 5.3 that, to which order GFRF the monomial contributes independently. For instance, consider a nonlinear parameter $c_{3,2}(.)$, which corresponds to the nonlinear term $\prod_{i=1}^{3}\frac{d^{k_i}y(t)}{dt^{k_i}}\prod_{i=4}^{5}\frac{d^{k_i}u(t)}{dt^{k_i}}$. It follows from Property 5.3 that Z=(3+2)−1+1=5. Thus this nonlinear term has an independent contribution to the fifth order GFRF $H_5(.)$ and affects all the GFRFs of order larger than 5. Moreover, it has no effect on the GFRFs less than the fifth order.

Property 5.4 If $1 \leq r_i$ and $1 \leq k$, then the elements of $CE\left(\prod_{i=1}^{k} H_{r_i}(\cdot)\right)$ are all *Z*-order complete, where $Z=\sum_{i=1}^{k} r_i - k + 1$, and are all *j*-order incomplete for *j*>*Z*, and have no effect on the GFRFs of order less than Z. Similarly, the elements of $\prod_{i=1}^{k_1} c_{p_i q_i}(\cdot) \otimes CE\left(\prod_{i=1}^{k_2} H_{r_i}(\cdot)\right)$ are all *Z*-order complete, where $Z=\sum_{i=1}^{k_1}(p_i+q_i)+\sum_{i=1}^{k_2} r_i - k_1 - k_2 + 1$, and are all *j*-order incomplete for *j*>*Z*, and have no effect on the GFRFs of order less than *Z*. □

The proof of Property 5.4 is given in Sect. 5.5. Obviously, this property is an extension of Property 5.3, which shows that some computation by "⊗" between some parameters and the parametric characteristics of some different order GFRFs may result in the same parametric characteristic.

Property 5.5 $CE(H_{n,p}(\cdot)) = CE(H_{n-p+1}(\cdot))$. □

The proof of Property 5.5 is given in Sect. 5.5. This property, together with Property 5.4, provides a simplified approach to the recursive computation of the parametric characteristic of the *n*th-order GFRF in Eqs. (5.16a–c), which is summarized in Corollary 5.1 as follows.

Corollary 5.1 The parametric characteristic of the *n*th-order GFRF for model (2.11) can be recursively determined as

$$\begin{aligned} &CE(H_n(j\omega_1,\cdots,j\omega_n)) \\ &\quad = C_{0,n} \oplus \bigoplus_{q=1}^{n-1} \left\{ C_{n-q,q} \oplus \left(\bigoplus_{p=1}^{n-q-1} C_{p,q} \otimes \chi_C\left(n,p,q,\left\lfloor \tfrac{n-q}{2} \right\rfloor\right)\right)\right\} \\ &\qquad \oplus \left\{ C_{n,0} \oplus \left(\bigoplus_{p=2}^{n-1} C_{p,0} \otimes \chi_C\left(n,p,0,\left\lfloor \tfrac{n+1}{2} \right\rfloor\right)\right)\right\} \end{aligned} \tag{5.17}$$

where $\lfloor \cdot \rfloor$ is to take the integer part, $\chi_C(n,p,q,\aleph) = \begin{cases} CE(H_{n-p-q+1}(\cdot)) & p \leq \aleph \\ C_{0,n-p-q+1} & p > \aleph \end{cases}$, and $\aleph$ is a positive integer.

Proof Using Property 5.5, (5.16a) can be written as (n>1)

$$CE(H_n(j\omega_1,\cdots,j\omega_n)) = C_{0,n} \oplus \left(\mathop{\oplus}_{q=1}^{n-1} \mathop{\oplus}_{p=1}^{n-q} C_{p,q} \otimes CE\left(H_{n-q-p+1}(\cdot)\right)\right) \oplus \left(\mathop{\oplus}_{p=2}^{n} C_{p,0} \otimes CE\left(H_{n-p+1}(\cdot)\right)\right) \quad (5.18)$$

Note from Property 5.4 that some computations in the second and third parts of the last equation are repetitive. For example, the monomials in $C_{n-2,1} \otimes CE(H_{n-n+2-1+1}(\cdot)) = C_{n-2,1} \otimes CE(H_2(\cdot))$ (n>2) are included in $C_{1,1} \otimes CE(H_{n-1}(\cdot)) \cup C_{2,0} \otimes CE(H_{n-1}(\cdot))$, except the monomials in $C_{n-2,1} \cdot C_{0,2}$. For this reason, (5.18) can be further written as

$$\begin{aligned} &CE(H_n(j\omega_1,\cdots,j\omega_n)) \\ &= C_{0,n} \oplus \mathop{\oplus}_{q=1}^{n-1} \left\{ C_{n-q,q} \oplus \left(\mathop{\oplus}_{p=1}^{\lfloor n-q/2 \rfloor} C_{p,q} \otimes CE\left(H_{n-q-p+1}(\cdot)\right)\right) \right. \\ &\left. \oplus \left(\mathop{\oplus}_{p=\lfloor n-q/2 \rfloor+1}^{n-q-1} C_{p,q} \otimes C_{0,n-q-p+1} \right) \right\} \\ &\oplus \left\{ C_{n,0} \oplus \left(\mathop{\oplus}_{p=2}^{\lfloor n+1/2 \rfloor} C_{p,0} \otimes CE\left(H_{n-p+1}(\cdot)\right)\right) \oplus \left(\mathop{\oplus}_{p=\lfloor n+1/2 \rfloor+1}^{n-1} C_{p,0} \otimes C_{0,n-p+1} \right) \right\} \end{aligned}$$

This produces Eq. (5.17). The proof is completed. □

Remark 5.3 Corollary 5.1 provides an alternative recursive way to determine the parametric characteristic of the *n*th-order GFRF. If there are only some nonlinear parameters in (5.13) of interest, then Eq. (5.17) and all the results above can still be used by taking other parameters as 1 if they are nonzero, or as zero if they are zero. Therefore, whatever nonlinear parameters (for instance x) are concerned, the parametric characteristic function with respect to x denoted by $H_n(x)_{(\omega_1,\cdots,\omega_n;C(n,K)\backslash x)}$ and the parametric characteristic $CE(H_n(j\omega_1,\cdots,j\omega_n))$ can all be derived by following the same method established above. □

The parametric characteristic analysis of this section can be used to demonstrate how the parameters of interest affect the GFRFs and consequently provide useful information for both the GFRF evaluation and system analysis. The following example provides an illustration for this.

Example 5.1 Consider the parametric characteristics of the following two cases:

Case 1: Suppose there is only one input nonlinear term $C_{0,3} \neq 0$, and all the other nonlinear parameters are zero in model (2.11). Then the parametric characteristics of the nth-order GFRF can be computed as

If n<3, it follows from Property 5.1 that $CE(H_n(j\omega_1,\cdots,j\omega_n)) = 0$.

If n=3, it also follows from Property 5.1 that the parameters in $C_{0,3}$ are all 3-order complete. Thus $CE(H_3(j\omega_1,\cdots,j\omega_3)) = C_{0,3}$.

If n>3, it follows from Property 5.2, $C_{0,3}$ should be n-order incomplete in this case. However, from the Definition 5.1, a complete monomial should have at least one $p \geq 1$. Since there are no other nonzero nonlinear parameters, $CE(H_n(j\omega_1, \cdots, j\omega_n)) = 0$ for this case.

Therefore, $CE(H_n(j\omega_1, \cdots, j\omega_n)) = 0$ for $n \neq 1$ and $n \neq 3$ in Case 1. That is, only $H_1(j\omega)$ and $H_3(j\omega_1, \cdots, j\omega_3)$ are nonzero in this case. Obviously, the computation of the parametric characteristics can provide guidance to the computation and analysis of the GFRFs from this case study.

Case 2: Suppose only $C_{0,3} \neq 0$ and $C_{2,0} \neq 0$, and all the other nonlinear parameters are zero. Then the parametric characteristics of the GFRFs can be simply determined as

$$CE(H_1(j\omega_1)) = 1, \;\; CE(H_2(j\omega_1, j\omega_2)) = C_{2,0}, \;\; CE(H_3(j\omega_1, \cdots, j\omega_3)) = C_{2,0}^2 \oplus C_{0,3}$$
$$CE(H_4(j\omega_1, \cdots, j\omega_4)) = C_{2,0}^3 \oplus C_{0,3} \otimes C_{2,0}, \;\; CE(H_5(j\omega_1, \cdots, j\omega_5)) = C_{2,0}^4 \oplus C_{0,3} \otimes C_{2,0}^2$$
$$CE(H_6(j\omega_1, \cdots, j\omega_6)) = C_{2,0}^6 \oplus C_{0,3} \otimes C_{2,0}^3 \oplus C_{0,3}^2 \otimes C_{2,0}$$

Especially, if only $C_{0,3}$ is of interest for analysis, then $C_{2,0}$ can be regarded as constant 1. In this case, the parametric characteristics of the GFRFs can be obtained as

$$CE(H_1(j\omega_1)) = CE(H_2(j\omega_1, j\omega_2)) = 1, \;\; CE(H_3(j\omega_1, \cdots, j\omega_3)) = C_{0,3}$$
$$CE(H_4(j\omega_1, \cdots, j\omega_4)) = C_{0,3}, \;\; CE(H_5(j\omega_1, \cdots, j\omega_5)) = C_{0,3}, \;\; CE(H_6(j\omega_1, \cdots, j\omega_6)) = C_{0,3} \oplus C_{0,3}^2$$

Note that different parametric characteristics of the GFRFs correspond to different polynomial functions with respect to the parameters of interest, which can demonstrate how the parameters of interest affect the GFRFs and thus provide some useful information for the system analysis. For example, from the parametric characteristics in Case 2, it can be seen that the sensitivity of the GFRFs for $n<6$ with respect to $C_{0,3}$ is a constant when $C_{2,0}$ and the linear parameters are constant. This may imply that in order to make the system less sensitive to the input nonlinear term with coefficient $C_{0,3}$, it needs only to adjust the parameters in $C_{2,0}$ and the linear parameters of model (2.11) to reduce the corresponding constants in Case 2 under certain conditions. □

The parametric characteristic and its properties developed in this section for the nth-order GFRF demonstrate what the parametric characteristics of the GFRFs are, and how the nonlinear parameters in $C(n,K)$ make contributions to the nth-order GFRF. As demonstrated in Example 5.1, these fundamental results can be used to reveal how the nonlinear parameters affect the GFRFs and how the frequency response functions of model (2.11) are constructed and thus dominated by the model parameters which define system nonlinearities. Based on these results, useful results can be developed and will be discussed in more details in the following sections and chapters.

5.3 Parametric Characteristics Based Analysis

Based on the parametric characteristics of the GFRFs established in the last section, many significant results can be obtained. The parametric characteristic analysis can provide an important insight into at least the following aspects:

(a) System nonlinear effects on frequency response functions (including the GFRFs and output spectrum)—mainly discussed in this section, Chaps. 6–10 and 12;
(b) The detailed polynomial structure of frequency response functions—mainly discussed in this section and Chaps. 6–10;
(c) Computations of the GFRFs and output spectrum—mainly discussed in Chap. 11;
(d) Understanding of nonlinear behaviour in the frequency domain—mainly discussed in Chaps. 4–6 and 12;
(e) Analysis and design of system output behaviour by using nonlinearities—mainly discussed in Chaps. 9 and 10.

In this section, some of these results are given, and more detailed results will be discussed later in the following chapters.

5.3.1 *Nonlinear Effect on the GFRFs from Different Nonlinear Parameters*

As mentioned before, the nonlinearities in model (2.10) or model (2.11) can be classified into three categories as follows:

(a) Pure input nonlinearities. This refers to the nonlinear parameters $c_{0,n}(.)$, which are the first term in the parametric characteristics in Eq. (5.17) or (5.18);
(b) Pure output nonlinearities. This refers to the nonlinear parameters $c_{n,0}(.)$, which are the last term in Eq. (5.17) or (5.18);
(c) Input-output cross nonlinearities, This refers to the nonlinear parameters $c_{p,q}(.)$, which are the second term in (5.17) or (5.18).

It is known that different nonlinearity has a different effect on system dynamics. Different nonlinear parameters correspond to different degree and category of nonlinearities. Hence, the frequency characteristics of frequency response functions and the effects of different nonlinear parameters on system output behaviour can be revealed by the parametric characteristic analysis of the corresponding frequency response functions. Since the GFRFs represent system frequency characteristics, the study on the nonlinear effect on the GFRFs from different categories of nonlinearities can provide an important insight into the relationship between the system frequency characteristics and physical model parameters. In this section, the parametric characteristics based analysis is investigated and discussed for the GFRFs in order to reveal how different model parameters have their effect on the frequency response functions for model (2.11), and therefore affect the system frequency characteristics. In what follows, the k+1 in monomial $C_{pq} \otimes C_{p_1q_1} \otimes C_{p_2q_2} \otimes \cdots \otimes C_{p_kq_k}$ is referred to as the power of the monomial.

A. Pure Input Nonlinearities

As mentioned, this category of nonlinearities correspond to the nonlinear parameters of the form $c_{0,q}(.)$ with $q>1$. If $n=q$, then from Property 5.1 the parametric characteristic of the nth-order GFRF with respect to the parameters in $C_{0,q}$ is

$$CE(H_n(j\omega_1,\cdots,j\omega_n))_{C_{0,q}} = C_{0,q} \tag{5.19a}$$

and if $n<q$,

$$CE(H_n(j\omega_1,\cdots,j\omega_n))_{C_{0,q}} = 1 \tag{5.19b}$$

For $n>q$, since there is at least one parameter $c_{p,q}(.)$ with $p>0$ for any complete monomials (except $c_{0,n}(.)$) in $CE(H_n(j\omega_1,\cdots,j\omega_n))_{c_{0,q}(\cdot)}$ from Proposition 5.1, thus $c_{0,q}(.)^{\rho}$ for any $\rho>0$ cannot be an independent entry in $CE(H_n(j\omega_1,\cdots,j\omega_n))_{c_{0,q}(\cdot)}$. The largest power ρ can only appear in the monomial $c_{0,q}(.)^{\rho}c_{p',q'}(.)$, where $c_{p',q'}(.)$ is nonzero, satisfies $p'\geq 1$ and $p'+q'\geq 2$ and has the smallest $p'+q'$. In this case, ρ can be computed from Property 5.3 as

$$\rho(n,0,q) = \frac{n-p'-q'}{q-1}$$

For example, if $p'+q'=2$, then

$$\rho(n,0,q) = \begin{cases} \left\lfloor \frac{n-1}{q-1} \right\rfloor & \text{if } \frac{n-1}{q-1} \text{ is not an integer} \\ \frac{n-q}{q-1} & \text{else} \end{cases}$$

Therefore, for $n>q$,

$$CE(H_n(j\omega_1,\cdots,j\omega_n))_{C_{0,q}} = \begin{bmatrix} 1 & C_{0,q} & C_{0,q}{}^2 & \cdots & C_{0,q}{}^{\rho(n,0,q)} \end{bmatrix} \tag{5.19c}$$

In particular, when all the other nonlinear parameters are zero except for $C_{0,q}$, then (n>1)

$$CE(H_n(j\omega_1,\cdots,j\omega_n))_{C_{0,q}} = \begin{cases} C_{0,q} & \text{if q = n} \\ 0 & \text{else} \end{cases} \tag{5.19d}$$

It can further be verified that the parametric characteristic $CE(H_n(j\omega_1,\cdots,j\omega_n))_{C_{0,q}}$ is the same as (5.19d) even when only all the other categories of nonlinear parameters are zero except for the input nonlinearity.

From the parametric characteristic analysis of the nth-order GFRF for the input nonlinearity, it can be concluded that,

(A1) The parametric characteristic function with respect to the input nonlinearity for the nth-order GFRF is a polynomial of the largest degree $\rho(n, 0, q)$, i.e.,

$$H_n(C_{0,q})_{(j\omega_1,\cdots,j\omega_n;C(n,K)\backslash C_{0,q})} = \begin{bmatrix} 1 & C_{0,q} & {C_{0,q}}^2 & \cdots & {C_{0,q}}^{\rho(n,0,q)} \end{bmatrix} \cdot f_n(j\omega_1,\cdots,j\omega_n; C(n,K)\backslash C_{0,q})$$

where $f_n(j\omega_1,\cdots,j\omega_n; C(n,K)\backslash C_{0,q})$ is an appropriate function vector.

(A2) The largest power for the input nonlinearity of an independent contribution in $CE(H_n(j\omega_1,\cdots,j\omega_n))$ is 1, which corresponds to the nonlinear parameters in $C_{0,n}$.

(A3) For comparison with the other categories of nonlinearities, considering the individual effect of pure input nonlinearity when there are no other categories of nonlinearities, *i.e.*, output nonlinearity and input-output cross nonlinearity, it can be seen from (5.19d) that the input nonlinearities have no auto-crossing effects on system dynamics. That is, each degree of the input nonlinearities has an independent contribution to the corresponding order GFRF and the largest power of a complete monomial from input nonlinearities is 1, i.e., the nth-order GFRF is simply $H_n(j\omega_1,\cdots,j\omega_n) = C_{0,q} \cdot f_n(j\omega_1,\cdots,j\omega_n)$ from Proposition 5.1. Obviously, if $C_{0,n}=0$, there will be no contribution from the input nonlinearities in the nth-order GFRF. It will be seen that these demonstrate a quite different property for the input nonlinearity from other categories of nonlinearities.

It is known that a difficulty in the analysis of Volterra systems is that the Volterra kernels in the time domain usually interact with each order due to the crossing nonlinear effects from different nonlinearities, and so are the GFRFs in the frequency domain. From the discussions above, this difficulty does not hold for the case that there are only input nonlinearities, *e.g.*, for the class of Volterra systems studied in Kotsios (1997). The parametric characteristic analysis for the input nonlinearities can also make light on the selection of different parameters for the energy transfer filter design in Billings and Lang (2002).

B. Pure Output Nonlinearities

This category of nonlinearities correspond to the nonlinear parameters of the form $c_{p,0}(.)$ with $p>1$. If $n=p$, then from Property 5.1

$$CE(H_n(j\omega_1,\cdots,j\omega_n))_{C_{p,0}} = C_{p,0} \tag{5.20a}$$

If $n<p$, also from Property 5.1

$$CE(H_n(j\omega_1,\cdots,j\omega_n))_{C_{p,0}} = 1 \tag{5.20b}$$

These are similar to the input nonlinearity. If $n>p$, then from Properties 5.1–5.3 $C_{p,0}$ will contribute to all the GFRFs of order lager than p. From Property 5, $c_{p,0}(.)^{\rho}$ for $\rho > 0$ is a complete monomial for the Zth-order GFRFs where $Z = (p-1)\rho + 1$. For the nth-order GFRF with $n>p$, the largest power ρ can be computed from Property 5.3 as

$$\rho(n,p,0) = \left\lfloor \frac{n-1}{p-1} \right\rfloor$$

Thus, for $n>p$,

$$CE(H_n(j\omega_1,\cdots,j\omega_n))_{C_{p,0}} = \begin{bmatrix} 1 & C_{p,0} & C_{p,0}{}^2 & \cdots & C_{p,0}{}^{\rho(n,p,0)} \end{bmatrix} \tag{5.20c}$$

Consider the particular case where all nonlinear parameters are zero except the parameters in $C_{p,0}$, then for $n>1$

$$CE(H_n(j\omega_1,\cdots,j\omega_n))_{C_{p,0}} = \begin{cases} 0 & \text{if } p > n \text{ or } \frac{n-1}{p-1} \text{ is not an integer} \\ C_{p,0}{}^{\rho(n,p,0)} & \text{else} \end{cases} \tag{5.20d}$$

However, when all other nonlinear parameters are zero except output nonlinear parameters, the parametric characteristic $CE(H_n(j\omega_1,\cdots,j\omega_n))_{C_{p,0}}$ for $n>p$ is the same as (5.20c).

From the parametric characteristic analysis of the nth-order GFRF for the pure output nonlinearity, it can be concluded that,

(B1) The parametric characteristic function with respect to the output nonlinearity for the nth-order GFRF is a polynomial of the largest degree $\rho(n,p,0)$, i.e.,

$$H_n(C_{p,0})_{(j\omega_1,\cdots,j\omega_n;C(n,K)\backslash C_{p,0})} = \begin{bmatrix} 1 & C_{p,0} & C_{p,0}{}^2 & \cdots & C_{p,0}{}^{\rho(n,p,0)} \end{bmatrix} \cdot f_n(j\omega_1,\cdots,j\omega_n; C(n,K)\backslash C_{p,0})$$

where $f_n(j\omega_1,\cdots,j\omega_n; C(n,K)\backslash C_{p,0})$ is an appropriate function vector. Note that $\rho(n,p,0) \geq \rho(n,0,q)$, which may imply that for the same nonlinear degree, output nonlinearity has a larger effect on the system than input nonlinearity.

(B2) The largest power for the output nonlinear parameter $C_{p,0}$ of an independent contribution in $CE(H_n(j\omega_1,\cdots,j\omega_n))$ is $\rho(n,p,0)$, which corresponds to the n-order complete monomial $C_{p,0}{}^{\rho(n,p,0)}$. However, the largest power for the output nonlinearity of a complete monomial in $CE(H_n(j\omega_1,\cdots,j\omega_n))$ is k, corresponding to the monomial $C_{p_1,0} \otimes C_{p_2,0} \otimes \cdots \otimes C_{p_k,0}$, where $k = p_1 + \cdots + p_k + 1 - n$. This is quite different from the input nonlinearity.

(B3) Considering the individual effect of pure output nonlinearity when there are no other categories of nonlinearities, *i.e.*, input nonlinearity and input-output cross nonlinearity, it can be seen from (5.20c) that the output nonlinearities have auto-crossing nonlinear effects on system dynamics. That is, different degree of output nonlinearities can form a complete monomial in the nth-order GFRF and the largest power of this kind of complete monomials from output nonlinearities is k as mentioned in (B2). Obviously, if the degree-n nonlinear parameter $C_{n,0}=0$, there are still contributions from the output nonlinearities in the nth-order GFRF if there are other nonzero output nonlinear parameters of degree less than n. These may imply that output nonlinearity has more complicated and larger effect on the system than input nonlinearity of the same order, which shows a property different from that of the input nonlinearity as mentioned in (A3).

(B4) It can be seen from (5.20c, d) that $C_{p,0}$ will contribute independently to the GFRFs whose orders are $(p-1)i+1$ for $i=1,2,3,\ldots$. It is known that for a Volterra system, the system nonlinear dynamics is usually dominated by the first several order GFRFs (Taylor 1999; Boyd and Chua 1985). This implies that the nonlinear terms with coefficient $C_{p,0}$ of smaller nonlinear degree, *e.g.*, 2 and 3, take much greater roles in the GFRFs than other pure output nonlinearities. This property is significant for the design of nonlinear feedback controller design, where a desired degree of nonlinearity should be determined for control objectives (Jing et al.2006; Van Moer et al. 2001). This will be further discussed in Chap. 9.

C. Input-Output Cross Nonlinearities

This category of nonlinearities corresponds to the nonlinear parameters of the form $c_{p,q}(.)$ with $p \geq 1$ and $q \geq 1$. It can be verified that the parametric characteristics of the GFRFs with respect to such nonlinearities are very similar to those for the pure output nonlinearities as shown in B, and the conclusions held for the output nonlinearity still hold for the input-output cross nonlinearity. Thus the detailed discussions are omitted here. For a summary, the following parametric characteristics hold for both of these two categories of nonlinearities

$$CE(H_n(j\omega_1,\cdots,j\omega_n))_{C_{p,q}} = \begin{cases} 1 & \text{if } n < p+q \\ \begin{bmatrix} 1 & C_{p,q} & C_{p,q}{}^2 & \cdots & C_{p,q}{}^{\rho(n,p,q)} \end{bmatrix} & \text{else} \end{cases} \tag{5.21}$$

where, n>1, $\rho(n,p,q) = \left\lfloor \frac{n-1}{p+q-1} \right\rfloor$, $p \geq 1$ and $p+q \geq 2$.

A difference between the input-output cross nonlinearity and the pure output nonlinearity may be that the output nonlinearity can be relatively easily realized by a nonlinear state or output feedback control in practice. A simple comparison is summarized in Table 5.1.

Remark 5.4 Based on the parametric characteristic of the nth-order GFRF with respect to nonlinear parameters in $C_{p,q}$, the sensitivity of the GFRFs with respect to these nonlinear parameters can also be studied. From Proposition 5.1, the sensitivity of $H_n(j\omega_1, \cdots, j\omega_n)$ with respect to a specific nonlinear parameter c can be computed as

$$\frac{\partial H_n(c)_{(\omega_1,\cdots,\omega_n;C(M,K,n)\backslash c)}}{\partial c} = \frac{\partial H_n(j\omega_1,\cdots,j\omega_n)}{\partial c} = \frac{\partial CE(H_n(j\omega_1,\cdots,j\omega_n))}{\partial c} \cdot f_n(j\omega_1,\cdots,j\omega_n) \quad (5.22)$$

Thus, the sensitivity of the nth-order GFRF with respect to any nonlinear parameter $c{=}c_{p,q}(.)$ with $p \geq 1$ and $p+q \geq 2$ can be obtained from (5.21) as:

$$\frac{\partial H_n(c)_{(\omega_1,\cdots,\omega_n;C(K,n)\backslash c)}}{\partial c} = \begin{bmatrix} 0 & 1 & 2c & \cdots & \rho(n,p,q)c^{\rho(n,p,q)-1} \end{bmatrix} \cdot \bar{f}_n(j\omega_1,\cdots,j\omega_n; C(K,n)\backslash c) \quad (5.23)$$

where $\bar{f}_n(j\omega_1,\cdots,j\omega_n; C(K,n)\backslash c)$ is an appropriate function vector defined in Proposition 5.1. Obviously, the sensitivity to a specific parameter is still an analytical polynomial function of the nonlinear parameter. From the parametric characteristics in (5.19a–5.21), it can be concluded that the sensitivity of the nth-order GFRF with respect to an input nonlinear parameter must be zero or constant when there are no other category of nonlinearities. However, this can only happen to the output nonlinear parameters and input-output cross nonlinear parameters if the nonlinear degree of the parameter of interest is n. Otherwise, the sensitivity function with respect to an output or an input-output cross nonlinear parameter is still an analytical polynomial function of the parameter of interest and some other nonzero parameters.

5.4 Conclusions

The parametric characteristic analysis discussed in Chap. 4 is used in this Chapter for the study of the parametric characteristics of the GFRFs of Volterra-type nonlinear systems described by the NDE model (2.11) or NARX model (2.10). Fundamental and significant results have been established for the parametric characteristics of the GFRFs of the nonlinear systems. The method has been shown to be of great significance in understanding the system's frequency response functions and the nonlinear influence incurred by different nonlinear terms. As mentioned in Sect. 5.3, the significance has at least five aspects, some of which have been demonstrated in this chapter and more will be discussed and investigated later.

From the results of this Chapter, it can be seen that, the parametric characteristics of the GFRFs can explicitly reveal the relationship between the time domain

Table 5.1 Comparisons between different nonlinearities for $H_n(.)$ (Jing et al. 2009a)

For $H_n(.)$	Input nonlinearities	Output nonlinearities	Input-output nonlinearities
The parametric characteristic function with respect to different nonlinearity	$H_n(C_{0,q})_{(j\omega_1,\cdots,j\omega_n;C(n,K)\backslash C_{0,q})} = \left[1 \quad C_{0,q} \quad {C_{0,q}}^2 \quad \cdots \quad {C_{0,q}}^{\rho(n,0,q)}\right] \cdot f_n(j\omega_1,\cdots,j\omega_n;C(n,K)\backslash C_{0,q})$ for $n \geq q$	$H_n(C_{p,0})_{(j\omega_1,\cdots,j\omega_n;C(n,K)\backslash C_{p,0})} = \left[1 \quad C_{p,0} \quad {C_{p,0}}^2 \quad \cdots \quad {C_{p,0}}^{\rho(n,p,0)}\right] \cdot f_n(j\omega_1,\cdots,j\omega_n;C(n,K)\backslash C_{p,0})$ for $n \geq p$	$CE(H_n(j\omega_1,\cdots,j\omega_n))_{C_{p,q}} = \begin{cases} 1 & \text{if } n < p+q \\ \left[1 \quad C_{p,q} \quad {C_{p,q}}^2 \quad \cdots \quad {C_{p,q}}^{\rho(n,p,q)}\right] & \text{else} \end{cases}$
The largest power for a specific nonlinear parameter in an independent contribution to $H_n(.)$	1 (not $\rho(n,0,q)$)	$\rho(n,p,0) = \left\lfloor \frac{n-1}{p-1} \right\rfloor$	$\rho(n,p,q) = \left\lfloor \frac{n-1}{p+q-1} \right\rfloor$
The largest power for a pure nonlinearity of a complete monomial in $CE(H_n(.))$	1, corresponding to the monomial $C_{0,n}$	$k = p_1 + \cdots + p_k + 1 - n$ corresponding to the monomial $C_{p_1,0}C_{p_2,0}\cdots C_{p_k,0}$	$k = p_1 + q_1 \cdots + p_k + q_k + 1 - n$, corresponding to the monomial $C_{p_1,q_1}C_{p_2,q_2}\cdots C_{p_k,q_k}$
When there are no other kinds of nonlinearities	No auto-crossing effects on system dynamics; have straightforward relationship with output frequencies; and parameter sensitivity is zero or constant	Have complicated auto-crossing nonlinear effects; can be realized in feedback control; and lower degree parameters have more contributions to $H_n(.)$	Have complicated auto-crossing nonlinear effects; and lower degree parameters have more contributions to $H_n(.)$

model parameters and the GFRFs and therefore provide a useful insight into the analysis and design of nonlinear systems in the frequency domain. By using the parametric characteristic analysis, system nonlinear frequency domain characteristics can be studied in terms of the time domain model parameters which define system nonlinearities, and the dependence of the frequency response functions of nonlinear systems on model parameters can be revealed. As it will be shown further in the following chapters, the analytical relationship between system output spectrum and model parameters can also be determined explicitly, and the nonlinear effect on the system output frequency response from different nonlinearities can be unveiled. This will facilitate the study of nonlinear behaviours in the frequency domain and unveil the effects of different categories of system nonlinearities on the output frequency response. All these results provide a novel insight to the frequency domain analysis of nonlinear systems, which may be difficult to address with other existing methods in the literature.

5.5 Proofs

Proof of Property 5.3 From Proposition 5.1, $CE(H_Z(\cdot))$ includes all non-repetitive monomial functions of the nonlinear parameters in model (2.11) of the form $C_{pq} \otimes C_{p_1q_1} \otimes C_{p_2q_2} \otimes \cdots \otimes C_{p_kq_k}$, where the subscripts satisfy

$$p+q+\sum_{i=1}^{k'}(p_i+q_i)=Z+k',\ 2\le p_i+q_i\le Z-k',\ 0\le k'\le Z-2,\ 2\le p+q\le Z-k',$$

and noting $1\le p\le Z-k'$, thus $\overset{k}{\underset{i=1}{\otimes}} C_{p_iq_i}$ is included in $CE(H_Z(\cdot))$. Moreover, substitute k by $k+x$, where $x>0$ is an integer, then $Z'=\sum_{i=1}^{k+x}(p_i+q_i)-k-x+1$, which further yields $Z'-Z=\sum_{i=1}^{x}(p_i+q_i)-x$. Note that $2\le p_i+q_i$, thus $Z'-Z\ge\sum_{i=1}^{x}2-x=x$. Therefore, $\overset{k}{\underset{i=1}{\otimes}} C_{p_iq_i}$ must appear in $CE(H_j(j\omega_1,\cdots,j\omega_j))$ for $j>Z$ and but must not appear in the GFRFs of order less than j. This completes the proof. □

Proof of Property 5.4 From Proposition 5.1, any element $c_{p_1,q_1}(\cdot)c_{p_2,q_2}(\cdot)\cdots c_{p_{k_{r_i}},q_{k_{r_i}}}(\cdot)$ in $CE(H_{r_i}(\cdot))$ with $r_i>1$ satisfy

$$r_i = \sum_{i=1}^{k_{r_i}} (p_i + q_i) - k_{r_i} + 1$$

Note that if $r_i=1$, then CE($H_{r_i}(\cdot)$)=1. In this case, suppose $(p_i+q_i)=1$ for consistence. Therefore,

$$\sum_{i=1}^{k} r_i - k + 1 = \left(\sum_{i=1}^{k}\sum_{j=1}^{k_{r_i}} (p_j + q_j) - \sum_{i=1}^{k} k_{r_i} + k\right) - k + 1$$
$$= \sum_{i=1}^{k}\sum_{j=1}^{k_{r_i}} (p_j + q_j) - \sum_{i=1}^{k} k_{r_i} + 1 = Z.$$

This proves the first part of this property. The second part follows from the first part and Property 5.3. □

Proof of Property 5.5 A different proof was given in Proposition 3 of Jing et al. (2006), but here presents a more concise proof based on the properties developed in Sect. 5.2. Applying the CE operator to Eq. (5.5), it can be obtained that

$$CE\big(H_{n,p}(j\omega_1,\cdots,j\omega_n)\big) = \bigoplus_{\substack{r_1\cdots r_p=1\\ \sum r_i = n}}^{n-p+1} \bigotimes_{i=1}^{p} CE(H_{r_i}(\cdot))$$
$$= CE\big(H_{n-p+1}(\cdot)\big) \oplus \left(\bigoplus_{\substack{r_1\cdots r_p=1\\ \sum r_i = n}}^{n-p} \bigotimes_{i=1}^{p} CE(H_{r_i}(\cdot))\right)$$

From Property 5.4, it follows that all the elements in $\bigoplus_{\substack{r_1\cdots r_p=1\\ \sum r_i = n}}^{n-p} \bigotimes_{i=1}^{p} CE(H_{r_i}(\cdot))$ should be Z-order complete, where $Z = \sum_{i=1}^{p} r_i - p + 1 = n - p + 1$. This completes the proof. □

Chapter 6
The Parametric Characteristics of Nonlinear Output Spectrum and Applications

6.1 Introduction

The parametric characteristics of system output spectrum are studied, especially with respect to specific nonlinear parameters of interest. The results are developed based on the GFRFs of the NDE model (2.11) but would be the same for the NARX model (2.10). These results establish the foundation for nonlinear analysis in the frequency domain based on nonlinear output spectrum. Some potential applications of these results are partially demonstrated in this chapter, and more will be developed in the following chapters including nonlinear output spectrum based analysis, nonlinear characteristic output spectrum and so on.

6.2 Parametric Characteristics of Nonlinear Output Spectrum

The nonlinear output spectrum has been discussed in Chaps. 2 and 3. For convenience, it is rewritten here as

$$Y(j\omega) = \sum_{n=1}^{N} Y_n(j\omega) \tag{6.1}$$

when subject to a general input $u(t)$, in (6.1)

$$Y_n(j\omega) = \frac{1}{\sqrt{n}(2\pi)^{n-1}} \int_{\omega_1+\cdots+\omega_n=\omega} H_n(j\omega_1,\cdots,j\omega_n)\prod_{i=1}^{n} U(j\omega_i)d\sigma_\omega \tag{6.2}$$

X. Jing, Z. Lang, *Frequency Domain Analysis and Design of Nonlinear Systems based on Volterra Series Expansion*, Understanding Complex Systems,
DOI 10.1007/978-3-319-12391-2_6

When the input is a specific multi-tone function, i.e.,

$$u(t) = \sum_{i=1}^{\overline{K}} |F_i| \cos(\omega_i t + \angle F_i)$$

in (6.1)

$$Y_n(j\omega) = \frac{1}{2^n} \sum_{\omega_{k_1}+\cdots+\omega_{k_n}=\omega} H_n(j\omega_{k_1}, \cdots, j\omega_{k_n}) F(\omega_{k_1}) \cdots F(\omega_{k_n}) \tag{6.3}$$

where

$$F(\omega_{k_i}) = \left|F_{|k_i|}\right| e^{j\angle F_{|k_i|} \cdot \mathrm{sgn1}(k_i)} \quad \text{for } k_i \in \{\pm 1, \cdots, \pm \overline{K}\}, \text{ and } \mathrm{sgn1}(a)$$
$$= \begin{cases} 1 & a > 0 \\ 0 & a = 0 \\ -1 & a < 0 \end{cases} \quad \text{for } a \in \mathbb{R} \tag{6.4}$$

Definition 6.1 A function $y(h;s)$ is homogeneous of degree d with respect to h if $y(ch;s)=c^d y(h;s)$, where c is a constant, s denotes the independent variables of $y(.)$, and h may be a parameter or a function of certain variables and parameters.

The detailed properties of the functions and variables in Definition 6.1 are not necessarily considered here. The definition of a homogeneous function can also be referred to Rugh (1981). From Definition 6.1, it can be verified that (6.2) and (6.3) are both 1-degree homogeneous with respect to the nth-order GFRF $H_n(\cdot)$. From this definition, the following lemma is obvious.

Lemma 6.1 If $y(h;s_1)$ is a homogeneous function of degree d, and $h(.)$ is a separable function with respect to parameter x whose parametric characteristic function can be written as $h(x)=g(x)f(s_2)$, then $y(h;s_1)$ is a separable function with respect to x and its parametric characteristic function can be written as $y(x)_s = g(x)\,)^{[d]} f_y(f(s_2);s_1)$, where s_1 denotes the un-separable or un-interested parameters or variables in $h(.)$, s_2 denotes some variables in $y(.)$, $f_y(f(s_2);s_1)$ is an appropriate function vector, and $g(x)^{[d]}$ is the d times reduced kronecker product of $g(x)$.

From Proposition 5.1, Lemma 6.1 and (6.1)–(6.2), the following result can be obtained for a homogeneous function $Y(H_n(.); s)$ of degree d, where $H_n(.)$ is the nth-order GFRF.

Proposition 6.1 $Y_n(H_n(j\omega_1, \ldots, j\omega_n); \omega_1, \cdots, \omega_n)$ is a homogeneous function of degree d with respect to the nth-order GFRF $H_n(j\omega_1, \ldots, j\omega_n)$. Then $Y_n(H_n(j\omega_1, \ldots, j\omega_n); \omega_1, \cdots, \omega_n)$ is a separable function with respect to the nonlinear parameters in (5.13), whose parametric characteristic function can be described by

$$Y_n(C(M,K))_{\omega_1,\cdots,\omega_n} = \mathrm{CE}(H_n(j\omega_1,\ \ldots,j\omega_n)))^{[d]}\widetilde{Y}_n(f_n(j\omega_1,\cdots,j\omega_n);\omega_1,\cdots,\omega_n) \tag{6.5}$$

The sensitivity of the homogeneous function with respect to a specific parameter c is

$$\frac{\partial Y_n(C(M,K))_{\omega_1,\cdots,\omega_n}}{\partial c} = \frac{\partial CE(H_n(j\omega_1,\cdots,j\omega_n))^{[d]}}{\partial c} \cdot Y_n(f_n(j\omega_1,\cdots,j\omega_n);\omega_1,\cdots,\omega_n) \tag{6.6}$$

where $\tilde{Y}_n(f_n(j\omega_1,\cdots,j\omega_n);\omega_1,\cdots,\omega_n)$ is an appropriate function vector, and when $d=1$

$$\widetilde{Y}_n(f_n(j\omega_1,\cdots,j\omega_n);\omega_1,\cdots,\omega_n) = Y_n(f_n(j\omega_1,\cdots,j\omega_n);\omega_1,\cdots,\omega_n) \tag{6.7}$$

Proof The results are straightforward from Proposition 5.1, Lemma 6.1 and (6.1)–(6.2).

The following result can be concluded directly from Proposition 6.1 for the output spectrum of system (2.1) described by the NDE or NARX model (2.10) and (2.11).

Corollary 6.1 The output frequency response function $Y(j\omega)$ in (6.1) is separable with respect to the nonlinear parameters in (5.13), whose parametric characteristic function can be described by

$$Y(C(M,K))_{\omega} = \sum_{n=1}^{N} CE(H_n(\cdot)) \cdot Y_n(f_n(\cdot);j\omega) \tag{6.8a}$$

and whose parametric characteristic is

$$CE(Y(j\omega)) = \mathop{\oplus}_{n=1}^{N} CE(H_n(\cdot)) \tag{6.8b}$$

The sensitivity of the output frequency response with respect to a specific parameter c is

$$\frac{\partial Y(j\omega)}{\partial c} = \sum_{n=1}^{N} \frac{\partial CE(H_n(\cdot))}{\partial c} \cdot Y_n(f_n(\cdot);j\omega) \tag{6.9}$$

where, if the input is a general function, then $\omega = \omega_1 + \cdots + \omega_n$,

$$
\begin{aligned}
Y_n(f_n(\cdot);j\omega) &= Y_n(f_n(j\omega_1,\cdots,j\omega_n);j\omega) \\
&= \frac{1}{\sqrt{n}(2\pi)^{n-1}} \int\limits_{\omega_1+\cdots+\omega_n=\omega} f_n(j\omega_1,\cdots,j\omega_n)\prod_{i=1}^{n} U(j\omega_i)d\sigma_\omega
\end{aligned}
\tag{6.10}
$$

if the input is the multi-tone function given in (3.2), then $\omega = \omega_{k_1} + \cdots + \omega_{k_n}$,

$$
\begin{aligned}
Y_n(f_n(\cdot);j\omega) &= Y_n(f_n(j\omega_{k_1},\cdots,j\omega_{k_n});j\omega) \\
&= \frac{1}{2^n} \sum_{\omega_{k_1}+\cdots+\omega_{k_n}=\omega} f_n(j\omega_{k_1},\cdots,j\omega_{k_n})F(\omega_{k_1})\cdots F(\omega_{k_n})
\end{aligned}
\tag{6.11}
$$

□

From these results, it is noted that the system output spectrum can also be expressed by a polynomial function of the nonlinear parameters in *C*(*M*,*K*) based on the parametric characteristics of the GFRFs, and the detailed structure of this polynomial function with respect to any parameters of interest is completely determined by its parametric characteristics. Therefore, how the nonlinear parameters affect the system output spectrum can be studied through the parametric characteristic analysis as discussed in Chap. 5.

Remark 6.1 Note that $CE(H_n(\cdot))$ can be derived from the system model parameters according to the results developed in Chap. 5. Given a specific system described by model (2.10) or model (2.11), $Y(C(M,K)_\omega$ can be obtained by the FFT of the time domain output data from simulations or experiments at frequency ω. Therefore, $Y_n(f_n(\cdot);j\omega)$ for $n=1,\ldots,N$ can be obtained by the Least Square method as mentioned in Remark 4.1. Then $Y_n(C(n,K))_{\omega=\omega_1+\cdots+\omega_n} = CE(H_n(\cdot)) \cdot Y_n(f_n(\cdot);j\omega)$ for $n=1,\ldots,N$ and the sensitivity (6.6) and (6.8a,b) can all be obtained. This provides a numerical method to compute the output spectrum and its each order component which are now determined as analytical polynomial functions of any interested nonlinear parameters. Thus the analysis and design of the output performance of nonlinear systems can now be conducted in terms of these model parameters. Compared with the direct computation by using (2.12)–(2.16) or (2.19)–(2.24) and (3.1)–(3.3) or (2.3)–(2.4), the computational complexity is reduced. Compared with the results in Lang et al. (2007), the parametric characteristic analysis of this study provides an explicit analytical expression for the relationship between system output spectrum and model parameters with detailed polynomial structure up to any order and each order output spectrum component can also be determined. Moreover, let

$$G_n\big(C(M,K,n)\big)_{\omega=\omega_1+\cdots+\omega_n} = \frac{CE(H_n(\cdot))\cdot Y_n(f_n(\cdot);j\omega)}{\frac{1}{\sqrt{n}(2\pi)^{n-1}}\int\limits_{\omega_1+\cdots+\omega_n=\omega}\prod_{i=1}^{n}U(j\omega_i)d\sigma_\omega} \tag{6.12}$$

This is the parametric characteristic function of the nth-order nonlinear output frequency response function defined in Lang and Billings (2005), which can be used for the fault diagnosis of engineering systems and structures. □

6.2.1 *Parametric Characteristics with Respect to Some Specific Parameters in* $\mathbf{C_{p,q}}$

As discussed before, the parametric characteristic vector $CE(H_n(\cdot))$ for all the model parameters of nonlinear degree >1 (referred to as nonlinear parameters) can be obtained according to Proposition 5.1 or (5.17)–(5.18) in Corollary 5.1, and if there are only some parameters of interest, the computation can be conducted by only replacing the other nonzero parameters with 1. In many cases, only several specific model parameters, for example parameters in $C_{p,q}$, are of interest for the analysis of a specific nonlinear system. Thus, the computation of the parametric characteristic vector in (5.17) and (6.8a,b) can be simplified greatly. This section provides some useful results for the computation of parametric characteristics with respect to one or more specific parameters in $C_{\mathrm{p,q}}$, which can effectively facilitate the determination of the OFRF and the analysis based on the OFRF that will be discussed later.

Let

$$\delta(p) = \begin{cases} 1 & \text{if } p=0 \\ 0 & \text{else} \end{cases}, \quad \text{and} \quad pos(x) = \begin{cases} 1 & \text{if } x>0 \\ 0 & \text{else} \end{cases} \tag{6.13}$$

Proposition 6.2 Consider only the nonlinear parameter $C_{p,q}=c$. The parametric characteristic vector of the nth-order GFRF with respect to the parameter c is

$$CE(H_n(j\omega_1,\cdots,j\omega_n)) = \begin{bmatrix} 1 & c & c^2 & \cdots & c^{\left\lfloor\frac{n-1}{p+q-1}\right\rfloor-\delta(p)\cdot pos(n-q)} \end{bmatrix} \tag{6.14}$$

where $\lfloor\cdot\rfloor$ is to get the integer part of (.). □

The Proof of Proposition 6.2 is given in Sect. 6.6. Note that here c may be one parameter or a vector of some parameters of the same nonlinear degree and type in C_{pq}. Also note that $c^n = \underbrace{c\otimes\cdots c\otimes c}_{n}$ and $\otimes$ is the reduced Kronecker product defined in Chap. 4, when c is a vector. Proposition 6.2 establishes a very useful result to study the effects on the output frequency response from a specific nonlinear degree and type of nonlinear parameters. Note also that if some other nonlinear parameters in model (2.10) or (2.11) are zero, only several terms in (6.14) take effect. The detailed form of $CE(H_n(j\omega_1,\cdots,j\omega_n))$ can be derived from Proposition 5.1 or (5.17) in Corollary 5.1. However, a direct use of (6.14) does not affect the final result.

Corollary 6.2 If all the other nonlinear parameters are zero except $C_{p,q}=c$. Then the parametric characteristic vector of the nth-order GFRF with respect to the parameter c is: if (n>p+q and p>0), or (n=p+q), and if additionally $\frac{n-1}{p+q-1}$ is an integer, then

$$CE(H_n(j\omega_1,\cdots,j\omega_n)) = c^{\frac{n-1}{p+q-1}}$$

else

$$CE(H_n(j\omega_1,\cdots,j\omega_n)) = 0$$

which can be summarized as

$$CE(H_n(j\omega_1,\cdots,j\omega_n)) = c^{\frac{n-1}{p+q-1}} \cdot \delta\left(\frac{n-1}{p+q-1} - \left\lfloor\frac{n-1}{p+q-1}\right\rfloor\right) \cdot (1-\delta(p)pos(n-q)) \tag{6.15}$$

Proof The results can be directly achieved from Propositions 5.1 and 6.2. □

Corollary 6.2 provides a more special case of Volterra-type nonlinear systems described by (2.10) or (2.11). There are only several nonlinear parameters of the same nonlinear type and degree in the considered system. This result will be demonstrated in the simulation studies in the next chapter. The following results can be obtained for the output frequency response.

Proposition 6.3 Consider only the nonlinear parameter $C_{p,q}=c$. The parametric characteristic vector of the output spectrum in (6.1) with respect to the parameter c can be written as

$$\begin{aligned} CE(Y(j\omega)) &= \mathop{\oplus}_{n=1}^{N} CE(H_n(\cdot)) \\ &= \left[1 \quad c \quad c^2 \quad \cdots \quad c^{\left\lfloor\frac{N-1}{p+q-1}\right\rfloor - \delta(p)\cdot pos(N-q)\cdot\delta\left(\frac{N-1}{p+q-1}-\left\lfloor\frac{N-1}{p+q-1}\right\rfloor\right)}\right] \end{aligned} \tag{6.16}$$

Then there exists a complex valued function vector $F(j\omega_1,\cdots,j\omega_n;C(M,K)\backslash c)$ with appropriate dimension such that

$$Y(c)_{\omega;C(M,K)\backslash c} = CE(Y(j\omega))\cdot F(j\omega_1,\cdots,j\omega_n;C(M,K)\backslash c) \tag{6.17}$$

If all the other nonlinear parameters are zero except that $C_{p,q}=c\neq 0$ (p+q> 1). Then the parametric characteristic vector of the output spectrum in (6.1) with respect to the parameter c is: if p=0

$$\begin{aligned} CE(Y(j\omega)) &= 1 \oplus CE(H_q(\cdot)) \cdot (1 - pos(q-N)) \\ &= [1 \quad c \cdot (1 - pos(q-N))] \end{aligned} \tag{6.18}$$

else

$$\begin{aligned} CE(Y(j\omega)) &= \bigoplus_{i=0}^{\lfloor N-1/_{p+q-1} \rfloor} CE\big(H_{(p+q-1)i+1}(\cdot)\big) \\ &= \left[1 \quad c \quad c^2 \quad \cdots \quad c^{\lfloor N-1/_{p+q-1} \rfloor}\right] \end{aligned} \tag{6.19}$$

□

The proof of Proposition 6.3 is given in Sect. 6.6. From Corollary 6.2 and Proposition 6.3, it can be seen that different nonlinearities will result in a quite different polynomial structure for the output spectrum, and thus affect the system output frequency response in a different way. By using the results established above, the effect from different nonlinearities on system output frequency characteristics can now be studied. This will be further studied in the following sections.

Moreover, the results above involve the computation of c^n. If c is an I-dimension vector, there will be many repetitive terms involved in c^n. To simplify the computation, the following lemma can be used.

Lemma 6.2 Let be $c=[c_1,c_2,\ldots,c_I]$ which can also be denoted by $c[1:I]$, and $c^n = \underbrace{c \otimes c \cdots \otimes c}_{n}$, "⊗" is the reduced Kronecker product defined in Chap. 4, $n \geq 1$ and $I \geq 1$. Then

$$c^n = \left[c^{n-1} \cdot c_1, \cdots, c^{n-1}\left[s(1)_n - s(i)_n + 1 : s(1)_n\right] \cdot c_i, \cdots, c^{n-1}\left[s(1)_n\right] \cdot c_I\right]$$

where $s(i)_n = \sum_{j=i}^{I} s(j)_{n-1}$, $s(.)_1=1$, and $1 \leq i \leq I$. Moreover, $DIM(c^n)=s(1)_{n+1}$, and the location of c_i^n in c^n is $s(1)_{n+1}-s(i)_{n+1}+1$. □

The Proof of Lemma 6.2 is given in Sect. 6.6.

6.2.2 An Example

To illustrate the results above and to introduce the basic idea of the parametric characteristics based output spectrum analysis that will be discussed further, an example is given in this section. Consider a nonlinear system,

$$a_1\ddot{x} = -a_2x - a_3\dot{x} - c_1\dot{x}^3 - c_2\dot{x}^2x - c_3x^3 + bu(t) \tag{6.20}$$

which is a simple case of model (2.11) with M=3, K=2, $c_{10}(2)=a_1$, $c_{10}(1)=a_3$, $C_{10}(0)=a_2$, $c_{30}(111)=c_1$, $c_{30}(110)=c_2$, $c_{30}(000)=c_3$, $c_{01}(0)=-b$, all other parameters are zero. The GFRFs for system (6.20) can be computed according to (2.19)–(2.24). In the following, the parametric characteristics of the GFRFs for system (4.20) are discussed firstly. As will be seen, the parametric characteristics of the GFRFs provide a useful guidance to the analysis and computation of system frequency response functions.

When all the other nonlinear parameters are zero except $C_{p,q}$, it can be obtained from Corollary 6.2 that the parametric characteristic of the nth-order GFRF with respect to $C_{p,q}$ is

$$CE(H_n(j\omega_1,\cdots,j\omega_n)) = C_{p,q}{}^{\frac{n-1}{p+q-1}} \cdot \delta\left(\frac{n-1}{p+q-1} - \left\lfloor\frac{n-1}{p+q-1}\right\rfloor\right) \cdot (1-\delta(p)pos(n-q)) \tag{6.21}$$

For system (6.20), note that a_1, a_2, a_3 and b are all linear parameters, and the nonzero nonlinear parameters are

$$C_{30} = [\, c_{30}(000) \quad c_{30}(110) \quad c_{30}(111) \,] = [\, c_3 \quad c_2 \quad c_1 \,].$$

Hence,

$$CE(H_n(j\omega_1,\cdots,j\omega_n)) = C_{3,0}{}^{i} = [\, c_3 \quad c_2 \quad c_1 \,]^{i} \quad \text{for n} = 2\text{i}+1,\ \text{i} = 1,2,3,\ldots,$$

else

$$CE(H_n(j\omega_1,\cdots,j\omega_n)) = 0. \tag{6.22}$$

It is easy to compute from (6.22) as follows:

For n=3, $CE(H_3(j\omega_1,\cdots,j\omega_3)) = [\, c_3 \quad c_2 \quad c_1 \,]$;

For n=5, $CE(H_5(j\omega_1,\cdots,j\omega_5)) = [\, c_3 \quad c_2 \quad c_1 \,]^2 = [\, c_3 \quad c_2 \quad c_1 \,] \otimes [\, c_3 \quad c_2 \quad c_1 \,]$
$= [c_3{}^2, c_3c_2, c_3c_1, c_2{}^2, c_2c_1, c_1{}^2]$;

For n=7,

$$CE(H_7(j\omega_1,\cdots,j\omega_7)) = [\, c_3 \quad c_2 \quad c_1 \,]^3 = [\, c_3 \quad c_2 \quad c_1 \,] \otimes [\, c_3 \quad c_2 \quad c_1 \,] \otimes [\, c_3 \quad c_2 \quad c_1 \,]$$
$$= [c_3{}^3, c_3{}^2c_2, c_3{}^2c_1, c_3c_2{}^2, c_3c_2c_1, c_3c_1{}^2, c_2{}^3, c_2{}^2c_1, c_2c_1{}^2, c_1{}^3]$$

From Proposition 5.1, there must exist a complex valued function vector $\bar{f}_n(j\omega_1,\cdots,j\omega_n)$ with appropriate dimension, such that for n=2i+1, i=1,2,3,. . .,

$$H_n(c_1,c_2,c_3)_{(\omega_1,\cdots,\omega_n)} = [c_3 \quad c_2 \quad c_1]^i \cdot \bar{f}_n(j\omega_1,\cdots,j\omega_n) \tag{6.23}$$

else

$$H_n(c_1,c_2,c_3)_{(\omega_1,\cdots,\omega_n)} = 0.$$

When there is only one parameter for example c_1 is of interest for analysis, the parametric characteristic can be obtained by simply letting $C_{3,0}=c_1$ in (6.22), i.e., the parametric characteristic vector is: for n=2i+1 and i=1,2,3,. . .

$$CE(H_n(j\omega_1,\cdots,j\omega_n)) = \begin{bmatrix}1 & c_1 & c_1{}^2 & \cdots & c_1{}^i\end{bmatrix} \tag{6.24}$$

else

$$CE(H_n(j\omega_1,\cdots,j\omega_n)) = 0 \tag{6.25}$$

Thus the parametric characteristic function with respect to the parameter c_1 is: for n=2i+1 and i=1,2,3,. . .

$$H_n(c_1)_{(\omega_1,\cdots,\omega_n;c_2,c_3)} = \begin{bmatrix}1 & c_1 & c_1{}^2 & \cdots & c_1{}^i\end{bmatrix} \cdot \bar{f}_n(j\omega_1,\cdots,j\omega_n;c_2,c_3) \tag{6.26}$$

else

$$H_n(c_1)_{(\omega_1,\cdots,\omega_n;c_2,c_3)} = 0 \tag{6.27}$$

where, $\bar{f}_n(j\omega_1,\cdots,j\omega_n;c_2,c_3)$ is a complex valued function vector with appropriate dimension. The sensitivity of the nth-order GFRFs for n=2i+1 and i=1,2,3,. . . with respect to the parameter c_1 can also be obtained as

$$\frac{\partial H_n(c_1)_{(\omega_1,\cdots,\omega_n;c_2,c_3)}}{\partial c_1} = \begin{bmatrix}0 & 1 & 2c_1 & \cdots & ic^{i-1}\end{bmatrix} \cdot \bar{f}_n(j\omega_1,\cdots,j\omega_n;c_2,c_3) \tag{6.28}$$

Consider the output spectrum of system (6.20). From Proposition 6.3,

$$CE(X(j\omega)) = \mathop{\oplus}_{i=0}^{\lfloor N-1/_2 \rfloor} CE(H_{2i+1}(\cdot)) = \mathop{\oplus}_{i=0}^{\lfloor N-1/_2 \rfloor} C_{30}{}^i \tag{6.29}$$

Suppose the output function of interest is

$$y = a_2x + a_3\dot{x} - c_1\dot{x}^3 - c_2\dot{x}^2x - c_3x^3 \tag{6.30}$$

It will be shown in Sect. 6.4 that

$$CE(Y(j\omega)) = CE(X(j\omega)) \tag{6.31}$$

Then from Proposition 6.3, the parametric characteristic function for the output frequency response $Y(j\omega)$ of system (6.20) with respect to nonlinear parameters c_1, c_2 and c_3 is

$$\begin{aligned} Y(c_1, c_2, c_3)_\omega &= \sum_{i=0}^{\lfloor N-1/2 \rfloor} C_{30}{}^{i} \cdot Y_i(f_i(\cdot); j\omega) \\ &= \left(\bigoplus_{i=0}^{\lfloor N-1/2 \rfloor} C_{30}{}^{i} \right) \cdot \left[Y_0(f_0(\cdot); j\omega) \quad Y_1(f_1(\cdot); j\omega)^T \quad \cdots \quad Y_{\lfloor N-1/2 \rfloor}\left(f_{\lfloor N-1/2 \rfloor}(\cdot); j\omega\right)^T \right]^T \end{aligned} \tag{6.32}$$

For convenience, consider a much simpler case. Let $c_2 = c_3 = 0$, then $C_{30} = c_{30}(111) = c_1$. Therefore the parametric characteristic function in this simple case is

$$\begin{aligned} Y(c_1)_\omega &= Y_0(f_0(\cdot); j\omega) + c_1 \cdot Y_1(f_1(\cdot); j\omega) + \cdots + c_1{}^{\lfloor N-1/2 \rfloor} \cdot Y_{\lfloor N-1/2 \rfloor}\left(f_{\lfloor N-1/2 \rfloor}(\cdot); j\omega\right) \\ &= \left[1 \quad c_1 \quad \cdots \quad c_1{}^{\lfloor N-1/2 \rfloor} \right] \cdot \left[Y_0(f_0(\cdot); j\omega) \quad Y_1(f_1(\cdot); j\omega) \quad \cdots \quad Y_{\lfloor N-1/2 \rfloor}\left(f_{\lfloor N-1/2 \rfloor}(\cdot); j\omega\right) \right]^T \end{aligned} \tag{6.33}$$

As mentioned in Remark 4.1 and Remark 6.1, $\left[Y_0(f_0(\cdot); j\omega) \quad Y_1(f_1(\cdot); j\omega) \quad \cdots \quad Y_{\lfloor N-1/2 \rfloor}\left(f_{\lfloor N-1/2 \rfloor}(\cdot); j\omega\right) \right]^T$ can be computed by a numerical method for a specific input u(t) and at a specific frequency ω. The idea is to obtain $\lfloor N-1/2 \rfloor + 1$ system output frequency responses from $\lfloor N-1/2 \rfloor + 1$ simulations or experimental tests on the system (6.20) under $\lfloor N-1/2 \rfloor + 1$ different values of the nonlinear parameter c_1 and the same input u(t), then yielding

$$\begin{bmatrix} Y(j\omega)_0 \\ Y(j\omega)_1 \\ \vdots \\ Y(j\omega)_{\lfloor N-1/2 \rfloor} \end{bmatrix} = \begin{bmatrix} 1 & c_1(0) & \cdots & c_1(0)^{\lfloor N-1/2 \rfloor} \\ 1 & c_1(1) & \cdots & c_1(1) \\ 1 & \vdots & \ddots & \vdots \\ 1 & c_1(\lfloor N-1/2 \rfloor) & \cdots & c_1(\lfloor N-1/2 \rfloor)^{\lfloor N-1/2 \rfloor} \end{bmatrix} \cdot \begin{bmatrix} Y_0(f_0(\cdot); j\omega) \\ Y_1(f_1(\cdot); j\omega) \\ \vdots \\ Y_{\lfloor N-1/2 \rfloor}\left(f_{\lfloor N-1/2 \rfloor}(\cdot); j\omega\right) \end{bmatrix} \tag{6.34}$$

Hence,

$$\begin{bmatrix} Y_0(f_0(\cdot);j\omega) \\ Y_1(f_1(\cdot);j\omega) \\ \vdots \\ Y_{\lfloor N-1/_2 \rfloor}\left(f_{\lfloor N-1/_2 \rfloor}(\cdot);j\omega\right) \end{bmatrix} = \begin{bmatrix} 1 & c_1(0) & \cdots & c_1(0)^{\lfloor N-1/_2 \rfloor} \\ 1 & c_1(1) & \cdots & c_1(1) \\ 1 & \vdots & \ddots & \vdots \\ 1 & c_1(\lfloor N-1/_2 \rfloor) & \cdots & c_1(\lfloor N-1/_2 \rfloor)^{\lfloor N-1/_2 \rfloor} \end{bmatrix}^{-1} \cdot \begin{bmatrix} Y(j\omega)_0 \\ Y(j\omega)_1 \\ \vdots \\ Y(j\omega)_{\lfloor N-1/_2 \rfloor} \end{bmatrix} \quad (6.35)$$

Then (6.33) is determined explicitly, which is an analytical function of the nonlinear parameter c_1. The system output frequency response can therefore be analyzed and optimized in terms of the nonlinear parameters. And also from (6.33), the sensitivity of the system output frequency response with respect to the nonlinear parameter, and the nonlinear output frequency response function defined in (6.12) can both be studied. For more complicated cases, a similar process can be followed to conduct a required analysis and design in terms of multiple nonlinear parameters for model (2.11). Compared with the results in Lang et al. (2007), since the detailed polynomial structure for the output spectrum up to any order can be explicitly determined, this can greatly reduce the simulation amount needed in the numerical method when multiple parameters are considered.

6.3 Parametric Characteristic Analysis of Nonlinear Effects on System Output Frequencies

As discussed in Sect. 4.3, different nonlinearities may result in different output spectrum characteristics for a system. It is known that nonlinear systems have more abundant output frequencies than the driving frequencies, which demonstrates an energy transferring phenomenon and usually is difficult to predict when and where an output will happen. In order to achieve a desired output spectrum at certain given frequencies, the system should be properly designed to include or exclude some appropriate nonlinearity. In this subsection, how the different nonlinear parameters affect the system output frequencies when subject to a harmonic input is studied to further demonstrate the usefulness of the parametric characteristic analysis method. Consider the NDE model (2.11) subject to a harmonic input

$$u(t) = F_d \sin(\Omega t) \quad (6.36)$$

In this case, $F(\omega_{k_1})$ in the nth-order output spectrum (3.2) is

$$F(\omega_{k_l}) = -jk_l F_d, \text{ for } k_l = \pm 1, \ \omega_{k_l} = k_l \Omega, \text{ and } l = 1, \cdots, n \quad (6.37)$$

Thus (3.3) can be written as

$$Y_n(j\omega) = \frac{1}{2^n} \sum_{\omega_{k_1}+\cdots+\omega_{k_n}=\omega} H_n(j\omega_{k_1},\cdots,j\omega_{k_n}) \cdot F_d^n \cdot (k_1\cdots k_n) \cdot (-j)^n$$
$$= \frac{C_F(n)}{2^n} \sum_{\omega_{k_1}+\cdots+\omega_{k_n}=\omega} H_n(j\omega_{k_1},\cdots,j\omega_{k_n}) \tag{6.38}$$

where

$$C_F(n) = F_d^n \cdot (k_1\cdots k_n) \cdot (-j)^n \tag{6.39}$$

From (6.38), it can be seen that the output frequency range for the nth-order output spectrum in this case is completely determined by $\omega_{k_1} + \cdots + \omega_{k_n} = \omega$ and the nth-order GFRF. Note that $\omega_{k_l} = \pm\Omega$ for $l=1,\cdots,n$. Thus the output frequency range for the nth-order output spectrum can only possibly be $\{k\Omega | k=0,1,2,\ldots\}$, which can be written as

$$W_n = \{\omega | \omega = (\omega_{k_1} + \cdots + \omega_{k_n}) \cdot (1 - \delta(H_n(\cdot))),\ \omega_{k_l} = \pm\Omega,\ 1 \le l \le n\} \tag{6.40}$$

Therefore the output frequency range for the system output spectrum is

$$W_O = \bigcup_{1\le n\le N} W_n \tag{6.41}$$

Obviously, when certain orders of the GFRFs are zero, it will lead to no output spectrum at certain frequencies. From Proposition 5.1, (6.40) can also be written as

$$W_n = \{\omega | \omega = (\omega_{k_1} + \cdots + \omega_{k_n}) \cdot (1 - \delta(CE(H_n(\cdot)))),\ \omega_{k_l} = \pm\Omega,\ 1 \le l \le n\} \tag{6.42}$$

This demonstrates the effects of nonlinear parameters on the system output frequency characteristics. From (6.42), the following lemma is straightforward. In what follows, $\omega = 2l\Omega$ for l=0,1,2,... are called even frequencies, and $\omega = (2l+1)\Omega$ for l=0,1,2,... are called odd frequencies.

Lemma 6.3 Consider the output frequencies of model (2.11) when subject to a harmonic input (6.36). W_n can only include even frequencies, when n is an even number, and W_n only includes odd frequencies, when n is an odd number. □

Consider the effects of different nonlinear parameters in the following cases.

(A) Pure Input Nonlinearities

When there are only input nonlinearity in model (2.11) and supposing $C_{0,n} \neq 0$, then from Sect. 5.3.1A, $CE(H_n(\cdot)) = C_{0,n}$. Equation (6.42) can be written as

$$W_n = \{\omega|\omega = (\omega_{k_1} + \cdots + \omega_{k_n}) \cdot (1 - \delta(C_{0,n})),\ \omega_{k_l} = \pm\Omega,\ 1 \le l \le n\} \quad (6.43)$$

which further yields

$$W_n = \begin{cases} \{2l\Omega \cdot (1 - \delta(C_{0,n}))\ |0 \le l \le n/2, n \ge 2\ \} & \text{if n is an even integer} \\ \{(2l+1)\Omega \cdot (1 - \delta(C_{0,n}))\ |0 \le l \le (n-1)/2, n \ge 3\ \} & \text{if n is an odd integer} \end{cases} \quad (6.44)$$

Obviously $W_1 = \{\Omega\}$, which represents the linear output frequency. Equation (6.44) shows that for any a pure input nonlinear term of nonlinear degree n, it will introduce a finite output frequency range from 0 to $n\Omega$. If the nonlinear degree n is an even integer, the introduced output spectrum will appear at even frequencies, and if the nonlinear degree n is an odd integer, it will appear at odd frequencies.

Example 6.1 To verify the result above, a simulation result is provided for system (6.20) with a_1=1, a_2=1, a_3=0.5, c_1=c_2=c_3=0, b=u(t)^4 or u(t)^5. See Fig. 6.1.

(B) Output Nonlinearity and Input–Output Cross Nonlinearity
When all the other nonlinear parameters are zero except for $C_{p,q}$, then from (5.20d) and Sect. 5.3.1C, the parametric characteristic of the nth-order GFRF can be summarized as for n>1

$$CE(H_n(j\omega_1,\cdots,j\omega_n))_{C_{p,q}} = C_{p,q}^{\frac{n-1}{p+q-1}} \cdot step(n-p-q) \cdot \delta\left(\frac{n-1}{p+q-1} - \left\lfloor\frac{n-1}{p+q-1}\right\rfloor\right) \quad (6.45)$$

where

$$step(x) = \begin{cases} 1 & \text{if } x \ge 0 \\ 0 & \text{else} \end{cases} \quad (6.46)$$

Equation (6.45) demonstrates the independent contribution from parameters in $C_{p,q}$ to the nth-order GFRF. Only when $\frac{n-1}{p+q-1}$ is an integer and $n \ge p+q$, then $CE(H_n(j\omega_1,\cdots,j\omega_n))_{C_{p,q}}$ is nonzero. In this case, it can be obtained from (6.42) that

for n=1,

$$W_1 = \{\Omega\} \quad (6.47)$$

for n>1, and if $\frac{n-1}{p+q-1}$ is an integer and $n \ge p+q$,

$$W_n = \left\{\omega\middle|\omega = (\omega_{k_1} + \cdots + \omega_{k_n}) \cdot \left(1 - \delta\left(C_{p,q}^{\frac{n-1}{p+q-1}}\right),\ \omega_{k_l} = \pm\Omega,\ 1 \le l \le n\right\} \quad (6.48)$$

else

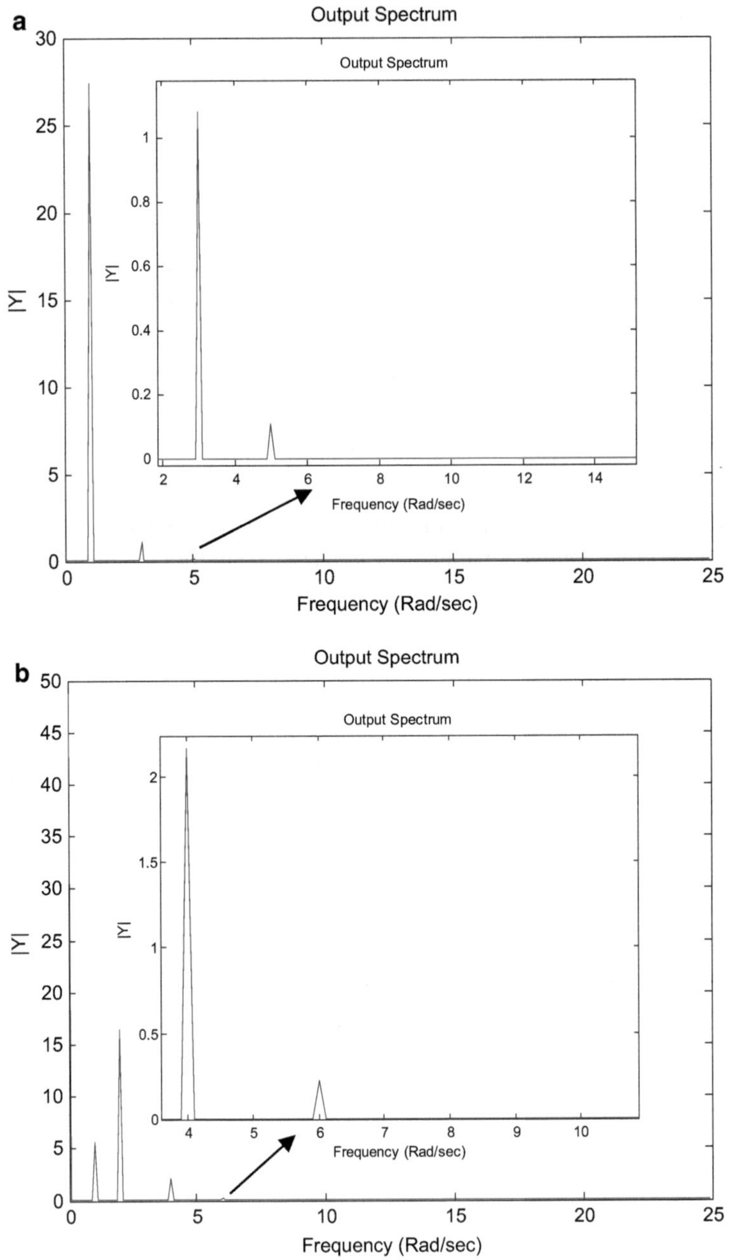

Fig. 6.1 Output spectrum of a nonlinear system with only one input nonlinear term having coefficient $c_{0,5}(.)$ (**a**) or $c_{0,6}(.)$ (**b**), subject to the input $u(t) = 10\sin(t)$ (Jing et al. 2009a)

$$W_n = \{\} \tag{6.49}$$

From (6.48), the nonlinear terms with coefficients $C_{p,q}$ will bring output spectrum at some frequencies only when $\frac{n-1}{p+q-1}$ is an integer and $n \geq p+q$. Let

$$\overline{p} = \frac{n-1}{p+q-1} \tag{6.50}$$

which is a positive integer and can go to infinity when n goes to infinity.

If $p+q=2l$ for $l=1,2,3,\ldots$, i.e., the nonlinear degree is an even integer, then from (6.50)

$$n = (2l-1)\overline{p} + 1 \tag{6.51}$$

In this case, (6.51) can not only give an even number but also an odd number. That is, for a nonlinear parameter with even nonlinear degree, it can make an independent contribution to both even and odd order of GFRFs. Therefore, from Lemma 1 an even degree nonlinear parameter will drive the system to have output spectrum at all the even and odd frequencies. Similarly, if $p+q=2l+1$ for $l=1,2,3,\ldots$, then from (6.50)

$$n = 2l\overline{p} + 1 \tag{6.52}$$

In this case, (6.52) must give an odd integer.

The following conclusion is straightforward from the discussions above.

Proposition 6.4 Consider the output frequencies of model (2.11) when subject to a harmonic input (6.36). For any nonlinear term with coefficient $c_{p,q}(.)$, where $p+q>1$ and $p>0$, if the nonlinear degree $p+q$ is an odd integer, it will bring super-harmonics to the system output spectrum only at these frequencies which are odd integer multiples of the input frequency; if $p+q$ is an even integer, it will introduce super-harmonics at all frequencies which are nonnegative integer multiples of the input frequency. □

These results can be verified by a simple example in simulation below.

Example 6.2 Consider system (6.20) subject to the harmonic input $u(t)=5\sin(t)$ with $a_1=1$, $a_2=1$, $a_3=0.5$, $c_3=0$, $b=1$ and two cases: $c_1=c_{3,0}(111)$, $c_2=0$ and $c_1=0$, $c_2=c_{6,0}(000111)\,x^2\dot{x}$. See Fig. 6.2, where the output frequency responses of the two cases are given. It can be seen that the harmonics happen only at odd frequencies when subject to the third degree nonlinearity; however, there are harmonics at all the integer frequencies when subject to the sixth degree nonlinearity.

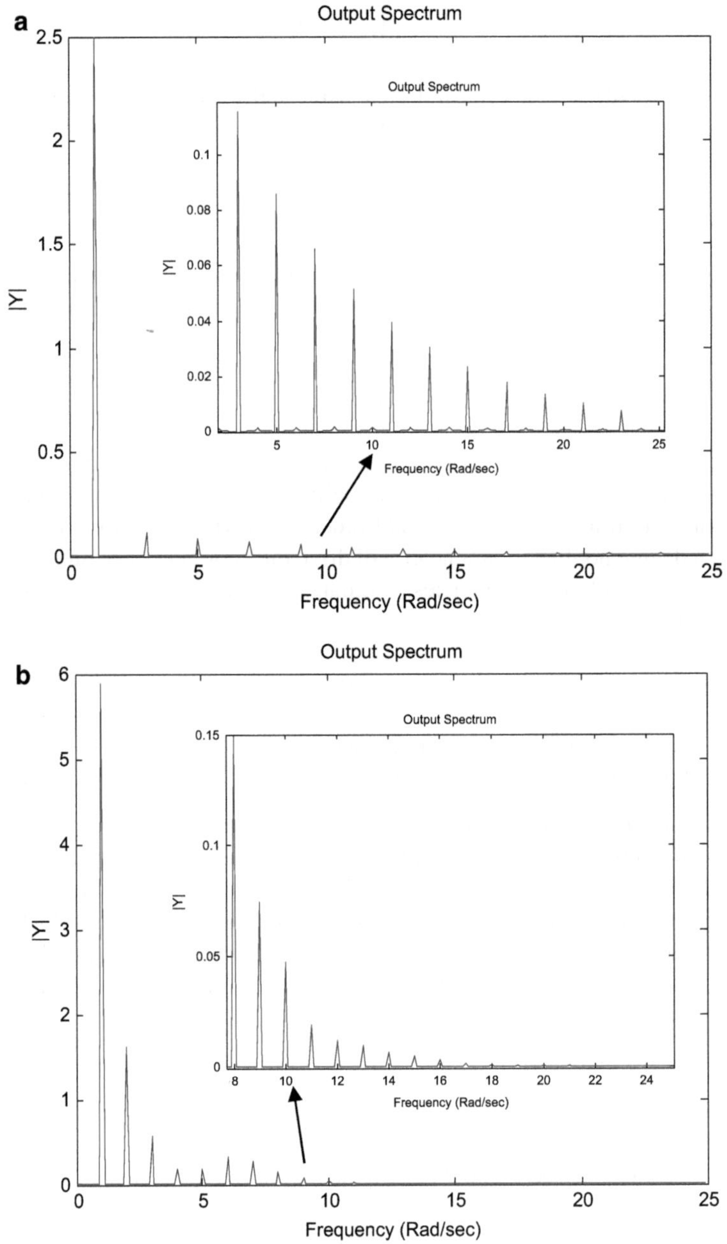

Fig. 6.2 Output spectrum of a nonlinear system with only one output nonlinear term having coefficient $c_{3,0}(.)$ (**a**) or $c_{6,0}(.)$ (**b**) (Jing et al. 2009a)

6.4 Parametric Characteristics of a Single Input Double Output Nonlinear System

Consider the SIDO nonlinear system in (2.28) or (2.34), which has a general nonlinear output, and can be frequently encountered in practice as mentioned before. The parametric characteristics of this kind of nonlinear systems are studied in this section. For convenience, the SIDO NARX system is written as follows,

$$x(t) = \sum_{m=1}^{M_1}\sum_{p=0}^{m}\sum_{k_1,k_m=0}^{K} \bar{c}_{p,m-p}(k_1,\cdots,k_m)\prod_{i=1}^{p} x(t-k_i)\prod_{i=p+1}^{m} u(t-k_i) \tag{6.53a}$$

$$y(t) = \sum_{m=1}^{M_2}\sum_{p=0}^{m}\sum_{k_1,k_m=0}^{K} \tilde{c}_{p,m-p}(k_1,\cdots,k_m)\prod_{i=1}^{p} x(t-k_i)\prod_{i=p+1}^{m} u(t-k_i) \tag{6.53b}$$

The notations and corresponding definitions can be referred to Sect. 2.3. The GFRFs of this system are given in (2.30)–(2.33). From the GFRFs of model (6.53a,b), the output frequency response of (6.53a,b) can also be derived readily by extending the results in (3.1) and (3.3). Regard x(t) and y(t) as two outputs actuated by the same input u(t), then

$$X(j\omega) = \sum_{n=1}^{N}\frac{1}{\sqrt{n}(2\pi)^{n-1}}\int_{\omega_1+\cdots+\omega_n=\omega} H_n^x(j\omega_1,\cdots,j\omega_n)\prod_{i=1}^{n} U(j\omega_i)d\sigma_\omega \tag{6.54a}$$

$$Y(j\omega) = \sum_{n=1}^{N}\frac{1}{\sqrt{n}(2\pi)^{n-1}}\int_{\omega_1+\cdots+\omega_n=\omega} H_n^y(j\omega_1,\cdots,j\omega_n)\prod_{i=1}^{n} U(j\omega_i)d\sigma_\omega \tag{6.54b}$$

When the system input is a multi-tone signal (3.2), then the system output frequency response can be similarly derived as:

$$X(j\omega) = \sum_{n=1}^{N}\frac{1}{2^n}\sum_{\omega_{k_1}+\cdots+\omega_{k_n}=\omega} H_n^x(j\omega_{k_1},\cdots,j\omega_{k_n})F(\omega_{k_1})\cdots F(\omega_{k_n}) \tag{6.55a}$$

$$Y(j\omega) = \sum_{n=1}^{N}\frac{1}{2^n}\sum_{\omega_{k_1}+\cdots+\omega_{k_n}=\omega} H_n^y(j\omega_{k_1},\cdots,j\omega_{k_n})F(\omega_{k_1})\cdots F(\omega_{k_n}) \tag{6.55b}$$

where $F(\omega) = \begin{cases} |F_i|e^{j\angle F_i} & \text{if } \omega \in \{\omega_k, k = \pm 1,\cdots,\pm K\} \\ 0 & \text{else} \end{cases}, \omega_{\pm k} = \pm\omega_k.$

It can be seen from the results above that the frequency response functions for nonlinear systems are quite different from those for linear systems. It is known that in a linear system, frequency response functions of different parts can be combined together via addition or multiplication. This is not the case for nonlinear systems.

For instance, if x(t) is only regarded as an input in (6.53b) independent of (6.53a), then the GFRFs $H_n^y(j\omega_1,\cdots,j\omega_n)$ and therefore the output spectrum $Y(j\omega)$ will all be changed completely for n>1, since in this case there are only input nonlinearities in (6.53b) and no output nonlinearities. Even so, it can also be seen from (6.54a,b) and (6.55a,b) that the output frequency ranges for both x(t) and y(t) are the same one, *i.e.*,

$$\bigcup_{n=1}^{N}\{\omega|\omega=\omega_1+\cdots+\omega_n,\omega_i\in R_\omega\} \tag{6.56}$$

where R_ω represents the input frequency range, for example $R_\omega=\{\omega_k,k=\pm1,\cdots,\pm K\}$ for the multi-tone signal (3.2).

As discussed before, the parametric characteristic analysis presented in Chaps. 4 and 5 can be used to reveal which model parameters contribute to and how these parameters affect system frequency response functions, and thus the explicit relationship between system frequency response and system time domain model parameters can be unveiled. In this section, the parameter characteristics of the output frequency response function related to the output y(t) of model (6.53a,b) with respect to nonlinear parameters are studied, and the nonlinear parameters in (6.53a) are focused since nonlinear parameters in (6.53b) has no effect on system dynamics. In what follows, let $\overline{C}(n)=\{\overline{c}_{p,q}(k_1\cdots k_{p+q})|1<p+q\le n,0\le k_i\le K,1\le i\le p+q\}$ denotes all the nonlinear parameters in (6.53a) with degree from 2 to n, and similarly denote all the parameters in (6.53b) with degree from 2 to n as: $\widetilde{C}(n)=\{\widetilde{c}_{p,q}(k_1\cdots k_{p+q})|1<p+q\le n,0\le k_i\le K,1\le i\le p+q\}$. All the $(p+q)$th degree nonlinear parameters in (6.53a,b) of form $c_{p,q}(.)$ construct a vector denoted by

$$C_{p,q}=\left[c_{p,q}(0,\cdots,0),c_{p,q}(0,\cdots,1),\cdots,c_{p,q}\left(\underbrace{K,\cdots,K}_{p+q}\right)\right]$$

In what follows, $CE(H_{CF})_\vartheta$ means to only extract the parameters in the set ϑ from H_{CF}, and without specialty $CE(H_{CF})$ means to extract all the nonlinear parameters (i.e., its nonlinearity degree >1) appearing in H_{CF}.

6.4.1 *Parametric Characteristic Analysis for* $\mathbf{H}_n^x(\mathbf{j}\omega_1,\cdots,\mathbf{j}\omega_n)$

Application of the CE operator to a complicated series for its parametric characteristics can be performed by directly replacing the addition and multiplication in the series by "⊕" and "⊗" respectively.

The parametric characteristic of the nth-order GFRF $H_n^x(j\omega_1,\cdots,j\omega_n)$ with respect to nonlinear parameters $\overline{C}(n)$ is

$$
\begin{aligned}
CE\left(H_n^x(j\omega_1,\cdots,j\omega_n)\right) &= CE\left(\frac{H_{n_u}^x(j\omega_1,\cdots,j\omega_n)+H_{n_{ux}}^x(j\omega_1,\cdots,j\omega_n)+H_{n_x}^x(j\omega_1,\cdots,j\omega_n)}{L_n(j(\omega_1+\cdots+\omega_n))}\right) \\
&= CE\left(H_{n_u}^x(j\omega_1,\cdots,j\omega_n)\right)\oplus CE\left(H_{n_{ux}}^x(j\omega_1,\cdots,j\omega_n)\right)\oplus CE\left(H_{n_x}^x(j\omega_1,\cdots,j\omega_n)\right) \\
&= \overline{C}_{0,n}\oplus\left(\mathop{\oplus}_{q=1}^{n-1}\mathop{\oplus}_{p=1}^{n-q}\overline{C}_{p,q}\otimes CE\left(H_{n-q,p}(\cdot)\right)\right)\oplus\left(\mathop{\oplus}_{p=2}^{n}\overline{C}_{p,0}\otimes CE\left(H_{n,p}(\cdot)\right)\right)
\end{aligned}
\tag{6.57}
$$

where

$$
\begin{aligned}
CE\left(H_{n,p}(\cdot)\right) &= \mathop{\oplus}_{i=1}^{n-p+1} CE\left(H_i^x(\cdot)\right)\otimes CE\left(H_{n-i,p-1}(\cdot)\right) \text{ or } CE\left(H_{n,p}(\cdot)\right) \\
&= \mathop{\oplus}_{\substack{r_1\cdots r_p=1 \\ \sum r_i=n}}^{n-p+1} \mathop{\otimes}_{i=1}^{p} CE\left(H_{r_i}^x(\cdot)\right)
\end{aligned}
\tag{6.58}
$$

$$
CE(H_{n,1}(\cdot)) = CE\left(H_n^x(\cdot)\right)) \tag{6.59}
$$

Note that in (6.57), $E(1/L_n(j(\omega_1+\cdots+\omega_n)))=1$ since there are no nonlinear parameters (in the set $\overline{C}(n)$) in $1/L_n(j(\omega_1+\cdots+\omega_n))$. It is shown in Chap. 5 that

$$
CE\left(H_{n,p}(\cdot)\right) = CE\left(H_{n-p+1}^x(\cdot)\right) \tag{6.60}
$$

and thus (6.57) is simplified as

$$
\begin{aligned}
&CE\left(H_n^x(j\omega_1,\cdots,j\omega_n)\right) \\
&\quad = \overline{C}_{0,n}\oplus\left(\mathop{\oplus}_{q=1}^{n-1}\mathop{\oplus}_{p=1}^{n-q}\overline{C}_{p,q}\otimes CE\left(H_{n-q-p+1}^x(\cdot)\right)\right) \\
&\quad \oplus\left(\overline{C}_{n,0}\oplus\mathop{\oplus}_{p=2}^{\lfloor n+1/2\rfloor}\overline{C}_{p,0}\otimes CE\left(H_{n-p+1}^x(\cdot)\right)\right)
\end{aligned}
\tag{6.61}
$$

From (6.61), $CE(H_n^x(j\omega_1,\cdots,j\omega_n))$ has no relationship with $\widetilde{C}(n)$. With the parametric characteristics (6.61), it can be concluded (referring to Chap. 5) that there must exist a complex valued function vector $f_n(j\omega_1,\cdots,j\omega_n)$ with appropriate dimension, such that

$$
H_n^x(j\omega_1,\cdots,j\omega_n) = CE\left(H_n^x(j\omega_1,\cdots,j\omega_n)\right)\cdot f_n(j\omega_1,\cdots,j\omega_n) \tag{6.62}
$$

Equation (6.62) provides an explicit expression for the relationship between nonlinear parameters $\overline{C}(n)$ and the nth-order GFRF from u(t) to x(t). For any parameter of interest, how its effect is on the GFRFs can be revealed by checking $CE(H_n^x(j\omega_1,\cdots,j\omega_n))$. From (6.62), $H_n^x(j\omega_1,\cdots,j\omega_n)$ is in fact a polynomial function of parameters in $\overline{C}(n)$ which define system nonlinearities, thus some qualitative properties of $H_n^x(j\omega_1,\cdots,j\omega_n)$ can also be indicated by $CE(H_n^x(j\omega_1,\cdots,j\omega_n))$. Moreover, using (6.62), (6.54a) can be written as

$$X(j\omega) = \sum_{n=1}^{N} CE\left(H_n^x(j\omega_1,\cdots,j\omega_n)\right) \cdot \overline{F}_n(j\omega) \tag{6.63}$$

where $\overline{F}_n(j\omega) = \frac{1}{\sqrt{n}(2\pi)^{n-1}} \int\limits_{\omega_1+\cdots+\omega_n=\omega} f_n(j\omega_1,\cdots,j\omega_n) \cdot \prod_{i=1}^{n} U(j\omega_i) d\sigma_\omega$. This is the parametric characteristic function expression for the output $X(j\omega)$. By using this expression, $X(j\omega)$ can be obtained by following a numerical method without complicated computation involved in (2.30)–(2.33).

6.4.2 *Parametric Characteristic Analysis for* $\mathbf{H}_n^y(\mathbf{j}\boldsymbol{\omega}_1,\cdots,\mathbf{j}\boldsymbol{\omega}_n)$

To study the parametric characteristic of the nth-order GFRF $H_n^y(j\omega_1,\cdots,j\omega_n)$ with respect to only model nonlinear parameters in $\overline{C}(n)$, the parametric characteristic with respect to model parameters in $\overline{C}(n)$ and $\widetilde{C}(n)$ are derived first and then the case with respect only to nonlinear parameters in $\overline{C}(n)$ is discussed.

Applying the CE operator to (2.32) yields,

$$\begin{aligned} CE(H_n^y(j\omega_1,\cdots,j\omega_n)) &= CE\left(H_{n_u}^y(j\omega_1,\cdots,j\omega_n)\right) \oplus CE\left(H_{n_{ux}}^y(j\omega_1,\cdots,j\omega_n)\right) \oplus CE\left(H_{n_x}^y(j\omega_1,\cdots,j\omega_n)\right) \\ &= \widetilde{C}_{0,n} \oplus \left(\mathop{\oplus}_{q=1}^{n-1} \mathop{\oplus}_{p=1}^{n-q} \widetilde{C}_{p,q} \otimes CE\left(H_{n-q,p}(j\omega_1,\cdots,j\omega_{n-q})\right)\right) \\ &\quad \oplus \left(\mathop{\oplus}_{p=1}^{n} \widetilde{C}_{p,0} \otimes CE\left(H_{n,p}(j\omega_1,\cdots,j\omega_n)\right)\right) \end{aligned}$$

using (6.60), which further gives

$$\begin{aligned} &CE\left(H_n^y(j\omega_1,\cdots,j\omega_n)\right) \\ &= \widetilde{C}_{0,n} \oplus \left(\mathop{\oplus}_{q=1}^{n-1} \mathop{\oplus}_{p=1}^{n-q} \widetilde{C}_{p,q} \otimes CE\left(H_{n-q-p+1}^x(j\omega_1,\cdots,j\omega_{n-q})\right)\right) \\ &\oplus \left(\mathop{\oplus}_{p=1}^{n} \widetilde{C}_{p,0} \otimes CE\left(H_{n-p+1}^x(j\omega_1,\cdots,j\omega_n)\right)\right) \end{aligned} \tag{6.64}$$

Thus the parametric characteristic of $H_n^y(j\omega_1,\cdots,j\omega_n)$ with respect to both nonlinear parameters in $\overline{C}(n)$ and $\widetilde{C}(n)$ is obtained.

Especially, if $\widetilde{C}(n)$ is independent of $\overline{C}(n)$, the parametric characteristic of $H_n^y(j\omega_1,\cdots,j\omega_n)$ with respect to nonlinear parameters in $\widetilde{C}(n)$ can be written as

$$CE\left(H_n^y(j\omega_1,\cdots,j\omega_n)\right)_{\widetilde{C}(n)} = \widetilde{C}_{0,n} \oplus \left(\mathop{\oplus}_{q=1}^{n-1} \mathop{\oplus}_{p=1}^{n-q} \widetilde{C}_{p,q}\right) \oplus \left(\mathop{\oplus}_{p=2}^{n} \widetilde{C}_{p,0}\right) \tag{6.65}$$

Therefore, in this case $H_n^y(j\omega_1,\cdots,j\omega_n)$ can be expressed as a polynomial function of $\widetilde{C}(n)$ as

$$H_n^y\left(j\omega_1,\cdots,j\omega_n;\widetilde{C}(n)\right) = CE\left(H_n^y(j\omega_1,\cdots,j\omega_n)\right)_{\widetilde{C}(n)} \cdot f_n\left(j\omega_1,\cdots,j\omega_n;\overline{C}(n)\right) \tag{6.66}$$

where $f_n\left(j\omega_1,\cdots,j\omega_n;\overline{C}(n)\right)$ is a complex valued function vector with an appropriate dimension, which is also a function of the parameters in $\overline{C}(n)$ in this case. From (6.65), it can be seen that $CE\left(H_n^y(j\omega_1,\cdots,j\omega_n)\right)_{\widetilde{C}(n)}$ is a vector which is composed of all the elements in $\widetilde{C}(n)$. That is, the nth-order GFRF is a polynomial function of all the parameters in $\widetilde{C}(n)$ if $\widetilde{C}(n)$ is independent of $\overline{C}(n)$. This conclusion is straightforward. The case where $\widetilde{C}(n)$ is dependent on $\overline{C}(n)$ will be discussed in the following section.

6.4.2.1 Parametric Characteristics of $H_n^y(j\omega_1,\cdots,j\omega_n)$ with Respect to $\overline{C}(n)$

What is of more interest is the parametric characteristic of $H_n^y(j\omega_1,\cdots,j\omega_n)$ with respect to nonlinear parameters in $\overline{C}(n)$ which define system nonlinear dynamics. Consider two cases as follows.

(1) $\widetilde{C}(n)$ has no relationship with $\overline{C}(n)$

In this case, it can be derived from (6.64) that

$$\begin{aligned} CE\left(H_n^y(j\omega_1,\cdots,j\omega_n)\right)_{\overline{C}(n)} = &\left(\mathop{\oplus}_{q=1}^{n-1} \mathop{\oplus}_{p=1}^{n-q} \left(1-\delta\left(\widetilde{C}_{p,q}\right)\right)\cdot CE\left(H_{n-q-p+1}^x(j\omega_1,\cdots,j\omega_{n-q})\right)\right) \\ &\otimes \oplus\left(\mathop{\oplus}_{p=1}^{n} \left(1-\delta\left(\widetilde{C}_{p,0}\right)\right)\cdot CE\left(H_{n-p+1}^x(j\omega_1,\cdots,j\omega_n)\right)\right) \end{aligned} \tag{6.67}$$

where $\delta(C_{p,q}) = \begin{cases} 0 & C_{p,q} \neq 0 \\ 1 & C_{p,q} = 0 \end{cases}$. From (6.67) it can be seen that $CE\left(H_n^y(j\omega_1,\cdots,j\omega_n)\right)_{\overline{C}(n)}$ is the summation by "$\oplus$" of parametric characteristics of some GFRFs for x(t) from the 1st order to the nth order. From the definition of operation "$\oplus$", the repetitive terms should not be counted. Therefore, (6.67) is simplified as

$$CE(H_n^y(j\omega_1,\cdots,j\omega_n))_{\overline{C}(n)} = \mathop{\oplus}_{p=1}^{n} \chi(n,p) \cdot CE\left(H_{n-p+1}^x(j\omega_1,\cdots,j\omega_n)\right)_{\overline{C}(n-p+1)} \tag{6.68}$$

where

$$\chi(n,p) = 1 - \delta\left(\sum_{\substack{0 \le q \le n-1,\, 1 \le p' \le n-q \\ p'+q=p}} \left(1 - \delta\left(\widetilde{C}_{p',q}\right)\right)\right) \tag{6.69}$$

Equation (6.69) means that if there is at least one nonzero $\widetilde{C}_{p',q}$ then the corresponding $CE(H_{n-q-p+1}^x(j\omega_1,\cdots,j\omega_{n-q}))$ will be counted in (6.68). According to Proposition 5.1 in Chap. 5, it follows from (6.68) that the *n*th-order GFRF for y(t) has relationship with all the nonlinear parameters in $\overline{C}(n)$ of degree from 2 to n' in this case, where $n' \le n$.

(2) $\widetilde{C}(n)$ *has linear relationship with* $\overline{C}(n)$ *by* $\widetilde{c}_{p,q}(\cdot) = \widetilde{\alpha} + \widetilde{\beta}\overline{c}_{p,q}(\cdot)$ *for some real number* α *and* β

Note that applying the CE operator to $\widetilde{c}_{p,q}(\cdot) = \widetilde{\alpha} + \widetilde{\beta}\overline{c}_{p,q}(\cdot)$ for the nonlinear parameter $\overline{c}_{p,q}(\cdot)$ gives $CE(\widetilde{c}_{p,q}(\cdot)) = CE\left(\widetilde{\alpha} + \widetilde{\beta}\overline{c}_{p,q}(\cdot)\right) = \overline{c}_{p,q}(\cdot)$, *i.e.*, $CE\left(\widetilde{C}_{p,q}\right) = \overline{C}_{p,q}$. Hence, in this case (6.64) should be

$$\begin{aligned} &CE(H_n^y(j\omega_1,\cdots,j\omega_n)) \\ &= \overline{C}_{0,n} \oplus \left(\mathop{\oplus}_{q=1}^{n-1} \mathop{\oplus}_{p=1}^{n-q} \overline{C}_{p,q} \otimes CE\left(H_{n-q-p+1}^x(j\omega_1,\cdots,j\omega_{n-q})\right)\right) \\ &\otimes \oplus \left(\mathop{\oplus}_{p=1}^{n} \overline{C}_{p,0} \otimes CE\left(H_{n-p+1}^x(j\omega_1,\cdots,j\omega_n)\right)\right) \end{aligned} \tag{6.70}$$

Equation (6.70) can be further written as

$$\begin{aligned} &CE(H_n^y(j\omega_1,\cdots,j\omega_n)) \\ &= \overline{C}_{0,n} \oplus \left(\mathop{\oplus}_{q=1}^{n-1} \mathop{\oplus}_{p=1}^{n-q} \overline{C}_{p,q} \otimes CE\left(H_{n-q-p+1}^x(j\omega_1,\cdots,j\omega_{n-q})\right)\right) \\ &\otimes \oplus \left(\mathop{\oplus}_{p=2}^{n} \overline{C}_{p,0} \otimes CE\left(H_{n-p+1}^x(j\omega_1,\cdots,j\omega_n)\right)\right) \\ &\otimes \oplus \overline{C}_{1,0} \otimes CE(H_n^x(j\omega_1,\cdots,j\omega_n)) \\ &= CE(H_n^x(j\omega_1,\cdots,j\omega_n)) \oplus \overline{C}_{1,0} \otimes CE(H_n^x(j\omega_1,\cdots,j\omega_n)) \end{aligned} \tag{6.71}$$

In the derivation of (6.71), (6.57) and (6.60) are used. Equation (6.71) can reveal that how the model parameters in (6.53a) affect system output frequency response. When only nonlinear parameters are considered under the assumption that linear parameters are fixed in the model, then (6.71) is simplified as

$$CE\left(H_n^y(j\omega_1,\cdots,j\omega_n)\right)_{\overline{C}(n)} = CE\left(H_n^x(j\omega_1,\cdots,j\omega_n)\right)_{\overline{C}(n)} \tag{6.72}$$

Equation (6.72) indicates that the parametric characteristics of the GFRFs for y(t) and x(t) are the same with respect to model nonlinear parameters in $\overline{C}(n)$. Note that (6.72) has a relationship with all the parameters in $\overline{C}(n)$ from degree 2 to n, which is different from (6.68). In this case both $X(j\omega)$ and $Y(j\omega)$ can be expressed as a polynomial function of model nonlinear parameters in $\overline{C}(n)$ with the same polynomial structure.

6.4.2.2 Some Further Results and Discussions

The following results can be summarized based on Sect. 6.4.2.1.

Proposition 6.5 Considering system (6.53a,b), there exists a complex valued function vector $\tilde{f}_n(j\omega_1,\cdots,j\omega_n)$ with appropriate dimension which is a function of linear parameters, such that

$$H_n^y(j\omega_1,\cdots,j\omega_n) = CE\left(H_n^y(j\omega_1,\cdots,j\omega_n)\right)_{\overline{C}(n)} \cdot \tilde{f}_n(j\omega_1,\cdots,j\omega_n) \tag{6.73}$$

and the output spectrum of system (6.53a,b) can be written as

$$Y\left(j\omega;\overline{C}(N)\right) = \sum_{n=1}^{N} CE\left(H_n^y(j\omega_1,\cdots,j\omega_n)\right)_{\overline{C}(n)} \cdot \tilde{F}_n(j\omega) \tag{6.74}$$

where $\tilde{F}_n(j\omega) = \frac{1}{\sqrt{n}(2\pi)^{n-1}} \int_{\omega_1+\cdots+\omega_n=\omega} \tilde{f}_n(j\omega_1,\cdots,j\omega_n) \cdot \prod_{i=1}^{n} U(j\omega_i) d\sigma_\omega$. If the input of system (6.53a,b) is the multi-tone signal (3.2), then the output spectrum of system (6.53a,b) can be expressed as

$$Y\left(j\omega;\overline{C}(N)\right) = \sum_{n=1}^{N} CE\left(H_n^y(j\omega_{k_1},\cdots,j\omega_{k_n})\right)_{\overline{C}(n)} \cdot \tilde{\tilde{F}}_n(j\omega) \tag{6.75}$$

where $\tilde{\tilde{F}}_n(j\omega) = \frac{1}{2^n} \sum_{\omega_{k_1}+\cdots+\omega_{k_n}=\omega} \tilde{f}_n(j\omega_{k_1},\cdots,j\omega_{k_n}) \cdot F(\omega_{k_1}) \cdots F(\omega_{k_n})$, and $CE\left(H_n^y(j\omega_1,\cdots,j\omega_n)\right)_{\overline{C}(n)}$ is given in (6.68) or (6.72).

Proof The results are straightforward from the discussions above and the results in Chaps. 4 and 5. □

Proposition 6.6 Under the same assumption as Proposition 6.5 for system (6.53a, b). If $\widetilde{C}(n)$ has either no relationship or linear relationship with $\overline{C}(n)$, then $CE\left(H_n^y(j\omega_1,\cdots,j\omega_n)\right)_{\overline{C}(n)}$ is given in (6.68) or (6.72), and the parametric characteristic vector for $Y(j\omega)$ can both be written as

$$CE(Y(j\omega))_{\overline{C}(N)} = \bigoplus_{n=1}^{N} CE\left(H_n^x(j\omega_1,\cdots,j\omega_n)\right)_{\overline{C}(n)} \tag{6.76}$$

That is, there exists a complex valued function vector $\hat{F}(j\omega_1,\cdots,j\omega_n)$ with appropriate dimension, which is a function of nonlinear parameters in $\widetilde{C}(N)$, linear parameters and the input, such that

$$Y\left(j\omega;\overline{C}(N)\right) = \left(\bigoplus_{n=1}^{N} CE\left(H_n^x(j\omega_1,\cdots,j\omega_n)\right)_{\overline{C}(n)}\right)\cdot\hat{F}(j\omega) \tag{6.77}$$

Proof See the proof in Sect. 6.6. □

From Proposition 6.6, both of the two mentioned cases have the same parametric characteristics for the output spectrum $Y(j\omega)$. If $\widetilde{C}(n)$ has no relationship with $\overline{C}(n)$, (6.76) may be conservative since some terms in (6.76) have no contribution. However, this does not affect the result of (6.77) because the corresponding terms in the complex valued vector will actually be zero after numerical identification. Once the parametric characteristics $CE\left(H_n^y(j\omega_1,\cdots,j\omega_n)\right)_{\overline{C}(n)}$ are derived, the polynomial structure of the parametric characteristic expression for $Y(j\omega)$ is determined, and then as mentioned above, (6.74) and (6.75) can be determined by using a numerical method. Therefore, analysis, design and optimization of system output frequency response can be conducted based on this explicit polynomial expression in terms of nonlinear parameters in $\overline{C}(N)$.

Example 6.3 Consider nonlinear system (2.37) again. Note that there are only two nonlinear parameters in $\overline{C}(n)$, *i.e.*, $\overline{c}_{2,0}(11) = -a_2/k, \overline{c}_{3,0}(111) = -a_3/k$, and the nonlinear parameters in $\widetilde{C}(n)$ are linear functions of the corresponding parameters in $\overline{C}(n)$. Let $\overline{C}_{2,0} = -a_2/k, \overline{C}_{3,0} = -a_3/k$. The GFRFs up to the fifth orders are computed according to (6.72) as follows,

$$CE\left(H_1^y(j\omega_1)\right) = 1 \tag{6.78}$$

$$CE\left(H_2^y(j\omega_1,j\omega_2)\right)_{\overline{C}(2)} = CE\left(H_2^x(j\omega_1,j\omega_2)\right)_{\overline{C}(2)} = \overline{C}_{2,0} \oplus \mathop{\oplus}_{p=2}^{\lfloor 2+1/2 \rfloor} \overline{C}_{p,0} \otimes CE\left(H_{2-p+1}^x(\cdot)\right)$$
$$= \overline{C}_{2,0} \oplus 0 = \overline{C}_{2,0} = -a_2/k \tag{6.79}$$

$$CE\left(H_3^y(j\omega_1,\cdots,j\omega_3)\right)_{\overline{C}(3)} = CE\left(H_3^x(j\omega_1,\cdots,j\omega_3)\right)_{\overline{C}(3)} = \overline{C}_{3,0} \oplus \mathop{\oplus}_{p=2}^{\lfloor 3+1/2 \rfloor} \overline{C}_{p,0} \otimes CE\left(H_{3-p+1}^x(\cdot)\right)$$
$$= \overline{C}_{3,0} \oplus \overline{C}_{2,0} \otimes CE\left(H_2^x(\cdot)\right) = \overline{C}_{3,0} \oplus \overline{C}_{2,0}{}^2 = \left[-\frac{a_3}{k}, \frac{a_2^2}{k^2}\right] \tag{6.80}$$

$$CE\left(H_4^y(j\omega_1,\cdots,j\omega_4)\right)_{\overline{C}(4)} = CE\left(H_4^x(j\omega_1,\cdots,j\omega_4)\right)_{\overline{C}(4)} = \overline{C}_{4,0} \oplus \mathop{\oplus}_{p=2}^{\lfloor 4+1/2 \rfloor} \overline{C}_{p,0} \otimes CE\left(H_{4-p+1}^x(\cdot)\right)$$
$$= 0 \oplus \overline{C}_{2,0} \otimes CE\left(H_3^x(\cdot)\right) = \overline{C}_{2,0} \otimes \left(\overline{C}_{3,0} \oplus \overline{C}_{2,0}{}^2\right)$$
$$= \overline{C}_{2,0} \otimes \overline{C}_{3,0} \oplus \overline{C}_{2,0}{}^3 = \left[\frac{a_2 a_3}{k^2}, -\frac{a_2^3}{k^3}\right] \tag{6.81}$$

$$CE\left(H_5^y(j\omega_1,\cdots,j\omega_5)\right)_{\overline{C}(5)} = CE\left(H_5^x(j\omega_1,\cdots,j\omega_5)\right)_{\overline{C}(5)} = \overline{C}_{5,0} \oplus \mathop{\oplus}_{p=2}^{\lfloor 5+1/2 \rfloor} \overline{C}_{p,0} \otimes CE\left(H_{5-p+1}^x(\cdot)\right)$$
$$= 0 \oplus \overline{C}_{2,0} \otimes CE\left(H_4^x(\cdot)\right) \oplus \overline{C}_{3,0} \otimes CE\left(H_3^x(\cdot)\right)$$
$$= \overline{C}_{2,0}{}^2 \otimes \overline{C}_{3,0} \oplus \overline{C}_{2,0}{}^4 \oplus \overline{C}_{3,0}{}^2 = \left[\frac{a_2^2 a_3}{k^3}, \frac{a_2^4}{k^4}, \frac{a_3^2}{k^2}\right] \tag{6.82}$$

The parametric characteristic of the output spectrum up to the fifth order can be obtained as

$$CE(Y(j\omega))_{\overline{C}(5)} = \mathop{\oplus}_{n=1}^{5} CE\left(H_n^y(j\omega_1,\cdots,j\omega_n)\right)_{\overline{C}(n)}$$
$$= \left[1, -\frac{a_2}{k}, -\frac{a_3}{k}, \frac{a_2^2}{k^2}, \frac{a_2 a_3}{k^2}, -\frac{a_2^3}{k^3}, \frac{a_2^2 a_3}{k^3}, \frac{a_2^4}{k^4}, \frac{a_3^2}{k^2}\right] \tag{6.83}$$

Then according to Proposition 6.6, there exists a complex valued function vector $\hat{F}(j\omega_1,\cdots,j\omega_5)$ such that

$$Y(j\omega; a_2, a_3) = \left[1, -\frac{a_2}{k}, -\frac{a_3}{k}, \frac{a_2^2}{k^2}, \frac{a_2 a_3}{k^2}, -\frac{a_2^3}{k^3}, \frac{a_2^2 a_3}{k^3}, \frac{a_2^4}{k^4}, \frac{a_3^2}{k^2}\right]$$
$$\cdot \hat{F}(j\omega_1,\cdots,j\omega_5) \tag{6.84}$$

It should be noted that the system output spectrum in (6.84) is only approximated up to the fifth order. In order to have a higher accuracy, higher order approximation might be needed in practice. To obtain the explicit relationship between system output spectrum and the nonlinear parameters a_2 and a_3 at a specific frequency of interest, $\hat{F}(j\omega_1, \cdots, j\omega_5)$ in (6.84) can be determined by using a numerical method as mentioned before. The idea is to obtain Z system output frequency responses from Z simulations or experimental tests on the system (2.37) under Z different values of the nonlinear parameters ($a_2 a_3$) and the same input u(t), then yielding

$$\begin{aligned} Y_Z &= \left[Y(j\omega; a_2, a_3)_1 \quad Y(j\omega; a_2, a_3)_2 \quad \cdots \quad Y(j\omega; a_2, a_3)_Z \right]^T \\ &= \Phi \cdot \hat{F}(j\omega_1, \cdots, j\omega_5) \end{aligned} \tag{6.85}$$

where

$$\Phi = \begin{bmatrix} 1, -\frac{a_2(1)}{k}, -\frac{a_3(1)}{k}, \frac{a_2^2(1)}{k^2}, \frac{a_2(1)a_3(1)}{k^2}, -\frac{a_2^3(1)}{k^3}, \frac{a_2^2(1)a_3(1)}{k^3}, \frac{a_2^4(1)}{k^4}, \frac{a_3^2(1)}{k^2} \\ 1, -\frac{a_2(2)}{k}, -\frac{a_3(2)}{k}, \frac{a_2^2(2)}{k^2}, \frac{a_2(2)a_3(2)}{k^2}, -\frac{a_2^3(2)}{k^3}, \frac{a_2^2(2)a_3(2)}{k^3}, \frac{a_2^4(2)}{k^4}, \frac{a_3^2(2)}{k^2} \\ \vdots \\ 1, -\frac{a_2(Z)}{k}, -\frac{a_3(Z)}{k}, \frac{a_2^2(Z)}{k^2}, \frac{a_2(Z)a_3(Z)}{k^2}, -\frac{a_2^3(Z)}{k^3}, \frac{a_2^2(Z)a_3(Z)}{k^3}, \frac{a_2^4(Z)}{k^4}, \frac{a_3^2(Z)}{k^2} \end{bmatrix} \tag{6.86}$$

Then

$$\hat{F}(j\omega_1, \cdots, j\omega_5) = \left(\Phi^T \Phi\right)^{-1} \Phi^T Y_Z \tag{6.87}$$

Therefore, (6.84) can be determined, which is an explicitly analytical function of the nonlinear parameters a_2 and a_3. By using this method, the system output frequency response can thus be analyzed and designed in terms of model nonlinear parameters of interest.

6.5 Conclusions

The parametric characteristic analysis is performed for nonlinear output spectrum of Volterra-type nonlinear systems described by NDE models or NARX models in this Chapter and some fundamental results for the parametric characteristics of nonlinear output spectrum are established and demonstrated, including parametric characteristics based analysis, parametric characteristic analysis of nonlinear effects on system output frequency, and parametric characteristics of SIDO nonlinear systems etc.

Based on these results, the parametric characteristics based output spectrum analysis for nonlinear systems will be further shown in the following two chapters, and the nonlinear characteristic output spectrum is thereafter developed in Chap. 9.

6.6 Proofs

Proof of Proposition 6.2

Regard all other nonlinear parameters as constants or 1. From Proposition 5.1 and Properties 5.1–5.5, if $p+q>n$ then the parameter has no contribution to CE(H_n(.)), in this case CE(H_n(.))=1 with respect to this parameter. Similarly, if $p+q=n$ then the parameter is an independent contribution in CE(H_n(.)), thus CE(H_n(.))=[1 c] with respect to this parameter in this case. If $p+q<n$ and $p>0$, then the independent contribution in CE(H_n(.)) for this parameter should be $c^{\left\lfloor\frac{n-1}{p+q-1}\right\rfloor}$, and the monomials c^x for $0 \leq x < \left\lfloor\frac{n-1}{p+q-1}\right\rfloor$ are all not independent contributions in CE(H_n(.)). Hence $CE(H_n(\cdot)) = \begin{bmatrix} 1 & c & c^2 & \cdots & c^{\left\lfloor\frac{n-1}{p+q-1}\right\rfloor} \end{bmatrix}$ for this case. The similar result is held for the case $p+q<n$ and $p=0$. However, since there should be at least one $p>0$ in a complete monomial, thus in this latter case c^x for any x are not complete, which follows $CE(H_n(\cdot)) = \begin{bmatrix} 1 & c & c^2 & \cdots & c^{\left\lfloor\frac{n-1}{p+q-1}\right\rfloor-1} \end{bmatrix}$. The parametric characteristic vector for the nonlinear parameter c for all the cases above can be summarized into

$$CE(H_n(\cdot)) = \begin{bmatrix} 1 & c & c^2 & \cdots & c^{\left\lfloor\frac{n-1}{p+q-1}\right\rfloor-\delta(p)\cdot pos(n-q)} \end{bmatrix}$$

This completes the proof. □

Proof of Proposition 6.3

Equation (6.16) is summarized from (5.19)–(5.21), and when all the other parameters are zero except $c=c_{p,q}(.)$, the following equation can also be summarized from (5.19)–(5.21)

$$CE(H_n(j\omega_1,\cdots,j\omega_n)) = c^{\frac{n-1}{p+q-1}} \cdot \delta\left(\frac{n-1}{p+q-1} - \left\lfloor\frac{n-1}{p+q-1}\right\rfloor\right) \cdot (1-\delta(p)pos(n-q))$$

Therefore, it can be shown that

$$\begin{aligned} CE(Y(j\omega)) &= \bigoplus_{i=0}^{\left\lfloor {N-1}/{p+q-1}\right\rfloor} CE\big(H_{(p+q-1)i+1}(\cdot)\big) \\ &= \bigoplus_{i=0}^{\left\lfloor {N-1}/{p+q-1}\right\rfloor} c^i \cdot \delta(i - \lfloor i \rfloor) \cdot (1-\delta(p)pos((p+q-1)i+1-q)) \\ &= \bigoplus_{i=0}^{\left\lfloor {N-1}/{p+q-1}\right\rfloor} c^i \cdot (1-\delta(p)pos((p+q-1)i+1-q)) \end{aligned}$$

If p=0, $1-\delta(p)pos((p+q-1)i+1-q) = 1-pos((q-1)i+1-q)$, which yields,

$$CE(Y(j\omega)) = [1 \quad c \cdot (1 - pos(q - N))]$$

else, $1 - \delta(p)pos((p+q-1)i+1-q) = 1$, which yields

$$CE(Y(j\omega)) = \bigoplus_{i=0}^{\lfloor N-1/_{p+q-1} \rfloor} CE(H_{(p+q-1)i+1}(\cdot)) = \begin{bmatrix} 1 & c & c^2 & \cdots & c^{\lfloor N-1/_{p+q-1} \rfloor} \end{bmatrix}$$

This completes the proof. □

Proof of Lemma 6.2

The lemma is summarized by the following observation. For clarity, let I=3.

$$c^n = \underbrace{c \otimes c \cdots \otimes c}_{n} \qquad\qquad s(i)_n = \sum_{j=i}^{I} s(j)_{n-1} \quad \text{for i} = 1,2,3$$

$$\begin{array}{llrrr}
\text{n} = 1 & [\text{c}_1 \ \text{c}_2 \ \text{c}_3] & 1 & 1 & 1 \\
\text{n} = 2 & [\text{c}_1 \ \text{c}_2 \ \text{c}_3] \otimes [\text{c}_1 \ \text{c}_2 \ \text{c}_3] & 3 & 2 & 1 \\
 & = [\text{c}_1{}^2 \ \text{c}_1\text{c}_2 \ \text{c}_1\text{c}_3 \ \text{c}_2{}^2 \ \text{c}_2\text{c}_3 \ \text{c}_3{}^2] & & & \\
\text{n} = 3 & [\text{c}_1{}^2 \ \text{c}_1\text{c}_2 \ \text{c}_1\text{c}_3 \ \text{c}_2{}^2 \ \text{c}_2\text{c}_3 \ \text{c}_3{}^2] \otimes [\text{c}_1 \ \text{c}_2 \ \text{c}_3] & 6 & 3 & 1 \\
 & = [\text{c}_1{}^3 \ \text{c}_1{}^2\text{c}_2 \ \text{c}_1{}^2\text{c}_3 \ \text{c}_1\text{c}_2{}^2 \ \text{c}_1\text{c}_2\text{c}_3 \ \text{c}_1\text{c}_3{}^2 \ \text{c}_2{}^3 \ \text{c}_2{}^2\text{c}_3 \ \text{c}_2\text{c}_3{}^2 \ \text{c}_3{}^3] & & & \\
\text{n} = 4 & [\text{c}_1{}^3 \ \text{c}_1{}^2\text{c}_2 \ \text{c}_1{}^2\text{c}_3 \ \text{c}_1\text{c}_2{}^2 \ \text{c}_1\text{c}_2\text{c}_3 \ \text{c}_1\text{c}_3{}^2 \ \text{c}_2{}^3 \ \text{c}_2{}^2\text{c}_3 \ \text{c}_2\text{c}_3{}^2 \ \text{c}_3{}^3] \otimes [\text{c}_1 \ \text{c}_2 \ \text{c}_3] & 10 & 4 & 1 \\
 & = [\text{c}_1{}^4 \ \text{c}_1{}^3\text{c}_2 \ \text{c}_1{}^3\text{c}_3 \ \text{c}_1{}^2\text{c}_2{}^2 \ \text{c}_1{}^2\text{c}_2\text{c}_3 \ \text{c}_1{}^2\text{c}_3{}^2 \ \text{c}_1\text{c}_2{}^3 \ \text{c}_1\text{c}_2{}^2\text{c}_3 \ \text{c}_1\text{c}_2\text{c}_3{}^2 \ \text{c}_1\text{c}_3{}^3 & & & \\
 & \quad \text{c}_2{}^4 \ \text{c}_2{}^3\text{c}_3 \ \text{c}_2{}^2\text{c}_3{}^2 \ \text{c}_2\text{c}_3{}^3 \ \text{c}_3{}^4] & & & \\
\text{n} = 5 & [\text{c}_1{}^4 \ \text{c}_1{}^3\text{c}_2 \ \text{c}_1{}^3\text{c}_3 \ \text{c}_1{}^2\text{c}_2{}^2 \ \text{c}_1{}^2\text{c}_2\text{c}_3 \ \text{c}_1{}^2\text{c}_3{}^2 \ \text{c}_1\text{c}_2{}^3 \ \text{c}_1\text{c}_2{}^2\text{c}_3 \ \text{c}_1\text{c}_2\text{c}_3{}^2 \ \text{c}_1\text{c}_3{}^3 & 15 & 5 & 1 \\
 & \quad \text{c}_2{}^4 \ \text{c}_2{}^3\text{c}_3 \ \text{c}_2{}^2\text{c}_3{}^2 \ \text{c}_2\text{c}_3{}^3 \ \text{c}_3{}^4] \otimes [\text{c}_1 \ \text{c}_2 \ \text{c}_3] & & & \\
 & = [\text{c}_1{}^5 \ \text{c}_1{}^4\text{c}_2 \ \text{c}_1{}^4\text{c}_3 \ \text{c}_1{}^3\text{c}_2{}^2 \ \text{c}_1{}^3\text{c}_2\text{c}_3 \ \text{c}_1{}^3\text{c}_3{}^2 \ \text{c}_1{}^2\text{c}_2{}^3 \ \text{c}_1{}^2\text{c}_2{}^2\text{c}_3 \ \text{c}_1{}^2\text{c}_2\text{c}_3{}^2 \ \text{c}_1{}^2\text{c}_3{}^3 & & & \\
 & \quad \text{c}_1\text{c}_2{}^4 \ \text{c}_1\text{c}_2{}^3\text{c}_3 \ \text{c}_1\text{c}_2{}^2\text{c}_3{}^2 \ \text{c}_1\text{c}_2\text{c}_3{}^3 \ \text{c}_1\text{c}_3{}^4 \ \text{c}_2{}^5 \ \text{c}_2{}^4\text{c}_3 \ \text{c}_2{}^3\text{c}_3{}^2 \ \text{c}_2{}^2\text{c}_3{}^3 \ \text{c}_2\text{c}_3{}^4 \ \text{c}_3{}^5] & & &
\end{array}$$

To complete the proof, the complete mathematical induction can be adopted. An outline for this proof is given here. Note that

$$c^n = \left[c^{n-1} \cdot c_1, \cdots, c^{n-1}\left[s(1)_n - s(i)_n + 1 : s(1)_n\right] \cdot c_i, \cdots, c^{n-1}\left[s(1)_n\right] \cdot c_I\right]$$

includes all the non-repetitive terms of form $c_1^{k_1} c_2^{k_2} \cdots c_I^{k_I}$ with $k_1 + k_2 + \cdots + k_I = n$ and $0 \le k_1, k_2, \cdots, k_I \le n$. These terms can be separated into I parts, the ith part of which, i.e., $c^{n-1}[s(1)_n - s(i)_n + 1 : s(1)_n] \cdot c_i$, includes all the non-repetitive terms of degree n which are obtained by the parameter c_i timing the components of degree $n-1$ in $c^{\text{n}-1}$ from $s(1)_n - s(i)_n + 1$ to $s(1)_n$. Assume that the lemma holds for step n. Then for the step n+1, the ith part of the components in $c^{\text{n}+1}$ must be $c^n[s(1)_{n+1} - (s(i)_n + \cdots + s(I)_n) + 1 : s(1)_{n+1}] \cdot c_i$ which is $c^n[s(1)_{n+1} - s(i)_{n+1} + 1 : s(1)_{n+1}] \cdot c_i$. This completes the proof of Lemma 6.2.

Proof of Proposition 6.6

From (6.74) and (6.75), the parametric characteristic vector for $Y(j\omega)$ is

$$CE(Y(j\omega))_{\overline{C}(N)} = \mathop{\oplus}_{n=1}^{N} CE\big(H_n^y(j\omega_1,\cdots,j\omega_n)\big)_{\overline{C}(n)} \tag{A1}$$

If $\widetilde{C}(n)$ has a linear relationship with $\overline{C}(n)$, then $CE\big(H_n^y(j\omega_1,\cdots,j\omega_n)\big)_{\overline{C}(n)}$ is given by (6.72). In this case, (6.76) is straightforward by substituting (6.72) into (A1). If $\widetilde{C}(n)$ has no relationship with $\overline{C}(n)$, then substituting (6.68) into (A1) yields

$$CE(Y(j\omega))_{\overline{C}(N)} = \mathop{\oplus}_{n=1}^{N} \left(\mathop{\oplus}_{p=1}^{n} \chi(n,p) \cdot CE\left(H_{n-p+1}^x(j\omega_1,\cdots,j\omega_n)_{\overline{C}(n)}\right)\right) \tag{A2}$$

By the definition of operation "$\oplus$", repetitive terms should be removed. Therefore, (A2) further gives

$$CE(Y(j\omega))_{\overline{C}(N)} = \mathop{\oplus}_{p=1}^{N} \chi(N,p) \cdot CE\left(H_{N-p+1}^x(j\omega_1,\cdots,j\omega_N)_{\overline{C}(N)}\right) \tag{A3}$$

Note that, all the elements in vector $\mathop{\oplus}_{p=1}^{N} \chi(N,p) \cdot CE\left(H_{N-p+1}^x(j\omega_1,\cdots,j\omega_N)_{\overline{C}(N)}\right)$ must be elements in vector $\mathop{\oplus}_{n=1}^{N} CE\left(H_n^x(j\omega_1,\cdots,j\omega_n)_{\overline{C}(n)}\right)$. Hence, the parametric characteristics in (A3) are all included in (6.76). Equation (6.77) is straightforward from Proposition 6.5. □

Chapter 7
The Parametric Characteristics Based Output Spectrum Analysis

7.1 Introduction

The parametric characteristic analysis provides a powerful tool for nonlinear analysis in the frequency domain, which can be used for many important issues related to analysis, design and understanding of nonlinear dynamics and influence, from the viewpoints of output frequency response of nonlinear systems and/or the GFRFs. In this chapter, the output frequency response or output spectrum based analysis method is demonstrated, which actually has already been discussed in remarks and examples in the previous chapters. The parametric characteristic analysis can provide obvious benefits for example in determination of the parametric structure and in reduction of computation cost, which will be theoretically addressed in the chapter thereafter. Based on these results, the nonlinear characteristic output spectrum based analysis is established in Chap. 9, which is a much improved version of the output frequency response function based analysis method in this chapter. Following these, the GFRFs based analysis with the parametric characteristic approach will be investigated more, including understanding of nonlinear influence in the frequency domain, and bound evaluation of output response of nonlinear systems etc.

7.2 The Parametric Characteristics Based Output Spectrum Analysis

Using the parametric characteristic result in Corollary 6.1, the output spectrum (6.8a) can be simply rewritten as

X. Jing, Z. Lang, *Frequency Domain Analysis and Design of Nonlinear Systems based on Volterra Series Expansion*, Understanding Complex Systems,
DOI 10.1007/978-3-319-12391-2_7

$$Y(j\omega) = \psi \cdot \Phi(j\omega)^T \tag{7.1}$$

Where $\psi = \mathop{\oplus}\limits_{n=1}^{N} CE(H_n(\cdot))$, $\Phi(j\omega) = [\,\phi_1(j\omega) \quad \phi_2(j\omega) \quad \cdots \quad \phi_N(j\omega)\,]$. Note that $\phi_1(j\omega){=}H_1(j\omega)$ is the first order GFRF, which represents the linear part of model (2.10) or (2.11), and $\phi_n(j\omega){=}Y_n(f_n(\cdot);j\omega)$. The function in (7.1) is referred to here as output frequency response function (OFRF) or nonlinear output spectrum with parametric characteristics.

As discussed before, (6.8) or (7.1) provide a more straightforward analytical expression for the relationship between system time-domain model parameters and system output frequency response. By using this explicit relationship, system output frequency response can therefore be analyzed in terms of any model parameters of interest. Hence, it can considerably facilitate the analysis and design of output frequency response characteristics of nonlinear systems in the frequency domain. As demonstrated in Sect. 6.2.2, the main idea for the parametric characteristics based output spectrum analysis in this Chapter is that, given the model of a nonlinear system in the form of model (2.10) or (2.11), $CE(H_n(\cdot))$ can be computed according to Proposition 5.1 or Corollary 5.1, and $\phi_n(j\omega)$ can be obtained according to a numerical method which is mentioned before and will be discussed in more details later. Thus the output spectrum of the nonlinear system subject to any specific input can be obtained, which is an analytical function of nonlinear parameters of the system model, and finally frequency domain analysis for the nonlinear system can be conducted in terms of specific model parameters of interest.

The parametric characteristics based output spectrum analysis for system (2.10) or (2.11) is discussed in Sect. 7.2.1. In order to conduct the parametric characteristics based output spectrum analysis, a general procedure is provided in Sect. 7.2.2, where several basic algorithms and related results are discussed.

7.2.1 A General Frequency Domain Method

The parametric characteristics based output spectrum analysis for nonlinear systems described by (2.10) or (2.11) is totally a new frequency domain method for nonlinear analysis. The most noticeable advantage of this method is that any system model parameters of interest can be directly related to the interested engineering analysis and design objective which is dependent on system output frequency response, and thus the system output frequency response can be analysed in terms of some model parameters of interest in an easily-manipulated manner. This method does not restrict to a specific input signal and can be used for a considerable larger class of nonlinear systems. These are the main differences of this method from the other existing methods such as Popov-theory based analysis, describing functions and harmonic balance methods as discussed in Chap. 1.

One important step of this method is to determine the output spectrum with parametric characteristics for the system under study. This will be discussed in more detailed in the following section. Once the system output spectrum is obtained, based on the result in Proposition 6.3 and (7.1), the output frequency response function with respect to a specific parameter c can be written as

$$Y(j\omega) = \overline{\varphi}_0(j\omega) + c\overline{\varphi}_1(j\omega) + c^2\overline{\varphi}_2(j\omega) + \cdots + c^\ell\overline{\varphi}_\ell(j\omega) + \cdots \tag{7.2a}$$

Since $Y(j\omega)$ is also a function of c, therefore, (7.2a) is rewritten more clearly as

$$Y(j\omega; c) = \overline{\varphi}_0(j\omega) + c\overline{\varphi}_1(j\omega) + c^2\overline{\varphi}_2(j\omega) + \cdots + c^\ell\overline{\varphi}_\ell(j\omega) + \cdots \tag{7.2b}$$

$Y(j\omega;c)$ is in fact a series of an infinite order, ℓ is a positive integer which can be determined by Proposition 6.3, $\overline{\varphi}_i(j\omega)$ is the complex valued function corresponding to the coefficient c^i in (7.2b). If all the other degree and type of nonlinear parameters are zero except that $C_{p,q}=c\neq 0$ (p+q>1), then $\overline{\varphi}_{i+1}(j\omega) = \varphi_i(j\omega)$ ($\varphi_i(j\omega)$ is defined in (6.8), (6.10), (6.11), or (7.1). Based on (7.2a,b), the following analysis can be conducted.

- **Sensitivity of the output frequency response to nonlinear parameters**
  Based on (7.2a,b), this can be obtained easily as

$$\frac{\partial Y(j\omega; c)}{\partial c} = \overline{\varphi}_1(j\omega) + 2c\overline{\varphi}_2(j\omega) + \cdots + \ell c^{\ell-1}\overline{\varphi}_l(j\omega) + \cdots \tag{7.3}$$

Similarly, the sensitivity of the magnitude of the output frequency response with respect to the nonlinear parameters can also be derived. Note that

$$\begin{aligned} |Y(j\omega; c)|^2 &= Y(j\omega; c)Y(-j\omega; c) = (\overline{\varphi}_0(j\omega) + c\overline{\varphi}_1(j\omega) + c^2\overline{\varphi}_2(j\omega) + \cdots) \\ &\quad \times (\overline{\varphi}_0(-j\omega) + c\overline{\varphi}_1(-j\omega) + c^2\overline{\varphi}_2(-j\omega) + \cdots) \\ &= \overline{\varphi}_0 \cdot \overline{\varphi}_0^* + \sum_{\ell=1}^{\infty}\left(c^\ell \sum_{i=0}^{\ell} \overline{\varphi}_i \cdot \overline{\varphi}_{\ell-i}^*\right) : \\ &= p_0 + cp_1 + c^2p_2 + \cdots + c^{2\ell}p_{2\ell} + \cdots \end{aligned} \tag{7.4}$$

It is obvious that the spectral density of the output frequency response is still a polynomial function of the parameter c. Equation (7.4) can also be directly derived by following Process C that will be discussed later (Sect. 7.2.2.2). Thus, the sensitivity of the magnitude of the output spectrum to the parameter c can be obtained as

$$\frac{\partial|Y(j\omega;c)|}{\partial c}=\frac{1}{2|Y(j\omega;c)|}\frac{\partial|Y(j\omega;c)|^2}{\partial c}$$
$$=\frac{1}{2|Y(j\omega;c)|}\sum_{\ell=1}^{\infty}\left(\ell c^{\ell-1}\sum_{i=0}^{\ell}\overline{\varphi}_i\cdot\overline{\varphi}_{\ell-1}^*)\right) \tag{7.5a}$$

Given (7.3), (7.5a) can also be computed as

$$\frac{\partial|Y(j\omega;c)|}{\partial c}=\frac{1}{2|Y(j\omega;c)|}\frac{\partial|Y(j\omega;c)|^2}{\partial c}$$
$$=\frac{1}{2|Y(j\omega;c)|}\left(\frac{\partial Y(j\omega;c)}{\partial c}Y(-j\omega;c)+Y(j\omega;c)\frac{\partial Y(-j\omega;c)}{\partial c}\right)$$
$$=\Re\left(\frac{\partial Y(j\omega;c)}{\partial c}\cdot\frac{Y(-j\omega;c)}{|Y(j\omega;c)|}\right) \tag{7.5b}$$

The sensitivity function for system output spectrum with respect to a nonlinear parameter provides a useful insight into the effect on system output performance of specific model parameters. This will be illustrated in Sect. 7.3. Another possible application of the sensitivity function is vibration suppression. In many engineering practice, the effect of vibrations should be considerably suppressed. From (7.5a,b), it can be seen that if $Y(j\omega,c)$ represents the spectrum of a vibration, in order to suppress the vibration, it should be ensured that $\frac{\partial|Y(j\omega;c)|}{\partial c}$ < 0 for some c. Consider (7.4), the following conclusion is obvious.

(a) $\frac{\partial|Y(j\omega;c)|}{\partial c}<0$ for some $c\Rightarrow$ $\exists$some $n>0$, $\sum_{i=0}^{n} sign(c^{n-1})\overline{\varphi}_i\cdot\overline{\varphi}_{n-i}^*<0$

(b) $p_1=\mathrm{Re}(\overline{\varphi}_0(j\omega)\cdot\overline{\varphi}_1(-j\omega))<0\Rightarrow$ there exists $\varepsilon>0$ such that $\frac{\partial|Y(j\omega;c)|}{\partial c}<0$ for $0<c<\varepsilon$ or $-\varepsilon<c<0$. Where $sign(x)=\begin{cases}1 & x\geq 0\\ -1 & \text{else}\end{cases}$, $\mathrm{Re}(\cdot)$ is to get the real part of (.). If a nonlinear parameter c satisfies $p_1=\mathrm{Re}(\overline{\varphi}_0(j\omega)\cdot\overline{\varphi}_1(-j\omega))<0$, then it can be utilized for the vibration suppression objective.

- **Evaluation of the radius of convergence for the output frequency response with respect to nonlinear parameters**
  It is followed from (7.2a,b) that the radius of convergence is given by

$$R=\lim_{\ell\to\infty}\left|\frac{\overline{\varphi}_{\ell-1}(j\omega)}{\overline{\varphi}_{\ell}(j\omega)}\right| \tag{7.6}$$

Obviously, if $|c| < R$, then the series is convergent. Define a Ratio Function

$$R(\ell; c) = \left| \frac{\overline{\varphi}_{\ell-1}(j\omega)_c}{\overline{\varphi}_{\ell}(j\omega)_c} \right| \tag{7.7}$$

which is a function of ℓ and also varies with different nonlinear parameters. It can be seen that, if

$$\frac{\Delta R(\ell; c_1)}{\Delta \ell} > \frac{\Delta R(\ell; c_2)}{\Delta \ell} \tag{7.8}$$

then the output spectrum has a larger radius of convergence with respect to c_1 than that with respect to c_2. Equation (7.7) and inequality (7.8) can be used as an evaluation of the effect on the convergence of the Volterra series expansion for the nonlinear system under study from a model parameter and the comparative advantage between different parameters. Note that divergence of the Volterra series expansion can sometimes imply the instability of the system under study or the nonexistence of a Volterra series expansion. Thus this analysis can provide some useful information for the design of system output frequency response in terms of different model parameters.

- **Optimization of the output frequency response in terms of nonlinear parameters**
  Importantly, given a desired magnitude of the output frequency response Y^*, an optimal c^* in ∂S_c can be found such that

$$\min_{c \in \partial S_c} (|Y(j\omega; c)| - Y^*) \tag{7.9}$$

A systematic method for this purpose is yet to be developed, which will be discussed in the future study.

7.2.2 Determination of Output Spectrum Based on Its Parametric Characteristics

As mentioned before, an important step for output spectrum analysis based on the parametric characteristics is to obtain the parametric characteristic function of system output spectrum in (7.1) (i.e., the OFRF). In this section, a general procedure for the determination of the OFRF for a given model (2.10) or (2.11) is provided, and useful algorithms and techniques are discussed.

7.2.2.1 Computation of Parametric Characteristics

This step is to derive $\psi = \overset{N}{\underset{n=1}{\oplus}} CE(H_n(\cdot))$ in (7.1).

- **Determination of the largest order$N$**
  To derive the parametric characteristics of OFRF, the first task is to compute the largest order, i.e., N, of the Volterra series expansion for the nonlinear system, which is basically determined by the significance of the truncation error in the Volterra series expansion of finite order. This can alternatively be done by evaluating the magnitude of the nth-order output frequency response $Y_n(j\omega)$. For example, the magnitude bound of $Y_n(j\omega)$ for the NARX model (2.10) can be evaluated by (Jing et al. 2007a, b), which will be discussed in Chap. 13,

$$|Y_n(j\omega)| \leq \alpha_n \cdot b_n \cdot \hbar_n^T \tag{7.10}$$

  where α_n,$\hbar_n$ are complex valued functions, and b_n is a function vector of the system model parameters. For the detailed definitions for α_n,b_n,$\hbar_n$ refer to Jing et al. 2007a, b. If the magnitude bound of a certain order of $Y_n(j\omega)$ is less than a predefined value (for instance 10^{-8}), then the largest order N is obtained. It should be noted that the magnitude bound is a function of the model nonlinear parameters. Therefore, the largest range of interest for each nonlinear parameter should be considered in the evaluation of $|Y_n(j\omega)|$. The truncation order selection issue will be further discussed in Chap. 9, where it is well addressed from a different point of view.

- **Determination of the parametric characteristics**
  Once the largest order N is determined, the next step is to derive the parametric characteristics of GFRFs for the nonlinear system, i.e., $CE(H_n(\cdot))$ from n=2 to N, which will be used in the computation of $\psi = \overset{N}{\underset{n=1}{\oplus}} CE(H_n(\cdot))$. Note that $CE(H_n(\cdot))$ is computed in terms of the parameter vectors $C_{p,q} = [c_{p,q}(0,\cdots,0), c_{p,q}(0,\cdots,1),\cdots,c_{p,q}(\underbrace{K,\cdots,K}_{p+q})]$ for some p,q in (5.17).

  Basically, for some specific parameters to be analysed for a system, $CE(H_n(\cdot))$ can be recursively computed by (5.17) with respect to these parameters of interest with other nonzero nonlinear parameters being 1. Alternatively, $CE(H_n(\cdot))$ can also be determined directly without recursive computations by using the results in Proposition 5.1. Based on Proposition 5.1, the parametric characteristic $CE(H_n(\cdot))$ can be obtained as follows, which is referred to as Process A: for $0 \leq k \leq n-2$,

  (A1) Generate all the combinations (r_0, r_1, r_2..., r_k) satisfying $r_0 + \sum_{i=1}^{k} r_i = n + k$ and $2 \leq r_i \leq n-k$ with respect to a specific value of k;

(A2) Generate all the possible combinations (p_i,q_i) with respect to each r_i satisfying $p_i+q_i=r_i$, and note that when it is for r_0, $1 \leq p_0 \leq n-k$;
(A3) All the possible combinations can now be generated based on Step (A1) and (A2), then remove all the repetitive terms;
(A4) $CE(H_n(\cdot))$ is obtained in terms of the parameter vectors $C_{p,q}$ for some p,q, which can be stored for any future usage. For a specific nonlinear system, $CE(H_n(\cdot))$ can be obtained only by replacing the corresponding parameter vector $C_{p,q}$ of interest with respect to the specific nonlinear system, and the other parameters in $CE(H_n(\cdot))$ are set to be zero if it is zero or set to be 1 if it is not of interest;
(A5) Achieve the final result by manipulating $CE(H_n(\cdot))$ according to the operation rules of "$\oplus$" and "$\otimes$" (See Chap. 4), and removing the repetitive terms.

By this process, the parametric characteristic $CE(H_n(\cdot))$ can be obtained without recursive computations. For a summary, the parametric characteristic vector $\psi = \bigoplus_{n=1}^{N} CE(H_n(\cdot))$ can be computed by following the process below, which is referred to as Process B:

(B1) Determine the set of the nonlinear parameters of interest, denoted by S_C;
(B2) Determine the largest possible ranges for the nonlinear parameters of interest, denoted by ∂S_C;
(B3) Determine the largest order N of the Volterra series expansion according to (7.10) and the discussions there.
(B4) Computation of $CE(H_n(\cdot))$ with respect to the parameters S_C of interest following Process A or (5.17) from n=2 to N.
(B5) Combine the final parametric characteristic vector $\psi = \bigoplus_{n=1}^{N} CE(H_n(\cdot))$.

Therefore, based on Process A and Process B, the parametric characteristics of the output frequency response with respect to any specific model parameters of interest, which are the coefficients of the polynomial function (7.1), can be determined. Thus the structure of the polynomial (7.1) is explicitly determined at this stage. Note that, the parametric characteristic vector $CE(H_n(\cdot))$ for all the model nonlinear parameters in (5.13) can be obtained according to (5.17) or Process A, and if there are only some parameters of interest, the computation can be conducted by only replacing other nonzero parameters with 1 as mentioned above.

7.2.2.2 A Numerical Method

This is to determine $\Phi(j\omega) = [\phi_1(j\omega) \quad \phi_2(j\omega) \quad \cdots \quad \phi_N(j\omega)]$ in (7.1), then the OFRF in (7.1) is obtained consequently. Since the system model is supposed to be known, the parametric characteristic vector $\psi = \bigoplus_{n=1}^{N} CE(H_n(\cdot))$ is achieved, and note that $\Phi(j\omega)$ is invariant with respect to $\psi = \underset{n=1}{\overset{N}{\oplus}} CE(H_n(\cdot))$, thus $\Phi(j\omega)$ can be derived with respect to any a specific input by following a numerical method as follows, which is referred to as Process C:

(C1) Choose a series of different values of the nonlinear parameters of interest, which are properly distributed in ∂S_C, to form a series of vectors $\psi_1 \cdots \psi_{\rho(N)}$, where $\rho(N) = |\psi|$ denotes the dimension of vector ψ, such that

$$\Psi = \left[\psi_1{}^T \cdots \psi_{\rho(N)}{}^T\right]^T \text{ is non-singular} \tag{7.11}$$

(C2) Given a frequency ω where the output frequency response of the nonlinear system is to be analysed or designed. Excite the system using the same input under different values of the nonlinear parameters $\psi_1 \cdots \psi_{\rho(N)}$; collect the time domain output $y(t)$ for each case, and evaluate the output frequency response $Y(j\omega)_1 \cdots Y(j\omega)_{\rho(N)}$ at the frequency ω by FFT technique.

(C3) Step 2 yields

$$\Psi \cdot \Phi(j\omega)^T = \begin{bmatrix} \psi_1 \\ \psi_2 \\ \vdots \\ \psi_{\rho(N)} \end{bmatrix} \cdot \begin{bmatrix} \varphi_1(j\omega) \\ \varphi_2(j\omega) \\ \vdots \\ \varphi_{\rho(N)}(j\omega) \end{bmatrix} = \begin{bmatrix} Y(j\omega)_1 \\ Y(j\omega)_2 \\ \vdots \\ Y(j\omega)_{\rho(N)} \end{bmatrix} =: YY(j\omega) \tag{7.12}$$

Hence,

$$\begin{aligned} \Phi(j\omega)^T &= [\phi_1(j\omega) \quad \phi_2(j\omega) \quad \cdots \quad \phi_N(j\omega)]^T \\ &= [\varphi_1(j\omega) \quad \varphi_2(j\omega) \quad \cdots \quad \varphi_{\rho(N)}(j\omega)]^T = \Psi^{-1} \cdot YY(j\omega) \end{aligned} \tag{7.13}$$

In Step C1, $\rho(N)$ different values of the parameter vector Ψ in the parameter space ∂S_C, such that $\det(\Psi) \neq 0$ can be obtained by choosing a grid of parameter values of the nonlinear parameters of interest properly spanned in ∂S_C, or using a stochastic-based searching method or other optimization search methods such as GA to generate a non-singular matrix Ψ. In practices, it is not difficult to find such a matrix with a proper inverse, which will be illustrated in Sect. 7.3. In Step C2, given the largest order N of the system output spectrum, it can be seen that this algorithm needs $\rho(N)$ simulations to obtain $\rho(N)$ output frequency responses under different parameter values. Note from Step C1 that $\rho(N) = |\psi| = \left|\underset{n=1}{\overset{N}{\oplus}} CE(H_n(\cdot))\right|$, which is not only a function of the largest order N but also dependent on the number of parameters of interest. This implies the simulation burden will become heavier if the number of the parameters of interest and the largest order N are becoming larger. In Step C3, since $\det(\Psi) \neq 0$, the complex valued function vector $\Phi(j\omega)$ in (7.13) is unique, which implies the result in (7.13) can sufficiently approximate their real values if the truncation error incurred by the largest order N of the Volterra series is sufficiently small.

Therefore, by following Process C, the complex valued function vector $\Phi(j\omega)$ can accurately be obtained for the specific input function used in Step C2 and at the given frequency ω. Consequently, the OFRF (7.1) subject to the specific input function is now explicitly determined by following the method discussed above for the nonlinear system of interest. Although the function vector $\Phi(j\omega)$ is obtained by using the numerical method above and consequently the obtained OFRF is not an analytical function of the frequencies and the input, the achieved relationship between the output spectrum and model nonlinear parameters is analytical and explicit for the specific input function at the given frequency ω. Moreover, note that since $CE(H_n(\cdot))$ is known, and $\Phi(j\omega) = [\varphi_1(j\omega) \quad \varphi_2(j\omega) \quad \cdots \quad \varphi_N(j\omega)]$ is determined, then $Y_n(j\omega)=CE(H_n(\cdot))\cdot\varphi_n(j\omega)^T$ is also determined, which represents the analytical function for the nth-order output frequency response of nonlinear systems.

It should also be noted that, the proposed method above as demonstrated in this section enable the OFRF to be obtained directly in a concise polynomial form as (7.1) without the complex integration in the high-dimensional super-plane $\omega=\omega_1+\cdots+\omega_n$ especially when the nonlinearity order n is high. By using the proposed method above, the OFRF can be determined up to a very high order with respect to any specific model parameters of interest and any specific input signal at any given frequency. The cost may lie in that the new method needs $\rho(N)$ simulations.

Once the OFRF is obtained, the analysis and design of nonlinear systems described by model (2.10) or (2.11) can be carried out in terms of the model parameters of interest which define system nonlinearities and may represent some structural and controllable factors of a practical engineering system. For example, the sensitivity of system output frequency response with respect to a nonlinear parameter can be studied based on the analytical expression (7.1). By using the link between the nonlinear terms of interest and the components of a practical engineering system and structure, the OFRF may provide a useful insight into the design of nonlinear components in the system to achieve a desired output performance. Therefore, the OFRF based analysis method provides a novel approach to the analysis and synthesis of a large class of nonlinear systems subject to any input signal in the frequency domain.

7.3 Simulations

To demonstrate the application of the new frequency domain analysis method discussed in this Chapter, a nonlinear spring-damping system is studied as shown in Fig. 7.1. The system has two nonlinear passive components and one nonlinear active unit. The active unit is described by $F=c_1\dot{x}^2x+c_2x\dot{x}^2$, the output property of the spring satisfies $F = \hat{K}x + c_3x^3$, and the damper $F=B\dot{x}+c_4\dot{x}^3$. $u(t)$ is the external input force. The system dynamics can be described by

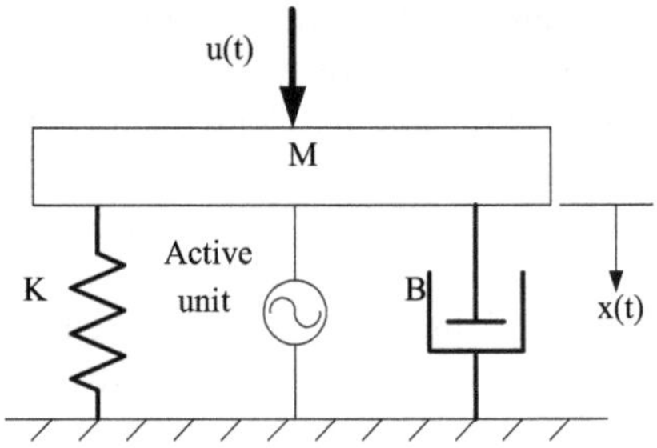

Fig. 7.1 A mechanical system

$$\hat{M}\ddot{x} = -\hat{K}x - B\dot{x} - c_1\dot{x}^2x - c_2\dot{x}x^2 - c_3x^3 - c_4\dot{x}^3 + u(t) \tag{7.14a}$$

Let the output be

$$y = \hat{K}x \tag{7.14b}$$

This may represent a vibration isolator system with nonlinear spring and damping characteristics. The task for this case study is to investigate how the nonlinear terms included both in passive and active unites affect the output and what the effect might be, and thus to provide a useful insight into the design of corresponding nonlinear parameters to achieve a desired output frequency response.

For clarity in discussion, let $\hat{M}$ = 240, $\hat{K}$ =16,000, and B=296, then (7.14a,b) can be rewritten as

$$240\ddot{x} = -16{,}000x - 296\dot{x} - c_1\dot{x}^2x - c_2\dot{x}x^2 - c_3x^3 - c_4\dot{x}^3 + u(t) \tag{7.15a}$$

$$y = 16{,}000x \tag{7.15b}$$

Equation (7.15a) is a simple case of the NDE model (2.11) with $\hat{M} = 3$, $\hat{K} = 2$, $c_{10}(2)=240$, $c_{10}(1)=296$, $c_{10}(0)=16000$, $c_{30}(111)=c_4$, $c_{30}(110)=c_1$, $c_{30}(100)=c_2$, $c_{30}(000)=c_3$, $c_{01}(0)=-1$, and all the other parameters are zero. Therefore, what is of interest for this study is to analyse the effect of the nonlinear terms with coefficients c_1, c_2, c_3 and c_4 on the system output frequency response. To achieve this objective, the procedure proposed in Sect. 7.2.2 are adopted to derive the OFRF of system (7.15a,b), and the results in Sect. 6.1 will be used for the computation of the parametric characteristic of the OFRF with respect to the nonlinear parameters c_1, c_2, c_3 and c_4. Moreover, though the method proposed in this paper is suitable for a general input function u(t), for convenience in discussion, the input of system (7.15a,b) is considered to be a sinusoidal function $u(t)=100\sin(8.1t)$. To illustrate the new results more clearly, first only the effect of parameter c_2 is considered and it is assumed that $c_1=c_3=c_4=0$. Then complicated cases where the effect of more than one nonlinear parameters is involved will also be investigated.

7.3.1 Determination of the Parametric Characteristics of OFRF

Note that all the parameters of interest belong to C_{30}, and the other degrees of nonlinear parameters are all zero. Thus Corollary 6.2 and Proposition 6.3 can be utilised directly. Therefore,

$$CE(H_n(j\omega_1,\cdots,j\omega_n)) = c^{\frac{n-1}{2}} \cdot \delta\left(\frac{n-1}{2} - \left\lfloor\frac{n-1}{2}\right\rfloor\right) \cdot (1-\delta(3)pos(n)) = c^{\frac{n-1}{2}} \cdot \delta\left(\frac{n-1}{2} - \left\lfloor\frac{n-1}{2}\right\rfloor\right)$$

$$\psi = CE(Y(j\omega)) = \bigoplus_{i=0}^{\lfloor N-1/_{p+q-1}\rfloor} CE\left(H_{(p+q-1)i+1}(\cdot)\right) = \left[1 \quad c \quad c^2 \quad \cdots \quad c^{\lfloor N-1/_{p+q-1}\rfloor}\right]$$
$$= \left[1 \quad c \quad c^2 \quad \cdots \quad c^{\lfloor N-1/_2\rfloor}\right] \tag{7.16}$$

where $c=c_2$. To derive the detailed form for ψ, the largest order N should be determined first according to Process B in Sect. 6.2.2. In order to have a larger range in which the parameters can vary, in this case let $c_2 \in (0, 10^8)$. The magnitude bound of $Y_n(j\omega)$ can then be evaluated as mentioned in Process B. However, for paper limitation, the detailed computation is omitted in this case. It can be verified that $N=23$ is enough for use in this case. Therefore,

$$\psi = \left[1 \quad c \quad c^2 \quad \cdots \quad c^{\lfloor 23-1/2\rfloor}\right] = \left[1,\ c_2,\ c_2^2,\ c_2^3,\ c_2^4,\ c_2^5,\ \ldots,\ c_2^{11}\right] \tag{7.17}$$

7.3.2 Determination of $\Phi(j\omega)$ for the OFRF

Following Process C, the matrix $\Psi=[\psi_1^T \cdots \psi_\rho^T]^T$ should be constructed first. In this case, for any 12 different values of c_2, the matrix Ψ is a Vandermonde matrix and thus non-singular. Note that in many cases, the parameters may be set to be some large values and cover a large range. This will make the element values in the matrix Ψ extraordinarily large. Then when the inverse of matrix Ψ is computed, there may be some computation error involved in Matlab. To overcome this problem, Ψ can be written as

$$\psi = \bigoplus_{i=0}^{\lfloor N-1/_{p+q-1}\rfloor} kCE\left(H_{(p+q-1)i+1}(\cdot)\right)/k = \left[1 \quad k(c/k) \quad k^2(c/k)^2 \quad \cdots \quad k^{\lfloor N-1/_2\rfloor}(c/k)^{\lfloor N-1/_2\rfloor}\right] \tag{7.18}$$

Then (7.1) can be written as

$$
\begin{aligned}
Y(j\omega;c) &= \psi \cdot \Phi(j\omega)^T \\
&= \begin{bmatrix} 1 & (c/k) & (c/k)^2 & \cdots & (c/k)^{\ell} \end{bmatrix} \begin{bmatrix} \varphi_1(j\omega) & k\varphi_2(j\omega) & \cdots & k^{\ell}\varphi_{\ell}(j\omega) \end{bmatrix}^T
\end{aligned} \tag{7.19}
$$

where $\ell = \lfloor N - ½ \rfloor$. Moreover, the range for each parameter can be divided into several sub-range, and the final result is the combination of these results obtained for each sub-range. In this study, let k=10^5, then $\overline{c}_2 = {}^{c_2}/_k \in [0, 1000]$. Choose $\overline{c}_2$ to be the following values to construct $\Psi = [{\psi_1}^T \cdots {\psi_\rho}^T]^T$, i.e., 0.1,1,50,65,80, 100,150,200,250,300,350,400,450,500,550,600,650,700,750,800,850,900,950,980, 1,000. The output frequency response

$$
YY(j\omega) = \begin{bmatrix} Y(j\omega)_1 & Y(j\omega)_2 & \cdots & Y(j\omega)_\rho \end{bmatrix} \tag{7.20}
$$

of system (7.15a,b) at ω=8.1 rad/s corresponding to different values of c_2 can be obtained through FFT of the time-domain output response. Then using (7.19), it can be derived from (7.13) that

$$
\begin{aligned}
\Phi(j\omega)^T &= \begin{bmatrix} \varphi_1(j\omega) & k\varphi_2(j\omega) & \cdots & k^{\ell}\varphi_{\ell}(j\omega) \end{bmatrix}^T \\
&= \left(\Psi^T\Psi\right)^{-1}\Psi^T \cdot YY(j\omega)
\end{aligned} \tag{7.21}
$$

Therefore, the output frequency response function of system (7.15a,b) with respect to nonlinear parameter c_2 in the case of $c_1=c_3=c_4=0$ is obtained as

$$
\begin{aligned}
Y(j\omega;c_2) = {} & (2.060893505718041\mathrm{e}+002 \; - 2.402014548824790\mathrm{e}+002\mathrm{i}) \\
& + \mathrm{k}^{-1}(\; - 5.14248529981906 + 5.35676372314361\mathrm{i})\, \mathrm{c}_2 \\
& + \mathrm{k}^{-2}(0.08589533966805 - \; 0.08827649204263\mathrm{i})\, {\mathrm{c}_2}^2 \\
& + \mathrm{k}^{-3}(-8.068953639113292\mathrm{e} - 004 \; + 8.248154776018186\mathrm{e} - 004\mathrm{i})\, {\mathrm{c}_2}^3 \\
& + \mathrm{k}^{-4}(4.598423724418538\mathrm{e} - 006 \; - 4.686570228695798\mathrm{e} - 006\mathrm{i})\, {\mathrm{c}_2}^4 \\
& + \mathrm{k}^{-5}(-1.679591261850433\mathrm{e} - 008 \; + 1.708497491564935\mathrm{e} - 008\mathrm{i})\, {\mathrm{c}_2}^5 \\
& + \mathrm{k}^{-6}(4.056287337706451\mathrm{e} - 011 \; - 4.120496550333245\mathrm{e} - 011\mathrm{i})\, {\mathrm{c}_2}^6 \\
& + \mathrm{k}^{-7}(-6.544911009113156\mathrm{e} - 014 \; + 6.641760366680977\mathrm{e} - 014\mathrm{i})\, {\mathrm{c}_2}^7 \\
& + \mathrm{k}^{-8}(6.976300614229155\mathrm{e} - 017 \; - 7.073928662624432\mathrm{e} - 017\mathrm{i})\, {\mathrm{c}_2}^8 \\
& + \mathrm{k}^{-9}(-4.713366512185836\mathrm{e} - 020 \; + 4.776287453573993\mathrm{e} - 020\mathrm{i})\, {\mathrm{c}_2}^9 \\
& + \mathrm{k}^{-10}(1.827866445826756\mathrm{e} - 023 \; - 1.851299290299388\mathrm{e} - 023\mathrm{i})\, {\mathrm{c}_2}^{10} \\
& + \mathrm{k}^{-11}(-3.098310700824303\mathrm{e} - 027 \; + 3.136656793561425\mathrm{e} - 027\mathrm{i})\, {\mathrm{c}_2}^{11}
\end{aligned} \tag{7.22}
$$

Based on this function, (7.4) can be further computed as

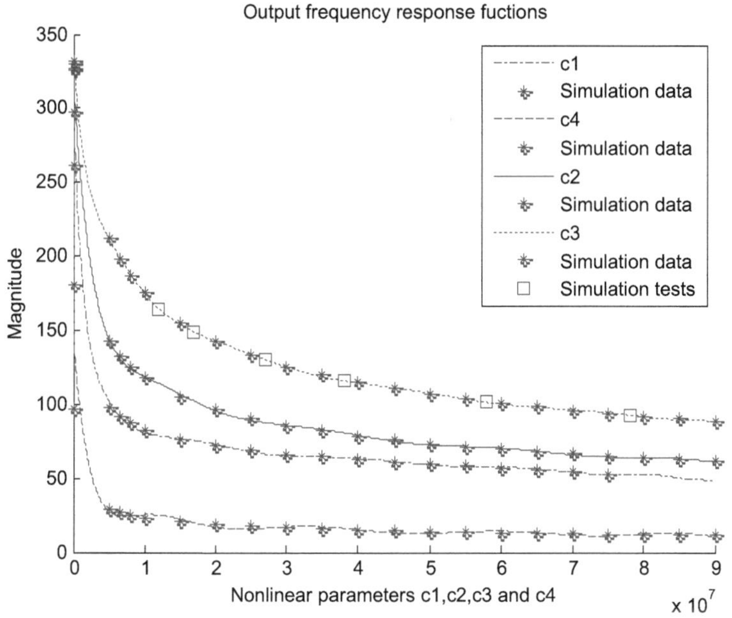

Fig. 7.2 Output frequency response functions with respect to c_1 to c_4 respectively (Jing et al. 2008d)

$$\begin{aligned}|Y(j\omega;c)|^2 &= p_0 + cp_1 + c^2p_2 + \cdots + c^{2\ell}p_{2\ell} + \cdots\cdots \\ &= (1.001695593467675e+005) + k^{-1}(-4.693027791051078e+003)c_2 \\ &\quad + k^{-2}(1.329525858242289e+002){c_2}^2 + k^{-3}(-2.55801250200731){c_2}^3 \\ &\quad + k^{-4}0.03645314106899{c_2}^4 + k^{-5}(-3.968756773045435e-004){c_2}^5 \\ &\quad + k^{-6}0.01517275811829{c_2}^6 + \ldots\end{aligned} \tag{7.23}$$

Note that this is an alternating series and it holds that $|p_i| > |p_{i+1}|$ and $|p_i| \to 0$. Hence the series may keep decreasing when c is going larger and within its radius of convergence. By following the similar method demonstrated above, the output frequency response functions of system (7.15a,b) with respect to nonlinear parameters c_1, c_2, c_3 and c_4 of different cases can all be obtained, for instance $Y(j\omega;c_1)$, $Y(j\omega;c_3)$, and $Y(j\omega;c_4)$ (The other nonlinear parameters are zero if not appearing in the function). The results are shown in Figs. 7.2, 7.3, and 7.4.

Figure 7.2 shows that the variation of the magnitude of the output frequency response functions with respect to each nonlinear parameter. It can be seen that there is a good matching between the theoretical computation results and the simulation results to which they have been fitted, and there is also a good match

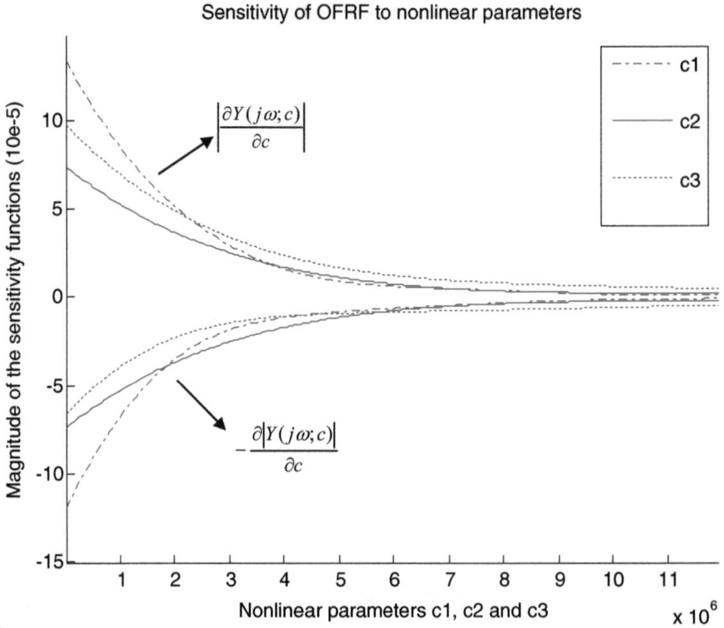

Fig. 7.3 Sensitivity function of the OFRFs with respect to c_1 to c_3 respectively (Jing et al. 2008d)

between the theoretical computation results and the simulation tests (for parameter c_3) which are independent of the fitted simulation results. From both Figs. 7.2 and 7.3 it can also be seen that the system output frequency response is much more sensitive to the variation of the nonlinear parameters when they are small. Once the value of a nonlinear parameter is sufficient large, then the sensitivity will tend to be zero. From the comparison of these four nonlinear terms, it can be concluded that the system output frequency response is more sensitive to the variation of the nonlinear parameter c_4 when the values are small; however when the values of each nonlinear parameters are sufficient large, the system output spectrum is more sensitive to the nonlinear parameter c_2. From Fig. 7.4 it can be seen that the convergence of the output frequency response functions are all very fast.. It is noted that the ratio functions of c_2 and c_3 go up much faster than that of c_1, especially c_2. This implies that the radius of convergence of the output spectrum corresponding to c_2 should be larger. Simulation tests verify that the system is still stable when $c_2 = 10^{17}$ where the magnitude of the output spectrum is 0.0216, while the system may tend to be unstable when c_1 tends to be larger than 10^8.

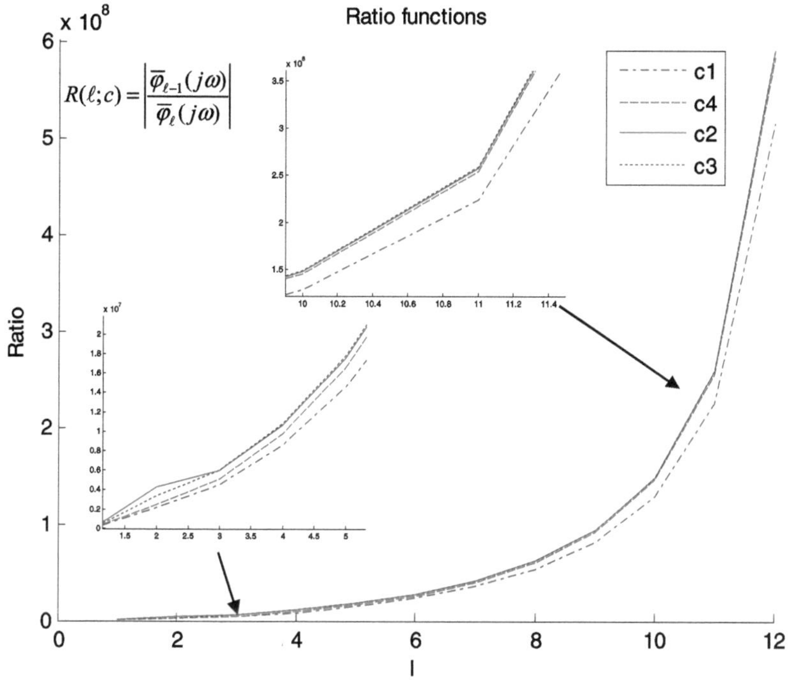

Fig. 7.4 Ratio functions with respect to c_1 to c_4 respectively (Jing et al. 2008d)

From the analysis above for the four nonlinear parameters of nonlinear degree 3, respectively, it can be seen that

- The computed system output spectrum has a larger radius of convergence with respect to c_2, c_3 and c_4.
- The system output spectrum is more sensitive to c_4 and less sensitive to c_3;
- If the output spectrum with respect to a nonlinear parameter is an alternating series satisfying $|p_i| > |p_{i+1}|$ and $|p_i| \to 0$, then the system output spectrum may be reduced to zero if additionally the radius of convergence for this parameter is sufficiently large.
- The magnitude of output spectrum decreases with the increase of the values of the nonlinear parameters. Thus an introduction of some simple nonlinear terms into a linear system may greatly improve the performance of output frequency response, and the stability of a nonlinear system is not necessarily deteriorated with increasing the values of nonlinear parameters; This also shows that a larger value of the parameter for a nonlinear term may not lead to a bad performance of a system.

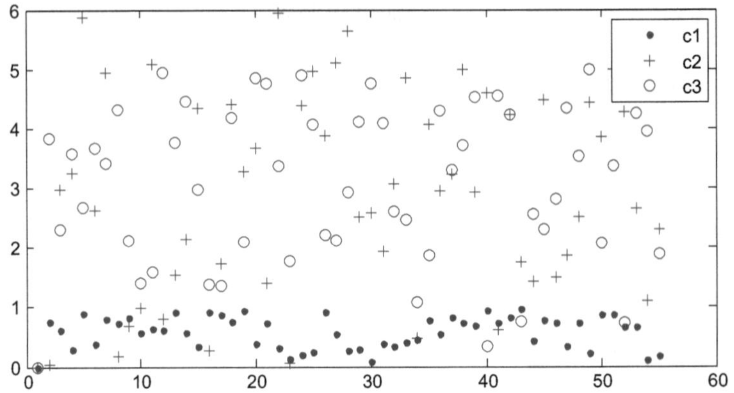

Fig. 7.5 A series of 55 points $c=[c_1,c_2,c_3]$ by random generation in {[0,1], [0,6], [0,5]} where the y-axis is the value of different parameters and the x-axis is the number of different point in the series (Jing et al. 2008d)

- For system (7.15a,b), the nonlinear parameters c_3 and c_4 can be designed to be large enough to achieve a sufficiently small transmitted force since they correspond to passive elements, and several nonlinear terms in the active part can work together to achieve a better performance.

To demonstrate further the advantage of the OFRF based analysis and to show more clearly the effect on the system output spectrum from several nonlinear parameters, the OFRF with respect to c_1, c_2 and c_3, i.e., $Y(j\omega; c_1, c_2, c_3)$ is derived. Let $c_1 \in [0, 10^5], c_2 \in [0, 6 \cdot 10^5], c_3 \in [0, 5 \cdot 10^5], c_4 = -500$, and the largest order N of the output spectrum is determined to be 11, then the parametric characteristic can be obtained as $(c=[c_1,c_2,c_3])$

$$\begin{aligned}\psi = \left[1 \quad c \quad c^2 \quad \cdots \quad c^{\lfloor 11-1/2 \rfloor}\right] = [&1, c_1, c_2, c_3, c_1{}^2, c_1c_2, c_1c_3, c_2{}^2, c_2c_3, c_3{}^2,\\ &c_1{}^3, c_1{}^2c_2, c_1{}^2c_3, c_1c_2{}^2, c_1c_2c_3, c_2{}^3, c_2{}^2c_3, c_2c_3{}^2, c_3{}^3, c_1{}^4, c_1{}^3c_2, c_1{}^3c_3, c_1{}^2c_2{}^2,\\ &c_1{}^2c_2c_3, c_1{}^2c_3{}^2, c_1c_2{}^3, c_1c_2{}^2c_3, c_1c_2c_3{}^2, c_1c_3{}^3, c_2{}^4, c_2{}^3c_3, c_2{}^2c_3{}^2, c_2c_3{}^3, c_3{}^4, c_1{}^5,\\ &c_1{}^4c_2, c_1{}^4c_3, c_1{}^3c_2{}^2, c_1{}^3c_2c_3, c_1{}^3c_3{}^2, c_1{}^2c_2{}^3, c_1{}^2c_2{}^2c_3, c_1{}^2c_2c_3{}^2, c_1{}^2c_3{}^3, c_1c_2{}^4,\\ &c_1c_2{}^3c_3, c_1c_2{}^2c_3{}^2, c_1c_2c_3{}^3, c_1c_3{}^4, c_2{}^5, c_2{}^4c_3, c_2{}^3c_3{}^2, c_2{}^2c_3{}^3, c_2c_3{}^4, c_3{}^5]\end{aligned} \tag{7.24}$$

In order to construct the non-singular matrix Ψ, the series of $\rho(N)=55$ different points $c=[c_1,c_2,c_3]$ in $\partial S_C=\{c=[c_1,c_2,c_3] | c_1 \in [0,1], c_2 \in [0,6], c_3 \in [0,5]\}$ can be obtained by using a simple stochastic-based searching method. In simulations, it is noticed that is easy to find such a series of points that $\det(\Psi) \neq 0$. For example, a series of points $c=[c_1,c_2,c_3]$ are illustrated in Fig. 7.5, and it can be obtained in this case that $\det(\Psi)=0.08608811188201$. It can be seen from simulations that it is easy to find a non-singular matrix Ψ with a proper inverse.

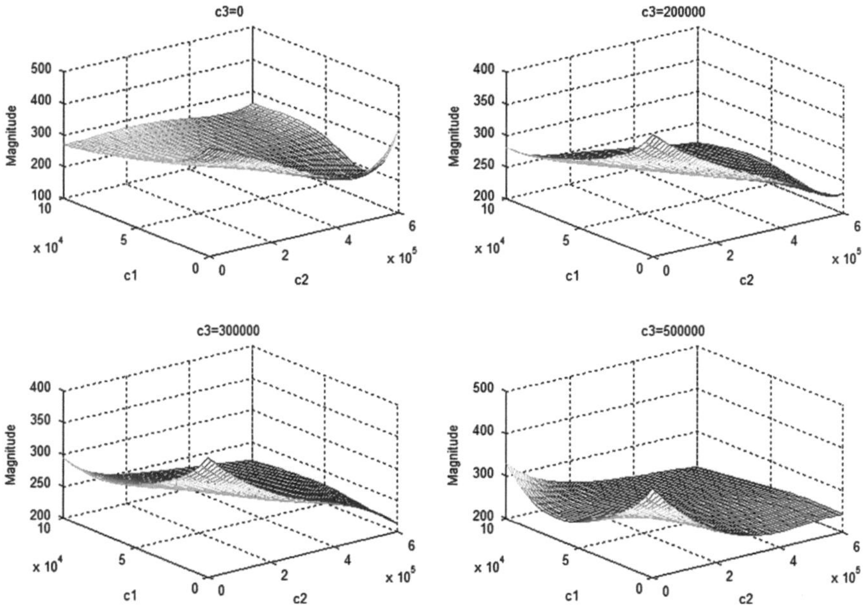

Fig. 7.6 Output spectrum with respect to c_1, c_2 and c_3 (Jing et al. 2008d)

Then following the same procedure as demonstrated above, the OFRF $Y(j\omega; c_1, c_2, c_3)$ in this case can be obtained. The results are shown in Figs. 7.6 and 7.7, and the following points can be summarized.

- By using the OFRF, the output spectrum can be plotted and analyzed under different combinations of the nonlinear parameters c_1, c_2 and c_3. This provides a straightforward understanding of the relationship between system output spectrum and model parameters which define nonlinearities.
- The OFRF varies with different values of c_1, c_2 and c_3. The effect on the output spectrum from any two nonlinear terms is not necessarily the simple combination of the contributions from each term respectively. Thus the parameters can be analyzed in order to obtain the best output frequency response performance. The OFRF provides a useful basis for this kind of analysis and optimization.

From the discussions above, it can be concluded that the OFRF based analysis provides a novel, effective and useful approach to the analysis and design of nonlinear systems in the frequency domain.

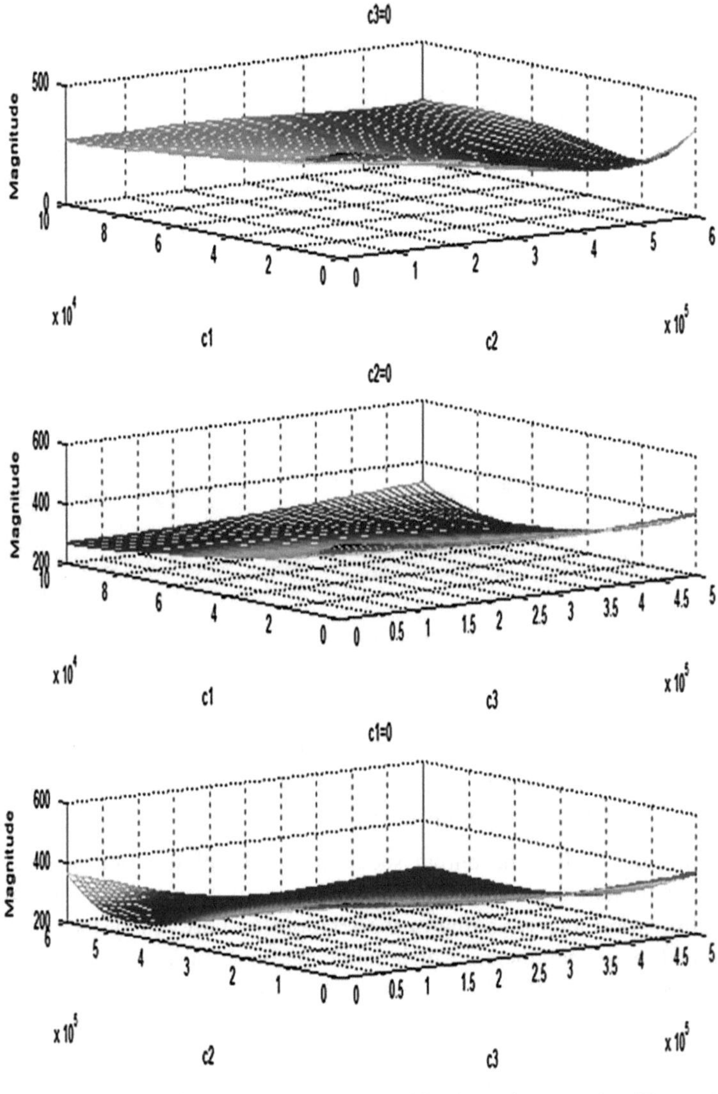

Fig. 7.7 Output spectrum with respect to any two combinations of c_1, c_2 and c_3 (Jing et al. 2008d)

7.4 Conclusions and Discussions

The OFRF based analysis for nonlinear Volterra systems is discussed and demonstrated in this chapter. The OFRF based analysis provides a novel and effective approach to the analysis and design of nonlinear systems in the frequency domain by using the explicit relationship between the system output frequency response and model parameters. The OFRF is characterized by its parametric characteristics multiplied with an associated complex valued frequency dependent function vector. Thus instead of the direct analytical computation of the OFRF, the proposed method simplifies the computation of the OFRF by splitting the computation procedure into two parts— the one is the computation of the parametric characteristics of the OFRF, which is analytical in the determination of the relationship between the output spectrum and model parameters, and simpler to be carried out, and the other is the determination of the complex valued frequency dependent function vectors, which are obtained by using the Least square method. Some fundamental results, techniques, and a general procedure for the determination of the OFRF for a given NDE or NARX model subjected to any specific input signal are provided. Although the proposed method needs $\rho(N)$ simulation data for the numerical method of Process C, and the OFRF obtained by the proposed method is not analytical with respect to the input signal and frequency variants at present, the case study for a simple mechanical system shows that the OFRF based analysis is a useful approach to the analysis and design of nonlinear systems in the frequency domain.

Chapter 8
Determination of Nonlinear Output Spectrum Based on Its Parametric Characteristics: Some Theoretical Issues

8.1 Introduction

Volterra-type nonlinear systems represent a considerably large class of nonlinear systems, and have been extensively applied in various engineering practice. As an important extension of traditional transfer function theory of linear systems, an important concept, referred to as the GFRF, initiates the frequency domain analysis and design of nonlinear systems. The GFRFs for a parametric nonlinear system described by a NDE or NARX model are given in Chap. 2, and nonlinear output frequency response are discussed in Chaps. 3, 6 and 7.

The output frequency response function (OFRF) of nonlinear systems is shown to be a polynomial function of model parameters (Chaps. 6 and 7). This reveals an explicit relationship between system output spectrum and model parameters, and consequently the system output frequency response can be studied in terms of any model parameters of interest subject to any input signals. This can greatly facilitate the analysis and design of nonlinear output response (or behavior) of nonlinear systems in the frequency domain. In order to perform an OFRF-based analysis for a nonlinear system, the OFRF can be analytically determined. Usually, this can be done by using the recursive algorithm in Chap. 2 to compute the GFRFs, then using the result in Chap. 6 (and also Chap. 11) to analytically obtain the output spectrum, and finally expressing the output spectrum to a polynomial form in terms of model parameters of concern. However, it can be seen that, the process above is very computationally intensive especially when the involved Volterra order is larger than 5.

To solve this problem, the detailed polynomial structure of the OFRF for up to any order in terms of any model parameters can be obtained by using the results in

X. Jing, Z. Lang, *Frequency Domain Analysis and Design of Nonlinear Systems based on Volterra Series Expansion*, Understanding Complex Systems,
DOI 10.1007/978-3-319-12391-2_8

Chaps. 6 and 7. Then if a series simulation or experimental data are collected, a numerical method may be adopted to determine the OFRF as discussed before. This can reduce the computational complexity as mentioned. However, the problem is, whether the analytical parametric relationship of the OFRF with respect to any model parameters can always be explicitly determined by using this numerical method with a possible specially-designed simulation or experimental data. To this aim, this study showed that, based on the parametric characteristics of the OFRF, the analytical parametric relationship of the OFRF up to any order and every specific order of the OFRF can all be determined accurately by using a simple Least Square method (when there is no data noise and measurement error). Practical methods to generate a special series of values for the parametric characteristic vector are discussed such that a unique solution can be obtained. This Chapter not only solves a fundamental problem for the OFRF-based method for nonlinear systems, but also provides a theoretical basis for the determination of the analytical parametric relationship of polynomial structures in dynamic systems. Theoretical analysis and simulations demonstrate the results.

8.2 The Problem

The input–output relationship of nonlinear systems can be approximated by a Volterra series up to a sufficiently high order N. Consider Volterra-type nonlinear systems described by the NDE model (2.11), i.e.,

$$\sum_{m=1}^{M}\sum_{p=0}^{m}\sum_{k_1,k_{p+q}=0}^{K} c_{p,q}(k_1,\cdots,k_{p+q})\prod_{i=1}^{p}\frac{d^{k_i}y(t)}{dt^{k_i}}\prod_{i=p+1}^{p+q}\frac{d^{k_i}u(t)}{dt^{k_i}}=0 \tag{8.1}$$

where the notations can be referred to Chap. 2.

The OFRF of system (8.1) can be expressed into a polynomial function of model parameters as studied in Chap. 6 (and will be discussed more in Chap. 11),

$$Y(j\omega)=\sum_{j_1=0}^{m_1}\cdots\sum_{j_{SN}=0}^{m_{SN}}\gamma_{j_1\cdots j_{SN}}(\omega;U(\omega))\cdot c_{1,1}^{j_1}(.)\cdots c_{SNSN}^{j_{SN}}(.) \tag{8.2}$$

$\gamma_{j_1\cdots j_{SN}}(\omega;U(\omega))$ are complex valued functions and $c_{1,1}^{j_1}(.)\cdots c_{SNSN}^{j_{SN}}(.)$ is a monomial function of model parameters, which also represents a combination among all the possible combinations consisting of model parameters from degree 0 to m_1+m_{SN}. Note that (8.2) includes many unnecessary terms $c_{1,1}^{j_1}(.)\cdots c_{SNSN}^{j_{SN}}(.)$ which do actually not appear in the OFRF. For this reason, the detailed parametric structure of this polynomial function can be revealed by using the method in Chaps. 4–6 as

$$Y(j\omega) = \sum_{n=1}^{N} Y_n(j\omega) \tag{8.3}$$

$$Y_n(j\omega) = CE(H_n(j\omega_1, \cdots, j\omega_n)) \cdot \hat{F}_n(j\omega; U(j\omega)) \tag{8.4}$$

where $\hat{F}_n(j\omega; U(j\omega))$ is a complex valued function vector and has the same dimension with $CE(H_n(j\omega_1, \cdots, j\omega_n))$. Note that $\hat{F}_n(j\omega; U(j\omega))$ is dependent on the system linear parameters and input $U(j\omega)$, which is thereafter denoted by $\hat{F}_n(j\omega)$ for convenience. $CE(H_n(j\omega_1, \cdots, j\omega_n))$ is referred to as the parametric characteristic of the nth-order GFRF for system (8.1), which is a vector whose elements are functions of model parameters, and can be recursively determined by

$$CE(H_n(j\omega_1, \cdots, j\omega_n)) = C_{0,n} \oplus \left(\mathop{\oplus}_{q=1}^{n-1} \mathop{\oplus}_{p=1}^{n-q} C_{p,q} \otimes CE\big(H_{n-q-p+1}(\cdot)\big) \right) \oplus \left(\mathop{\oplus}_{p=2}^{n} C_{p,0} \otimes CE\big(H_{n-p+1}(\cdot)\big) \right) \tag{8.5}$$

with terminating condition $CE(H_1(j\omega_i)) = 1$ or 0. Note that CE is a new operator with two operations "$\otimes$" and "$\otimes$" defined in Chap. 4, and $C_{p,q}$ represents the $p + q$th degree nonlinear parameter vector, i.e.,

$$C_{p,q} = \big[c_{p,q}(0, \cdots, 0), c_{p,q}(0, \cdots, 1), \cdots, c_{p,q}\underbrace{(K, \cdots, K)}_{p+q=m}\big]$$

For convenience, (8.3 and 8.4) can be written as

$$Y(j\omega) = \psi \cdot \Phi(j\omega)^T \tag{8.6a}$$

where

$$\psi = \mathop{\oplus}_{n=1}^{N} CE(H_n(j\omega_1, \cdots, j\omega_n)) \tag{8.6b}$$

$$\Phi(j\omega) = \left[\hat{F}_1(j\omega)^T \quad \hat{F}_2(j\omega)^T \quad \cdots \quad \hat{F}_N(j\omega)^T\right] \tag{8.6c}$$

At this stage, in order to obtain the analytical parametric relationship of the OFRF described by (8.6a–c) with the known polynomial structure in terms of any model parameters for system (8.1) under any specific input, the complex-valued frequency function $\Phi(j\omega)$ should be determined. As mentioned, the analytical computation of $\Phi(j\omega)$ can be conducted by using the results in Chap. 3–6, Jing et al. (2008e) and Jing and Lang (2009a), but it is very computationally intensive. However, this can alternatively be achieved by using the following method as discussed in Chap. 7, referred to here as Algorithm A:

(A1) Choose ρ series of different values of the model parameters to form a series of vectors $\psi_1 \cdots \psi_\rho$;

(A2) At a given frequency ω, actuate the system using the same input under the different values of the nonlinear parameters $\psi_1 \cdots \psi_\rho$, then collect the time domain output $y(t)$ for each case. Finally, obtain a series of output frequency response $Y(j\omega)_1 \cdots Y(j\omega)_\rho$ at the frequency ω by the FFT.

(A3) It follows from Step (A2)

$$\Psi \cdot \Phi(j\omega)^T = \begin{bmatrix} \psi_1 \\ \psi_2 \\ \vdots \\ \psi_\rho \end{bmatrix} \cdot \Phi(j\omega)^T = \begin{bmatrix} Y(j\omega)_1 \\ Y(j\omega)_2 \\ \vdots \\ Y(j\omega)_\rho \end{bmatrix} =: YY(j\omega)$$

Hence,

$$\Phi(j\omega)^T = \left(\Psi^T\Psi\right)^{-1} \cdot YY(j\omega) \tag{8.7}$$

From the algorithm above, it can be seen that ρ should at least be equal to the dimension of ψ_i and $\Psi = [\psi_1^T, \cdots, \psi_\rho^T]^T$ should be non-singular in order to achieve a unique and accurate evaluation for $\Phi(j\omega)$. When all the possible combinations of parametric monomials involved in (8.2) are considered, it is solvable for this problem by the algorithm above for any series of different parameter values (See the detailed in Chap. 7). However, the true polynomial coefficients of the OFRF are determined by the parametric characteristics $\bigoplus_{n=1}^{N} CE(H_n(\cdot))$, which is only a part of the monomials appearing in (8.2) in terms of the model parameters. Hence, the existence of the solution is not necessarily possible and how to generate a series of different parameter values is also yet to be solved.

For example, considering a polynomial $Y = y_0 + c_1c_2y_1 + c_1c_3y_2$ (c_i's are parameters and y_i's are yet to be determined), it needs at least two different values of (c_1, c_2, c_3) to obtain y_1 and y_2, i.e.,

$$\begin{bmatrix} Y(1) - y_0 \\ Y(2) - y_0 \end{bmatrix} = \begin{bmatrix} c_1(1)c_2(1) & c_1(1)c_3(1) \\ c_1(2)c_2(2) & c_1(2)c_2(2) \end{bmatrix} \begin{bmatrix} y_1 \\ y_2 \end{bmatrix}$$

If let (c_1, c_2, c_3) = (0,1,2) and (0,2,1), then the coefficient matrix is $\begin{bmatrix} 0 & 2 \\ 0 & 2 \end{bmatrix}$. In this case, the solution is not unique for y_1 and y_2.

This problem may be solved if the number ρ [in Step (A1)] of different values of (c_1, c_2, c_3) is increased. However, this does not guarantee the existence of the solution and will increase the simulation or experimental burden in Step (A2). Therefore, the problems are, given the detailed polynomial structure in terms of any model parameters, whether the polynomial (8.6a–c) can be solved by the algorithm above when ρ equals the dimension of ψ_i, and whether the complex valued function vectors $\hat{F}_n(j\omega)$ for $n=1$ to N can accurately be obtained and every specific component of the OFRF, i.e., $Y_n(j\omega)$ for $n=1$ to N, can also be determined from these complex valued function vectors. These are motivations of this study.

It shall be noted that the accurate determination of the polynomial structure of the OFRF in terms of any model parameters can effectively reduce the computation and simulation (experimental) burden in the determination process for the OFRF. This will be further discussed in the following section. Regarding the parametric characteristics of the OFRF, consider a special but frequently encountered case in practice for system (8.1) as follows, which can further simplify the determination of the OFRF structure.

Proposition 8.1 Consider the input function for system (8.1) to be $u(t)=F_d\sin(\Omega t)$. The parametric characteristics of the system OFRF at the driving frequency Ω is

$$CE(Y(j\Omega)) = \mathop{\oplus}_{n=0}^{\lfloor N-1/2 \rfloor} CE(Y_{2n+1}(j\Omega))$$

$$CE(Y_{2n+1}(j\Omega)) = CE(H_{2n+1}(\cdot)) \tag{8.8}$$

where $\lfloor \cdot \rfloor$ is to take the integer part.

Proof See the proof in Sect. 8.6A. □

In the practical analysis of a nonlinear system, a harmonic excitation like $u(t)=F_d\sin(\Omega t)$ is often adopted. In these cases, Proposition 8.1 provides a useful guidance for the accurate computation of the OFRF structure.

8.3 Solution Existence Theorem

In order to solve the problems mentioned above, some preliminary results are discussed first, which are summarized in Lemmas 8.1–8.5 below and demonstrate some important properties for the parametric characteristics of the OFRF and Algorithm A. The following Lemma 8.1 is an important result about the parametric characteristics of the GFRFs, which is Proposition 5.1 in Chap. 5.

Lemma 8.1 The elements of $CE(H_n(j\omega_1,\cdots,j\omega_n))$ include and only include the nonlinear parameters in C_{0n} and all the nonlinear parameter monomials in $C_{p,q} \otimes C_{p_1,q_1} \otimes C_{p_2,q_2} \otimes \cdots \otimes C_{p_k,q_k}$ for $0 \le k \le n-2$, where the subscripts satisfy

$$\left.\begin{array}{c} p+q+\sum_{i=1}^{k}(p_i+q_i)=n+k \\ 1\leq p\leq n-k,\ \ 2\leq p+q\leq n-k,\ \ 2\leq p_i+q_i\leq n-k \end{array}\right\} \tag{8.9}$$

According to Lemma 8.1, for example, a parameter monomial like $(c_{1,1}(\cdot))^2c_{2,0}(\cdot)$ $c_{0,2}(\cdot)$ must appear in the zth-order GFRF, where $Z=2\cdot(1+1)+2+2-3=5$. *CE* $(H_n(j\omega_1,\cdots,j\omega_n))$ can be obtained directly from model parameters according to the Lemma 8.1 without recursive computation. This can be carried out by counting k from 0 to $n-2$, then write out all the monomials satisfying the corresponding conditions in Lemma 8.1 and remove all the repetitive terms (see Definition 8.1 below). Based on these results, the following results can be obtained.

Lemma 8.2 (1) $CE(H_n(j\omega_1,\cdots,j\omega_n))$ includes and only includes all the nonlinear parameters of degree from 2 to n. (2) If $p>0$, $(c_{p,q}(\cdot))^{k+1}$ is an element of *CE* $(H_n(j\omega_1,\cdots,j\omega_n))$ with $k=\frac{n-p-q}{p+q-1}$.

Proof See the proof in Sect. 8.6B. □

Lemma 8.2 shows which degree of nonlinear parameters have contribution to $H_n(j\omega_1,\cdots,j\omega_n)$. From Lemma 8.2, it can also be seen that for the case that only one parameter $c_{pq}(.)\neq 0$ and all the other nonlinear parameters are zero for model (8.1), the parametric characteristic of the nth-order GFRF is $CE(H_n(j\omega_1,\cdots,j\omega_n))=\left(c_{p,q}(\cdot)\right)^{\frac{n-1}{p+q-1}}$ if ($n>p+q$ and $p>0$ and $(n-1)/(p+q-1)$ is an integer) or $(n=p+q)$, else $CE(H_n(j\omega_1,\cdots,j\omega_n))=0$. This will be used later.

For convenience, let

$$\begin{aligned}\mathrm{int}(a_1,b_1,a_2,b_2,\cdots,a_k,b_k)&=a_1 10^{2k}+b_1 10^{2k-1}+a_2 10^{2k-2}+b_2 10^{2k-3}\\ &\quad+\cdots+a_k 10^2+b_k\end{aligned} \tag{8.10}$$

where $a_1,b_1,a_2,b_2\cdots a_k,b_k$ are some non-negative integer numbers.

Definition 8.1 Consider monomials $c_{p_1,q_1}\left(k_1\cdots k_{p_1+q_1}\right)\cdot c_{p_2,q_2}\left(k_1\cdots k_{p_2q_2}\right)\cdots c_{p_k,q_k}$ $\left(k_1\cdots k_{p_kq_k}\right)$ and $c_{a_1,b_1}\left(k_1\cdots k_{a_1+b_1}\right)\otimes c_{a_2,b_2}\left(k_1\cdots k_{a_2+b_2}\right)\cdots c_{a_k,b_k}\left(k_1\cdots k_{a_k+b_k}\right)$. If there exists a permutation for the subscripts of $c_{p_1,q_1}(\cdot)\otimes c_{p_2,q_2}(\cdot)\cdots c_{p_k,q_k}(\cdot)$, i.e., $(p_1',q_1')(p_2',q_2')\cdots(p_k',q_k')$, such that

$$\frac{\mathrm{int}\left(p_1'q_1'p_2'q_2'\cdots p_k'q_k'\right)}{\mathrm{int}(a_1b_1a_2b_2\cdots a_kb_k)}=1 \text{ and } \frac{\mathrm{int}\left(k_1\cdots k_{p_1'+q_1'}k_1\cdots k_{p_2'q_2'}\cdots k_1\cdots k_{p_k'q_k'}\right)}{\mathrm{int}\left(k_1\cdots k_{a_1+b_1}k_1\cdots k_{a_2+b_2}\cdots k_1\cdots k_{a_k+b_k}\right)}=1$$

then the two monomials are repetitive, otherwise non-repetitive.

Remark 8.1 According to Definition 8.1, $c_{1,1}(1,1)\cdot c_{2,0}(1,1)$ and $c_{2,0}(1,1)$ $c_{1,1}(1,1)$ are repetitive, but $c_{1,1}(1,1)\cdot c_{2,0}(1,1)$ and $c_{2,0}(1,1)c_{1,1}(1,1)^2$ are non-repetitive. By the definition of the CE operator, there are no repetitive terms in the parametric characteristic $CE(H_n(j\omega_1,\cdots,j\omega_n))$.

Lemma 8.3 There are no repetitive elements between the parametric characteristics of any two different order GFRFs when all the nonlinear parameters are considered. This is denoted by

$$CE(H_n(j\omega_1,\cdots,j\omega_n)) \wedge CE(H_m(j\omega_1,\cdots,j\omega_m)) = 0 \text{ for } m \neq n.$$

Proof See the proof in Sect. 8.6C. □

Remark 8.2 Although there are always no repetitive terms in the parametric characteristics of the same order OFRF, there may be repetitive terms between the parametric characteristics of different order GFRFs in practices when there are only part of the model parameters are interested for an OFRF analysis. Lemma 8.3 shows that when all the nonlinear parameters are interested, then there must be no repetitive elements between the parametric characteristics of different order GFRFs. When there are no repetitive terms between different order GFRFs, (8.4) can be used to determine every specific component of the OFRF, i.e., $Y_n(j\omega)$ for $n=1,\ldots,n$.

The following lemma is a fundamental result for the proof of Theorem 8.1 below.

Lemma 8.4 Consider equation $e_Y = \zeta \cdot \varphi^T$, where $\zeta \in \Re^n$ whose elements are monomials of parameters $c_1, c_2, \ldots, c_m$ taking values in a parameter space S_C which is a subspace of $\Re^m$, φ is a nonzero complex-valued vector in $\boldsymbol{C}^n$ and is also independent of ζ. If there exist n points $(c_1(1), c_2(1), \ldots, c_m(1)] \ldots (c_1(n), c_2(n), \ldots, c_m(n))$ in S_C such that

$$\zeta|_{(c_1(i), c_2(i), \ldots, c_m(i))} \cdot \varphi^T = 0 \text{ for } i = 1 \text{ to } n$$

and

$$\zeta|_{(c_1(i), c_2(i), \ldots, c_m(i))} \text{ for } i = 1 \text{ to } n \text{ is a base of } \Re^n$$

then (p1) $\varphi = 0$;

(p2) $\zeta \cdot \varphi^T = 0$ for any $\zeta \in \Re^n$;

(p3) $\zeta \cdot \varphi^T = 0$ for any point in the parameter space S_C.

Proof See the proof in Sect. 8.6D. □

All the nonlinear parameters from degree 2 to N of the model (8.1) form a parameter vector C in $\Re^{\sigma_1}$, where σ_1 denotes the dimension of C which is a function of N. Let S_C denote a subspace of $\Re^{\sigma_1}$ around the zero point and be the definition domain of C. Recalling (8.6b), it is from Lemma 8.2 that elements of ψ are monomial functions of elements of C. Let σ_2 denotes the dimension of ψ. It should be noted from Remark 8.1 and Lemma 8.3 that there are no repetitive elements in ψ. That is, each element in ψ is a non-repetitive monomial function of some nonlinear parameters in C. The following lemma can be obtained, which is an important result for the accurate and unique determination of the OFRF by using Algorithm A and shows that there exists a series points in the parameter space S_C for the parametric characteristic vector of the OFRF such that a non-singular matrix $\Psi=[\psi_1^T,\cdots,\psi_\rho^T]^T$ required in Algorithm A can be generated.

Lemma 8.5 There exist σ_2 points $C(1)...C(\sigma_2)$ in S_C, such that $\psi|_{C(i)}$ for $i=1$ to σ_2 form a basis of $\Re^{\sigma_2}$.

Proof See the proof in Sect. 8.6E. □

Now consider (8.6a–c) and Algorithm A. Note from Jing et al. (2008e) and Jing and Lang (2009a) that for $n=1, \hat{F}_1(j\omega)$ in (8.6c) represents the frequency response of the linear part of the system, i.e., $\hat{F}_1(j\omega)=H_1(j\omega)U(j\omega)$ or $\frac{F_d}{2}H_1(j\omega_1)$ for the general input or multi-tone input respectively. Based on Lemmas 8.1–8.5, the following theorem can address the solution existence problem of Algorithm A.

Theorem 8.1 Consider Volterra systems described by NDE model (8.1) which has a parameter space S_C and subject to a specific input function u(t). The maximum order of the Volterra series is N, and the truncation error is denoted by $o(N+1)$. Suppose $o(N+1)$ =0, then there exist a series of points in S_C, i.e., $C(1), C(2)\ldots C(\sigma_2)$, such that the analytically parametric relationship for the system OFRF can be determined as

$$\widetilde{Y}(j\omega)=\psi\cdot\widetilde{\Phi}(j\omega)^T \tag{8.11a}$$

with zero error in S_C, and in case that S_C includes all the nonlinear parameters of model (8.1)

$$\widetilde{Y}_n(j\omega)=CE(H_n(\cdot))\cdot\widetilde{F}_n(j\omega)^T \tag{8.11b}$$

with zero error in S_C, where

$$\begin{aligned}\widetilde{\Phi}(j\omega)^T&=\left[\widetilde{F}_1(j\omega)\quad\widetilde{F}_2(j\omega)\quad\cdots\quad\widetilde{F}_N(j\omega)\right]^T=\Phi(j\omega)^T\\&=\left[\psi\big|_{C(1)}{}^T\quad\psi\big|_{C(2)}{}^T\quad\cdots\quad\psi\big|_{C(\sigma_2)}{}^T\right]^{-T}\left[Y(j\omega)\big|_{C(1)}\right.\\&\quad\left.Y(j\omega)\big|_{C(2)}\cdots Y(j\omega)\big|_{C(\sigma_2)}\right]^T\end{aligned} \tag{8.11c}$$

$Y(j\omega)|_{C(i)}$ is the output frequency response obtained by a simulation or experiment when the model parameter vector is $C(i)$ and actuated by the specific input $u(t)$. Considering the truncation error $o(N+1)\neq 0$, then

$$\begin{aligned}e\left(\widetilde{\Phi}(j\omega)-\Phi(j\omega)\right)&=\left\|\widetilde{\Phi}(j\omega)^T-\Phi(j\omega)^T\right\|\\&=\left\|\Psi^{-1}\big|_{C(1\cdots\sigma_2)}\cdot o_{N+1}\big|_{C(1\cdots\sigma_2)}\right\|\end{aligned} \tag{8.11d}$$

$$
\begin{aligned}
e\left(\widetilde{Y}(j\omega) - Y(j\omega)\right) &= \left\|\widetilde{Y}(j\omega) - Y(j\omega)\right\| \\
&= \left\|\psi \cdot \Psi^{-1}\big|_{C(1\cdots\sigma_2)} \cdot o_{N+1}\big|_{C(1\cdots\sigma_2)} - o(N+1)\right\| \qquad (8.11e)
\end{aligned}
$$

where, $\Psi\big|_{C(1\cdots\sigma_2)} = \left[\psi\big|_{C(1)}{}^T \quad \psi\big|_{C(2)}{}^T \quad \cdots \quad \psi\big|_{C(\sigma_2)}{}^T\right]^T$,

$$
o_{N+1}\big|_{C(1\cdots\sigma_2)} = \left[o(N+1)\big|_{C(1)} \quad o(N+1)\big|_{C(2)} \quad \cdots \quad o(N+1)\big|_{C(\sigma_2)}\right]^T
$$

Proof See the proof in Sect. 8.6F. □

From Theorem 8.1, it can be seen that, $\det\left(\Psi\big|_{C(1\cdots\sigma_2)}\right)$ is larger, the error of the algorithm will be smaller. Theorem 8.1 provides a fundamental result for the accurate numerical determination of the analytically parametric relationship for the OFRF and its every specific component. Given the model of a nonlinear system, to determine the analytically parametric relationship of the system OFRF based on Theorem 8.1, the following procedure can be followed (Algorithm B):

(B1) Determine the largest nonlinearity order N. Given the system model, the variation domain S_C of the model parameters of interest, the largest nonlinearity order N needed for an accurate Volterra series approximation can be obtained by evaluating the truncation error of the series. This can be done by following the bound evaluation method in Jing et al. (2007a).
(B2) Compute the parametric characteristics of the GFRFs $CE(H_n(j\omega_1,\cdots,j\omega_n))$ from 2 to N according to (8.5) or Lemma 8.1 to obtain $\psi = \bigoplus_{n=1}^{N} CE(H_n(\cdot))$.
(B3) Choose a series of points in S_C for the parameter vector C which consists of all the parameters of interest, such that $\psi|_{C(i)}$ for $i=1$ to σ_2 is a base of $\mathfrak{R}^{\sigma_2}$.
(B4) Using a specific input, actuate the system in simulations under different model parameters C(i) to obtain $Y(j\omega)|_{C(i)}$ for $i=1$ to σ_2.
(B5) Then the analytical parametric relationship for the OFRF and its different components can all be determined according to (8.11a–e) with respect to the specific input.

Remark 8.3 After the simulation (or experimental) data are collected according to the procedure above, the computation burden are only those in (8.11c). Compared with the analytical determination of the OFRF structure by using the recursive algorithm in Jing et al. (2008e) or Chap. 11, the parametric characteristic analysis facilitates the determination of the parametric relationship for the OFRF. Moreover, it can be seen that there are only σ_2 simulations needed for the collection of $Y(j\omega)|_{C(i)}$ in this algorithm. Thus the simulation (or experimental) burden is also greatly reduced. For example, suppose the largest nonlinearity order $N=3$ and only $c_{p,q}(1\cdots1)$ is nonzero in $C_{p,q}$, then according to Lemma 8.1 or (8.5), it can be obtained that

$$CE(H_1(j\omega_1)) = 1, CE(H_2(j\omega_1, j\omega_2)) = C_{0,2} \oplus C_{1,1} \oplus C_{2,0}$$

$$\begin{aligned} &CE(H_3(j\omega_1, \cdots, j\omega_3)) \\ &\quad = C_{0,3} \oplus C_{1,1} \otimes C_{0,2} \oplus C_{1,1}{}^2 \oplus C_{1,1} \otimes C_{2,0} \oplus C_{2,1} \oplus \\ &\quad C_{1,2} \oplus C_{2,0} \otimes C_{0,2} \oplus C_{2,0}{}^2 \oplus C_{3,0} \\ &\quad = C_{0,3} \oplus C_{1,1} \cdot C_{0,2} \oplus C_{1,1}{}^2 \oplus C_{1,1} \cdot C_{2,0} \oplus C_{2,1} \oplus \\ &\quad C_{1,2} \oplus C_{2,0} \cdot C_{0,2} \oplus C_{2,0}{}^2 \oplus C_{3,0} \end{aligned}$$

Therefore, $\sigma_2 = DIM(\psi) = DIM\left(\mathop{\oplus}\limits_{n=1}^{N} CE(H_n(\cdot))\right) = 13$, that is, only 13 simulations are needed. According to the method in Jing et al. (2008e) or Chap. 11, all the parameters from power 0 to 2 should be counted. Note that there are seven different parameters, thus there are totally 3^7 cases, which means that there should be 3^7 simulations needed. Especially, based on the parametric characteristics, every specific component of the OFRF can be determined readily after the OFRF is obtained, while this cannot be obtained by analytical computation. Therefore, the results developed in this paper facilitate the application of the OFRF based method for the frequency domain analysis of nonlinear systems.

Remark 8.4 To conduct the procedure in Algorithm B in order to determine the OFRF, a problem may be: how to find a proper series of the parameter vector in S_C, i.e., $C(1), C(2)\ldots C(\sigma_2)$, such that $\left[\psi|_{C(1)}{}^T \quad \psi|_{C(2)}{}^T \quad \cdots \quad \psi|_{C(\sigma_2)}{}^T\right]$ is non-singular. An improper series may result in the matrix to be ill-conditioned or even singular. To solve this problem, a simple stochastic searching method as given in the following or other searching methods such as GA can be used since the series of different values of the parameter vector exists from Lemma 8.5. For example (Algorithm C),

(C1) $C(1), C(2)\ldots C(\sigma_2)$ can be generated randomly in S_C or a smaller subspace $\overline{S}_C$ (where $\overline{S}_C \subseteq S_C$) according to a distribution function, or each parameter in C can be generated randomly in its own variation domain (or a sub-domain) according to a distribution function;

(C2) After a series of points are obtained, the determinate of the matrix $\left[\psi|_{C(1)}{}^T \quad \psi|_{C(2)}{}^T \quad \cdots \quad \psi|_{C(\sigma_2)}{}^T\right]$ can then be computed;

(C3) Repeat this process until find a series such that the determinate of $\left[\psi|_{C(1)}{}^T \quad \psi|_{C(2)}{}^T \quad \cdots \quad \psi|_{C(\sigma_2)}{}^T\right]$ is a satisfactory value.

This will be demonstrated in the next section.

8.4 Simulations

In this section, an example is provided to demonstrate the theoretical results above. Consider a nonlinear system (Fig. 8.1)

$$240\ddot{x} = -16,000x - f(x,\dot{x})\dot{x} + u(t) \tag{8.12a}$$

where $u(t) = 100\sin(8.1t)$, and $f(x,\dot{x}) = 296 + c_1\dot{x}^2 + c_2\dot{x}x$. The output is

$$y = 16,000x + f(x,\dot{x})\dot{x} \tag{8.12b}$$

Equation (8.12a) represents the transmitted force from u(t) to the ground, and is a simple case of system (8.1) with $M = 3$, $K = 2$, $c_{10}(2) = 240, c_{10}(1) = 296\, C_{10}(0) = 16,000\, c_{30}(111) = c_1, c_{30}(110) = c_2, c_{01}(0) = -1$, and all the other parameters are zero. This is a model of the following spring-damping system with nonlinear damping $f(x,\dot{x}) = 296 + c_1\dot{x}^2 + c_2\dot{x}x$.

In system (8.12a,b), only nonlinear parameters in C_{30} are not zero, i.e.,

$$C_{30} = \begin{bmatrix} c_{30}(110) & c_{30}(111) \end{bmatrix} = \begin{bmatrix} c_2 & c_1 \end{bmatrix}$$

In this case, it can be shown from (8.5) that

$CE(H_{2k}(j\omega_1,\cdots,j\omega_n)) = 0$ and $CE(H_{2k+1}(j\omega_1,\cdots,j\omega_n)) = C_{30}^k$, for k = 1,2, 3,...
This can also directly be obtained from Lemma 8.2. From Proposition 8.1,

$$\begin{aligned} CE(X(j\omega)) &= \bigoplus_{n=0}^{\lfloor N-1/2 \rfloor} CE(H_{2n+1}(j\omega_1,\cdots, j\omega_n)) \\ &= 1 \oplus C_{30} \oplus C_{30}^2 \oplus C_{30}^3 \oplus \cdots \oplus C_{30}^{\lfloor N-1/2 \rfloor} \end{aligned} \tag{8.13a}$$

That is

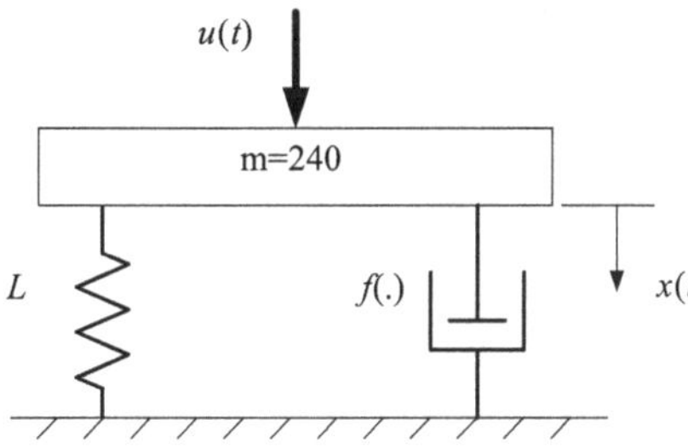

Fig. 8.1 A spring-damping system with nonlinear damping

$$X(j\omega) = \left(1 \oplus C_{30} \oplus C_{30}^2 \oplus C_{30}^3 \oplus \cdots \oplus C_{30}^{\lfloor N-1/2 \rfloor}\right)$$
$$\cdot \left[\overline{\overline{F}}_1(j\omega)^T \quad \overline{\overline{F}}_2(j\omega)^T \quad \cdots \quad \overline{\overline{F}}_N(j\omega)^T\right]^T \tag{8.13b}$$

$CE(X(j\omega))$can readily be computed according to (8.13a). For example, for N = 5

$$CE(X(j\omega)) = 1 \oplus C_{30} \oplus C_{30}^2 \oplus C_{30}^3 \oplus \cdots \oplus C_{30}^{\lfloor N-1/2 \rfloor}$$
$$= \left[1, c_2, c_1, {c_2}^2, c_2c_1, {c_1}^2, {c_2}^3, {c_2}^2c_1, c_2{c_1}^2, {c_1}^3, {c_2}^4, {c_2}^3c_1, {c_2}^2{c_1}^2,\right.$$
$$\left. c_2{c_1}^3, {c_1}^4, {c_2}^5, {c_2}^4c_1, {c_2}^3{c_1}^2, {c_2}^2{c_1}^3, c_2{c_1}^4, {c_1}^5\right] \tag{8.13c}$$

Therefore an explicit analytical expression for the OFRF $X(j\omega)$ for up to the fifth order in terms of the system nonlinear parameters c_1 and c_2 are obtained as given by (8.13b,c). It can be shown that $CE(Y(j\omega)) = CE(X(j\omega))CE(Y(j\omega)) = CE(X(j\omega))$ (Chap. 6). Therefore

$$CE(Y(j\omega)) = \mathop{\oplus}_{n=1}^{N} CE(H_n(j\omega_1, \cdots, j\omega_n))$$
$$= 1 \oplus C_{30} \oplus C_{30}^2 \oplus C_{30}^3 \oplus \cdots \oplus C_{30}^{\lfloor N-1/2 \rfloor} =: \psi \tag{8.14a}$$

and

$$Y(j\omega) = \psi \cdot \left[\overline{\overline{F}}_1(j\omega)^T \quad \overline{\overline{F}}_2(j\omega)^T \quad \cdots \quad \overline{\overline{F}}_N(j\omega)^T\right]^T \tag{8.14b}$$

To find a proper series of the points $CE(Y(j\omega))$ in S_C, for example $0 \le c_1, c_2 \le 5$, the Algorithm C mentioned in Remark 8.4 can be used. In simulations, it is easy to find a proper series. This verifies the result of Theorem 8.1. Locations of a series of the points $C[i] = (c_1[i], c_2[i])$ from $i = 1$ to 21 is demonstrated in Fig. 8.2, which are generated according to a uniform distribution. In this case, the determinate of the matrix $\left[\psi|_{C(1)}{}^T \quad \psi|_{C(2)}{}^T \quad \cdots \quad \psi|_{C(21)}{}^T\right]^{-1} = 0.9321875125788$.

For clarity of illustration, consider a much simpler case of $c_2 = 0$, i.e., $C_{30} = c_1$ (More complicated cases can be referred to Chap. 7). When $N = 21$, it can be obtained from (8.14a,b) that $CE(Y(j\omega)) = [1 \quad c_1 \quad c_1^2 \quad c_1^3 \quad \cdots \quad c_1^{10}]$, and consequently

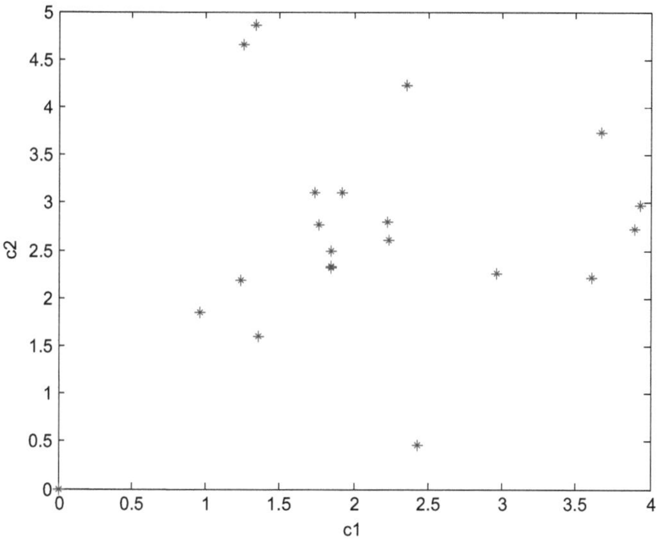

Fig. 8.2 A series of points (c_1[i],c_2[i]) from $i = 1$ to 21 (Jing et al. 2009d)

$$Y(j\omega) = \begin{bmatrix} 1 & c_1 & c_1^2 & c_1^3 & \cdots & c_1^{10} \end{bmatrix} \cdot \begin{bmatrix} \overline{\overline{F}}_1(j\omega) & \overline{\overline{F}}_2(j\omega) & \cdots & \overline{\overline{F}}_{11}(j\omega) \end{bmatrix}^T$$

Choose 11 different values of c_1, $\overline{\overline{F}}_i(j\omega)$ can be obtained according to Theorem 8.1 as

$$\overline{\overline{F}}(j\omega) = \begin{bmatrix} 1 & c_1(1) & \cdots & c_1^{10}(1) \\ 1 & c_1(2) & \cdots & c_1^{10}(2) \\ \vdots & \vdots & \ddots & \vdots \\ 1 & c_1(11) & & c_1^{10}(11) \end{bmatrix}^{-1} \cdot \begin{bmatrix} Y(j\omega)|_{c_1(1)} \\ Y(j\omega)|_{c_1(2)} \\ \vdots \\ Y(j\omega)|_{c_1(11)} \end{bmatrix} \tag{8.15}$$

It can be seen that the parameter matrix is a Vandermonde matrix. Thus if $c_1(i) \neq c_1(j)$ for $i \neq j$, it is non-singular. In order to determine $\overline{\overline{F}}_i(j\omega)$ in the above equation, simulation studies are carried out for 11 different values of c_1 as $c_1 = 0.5$, 50, 100, 500, 800, 1,200, 1,800, 2,600, 3,500, 4,500, 5,000, to produce 11 corresponding output responses. The FFT results of these responses at the system driving frequency $\omega_0 = 8.1$ rad/s were obtained as

$$Y_Y = \big[(3.355387229685395\mathrm{e}+002) - 9.144123368552089\mathrm{e}+000\mathrm{i},\\ (3.311400634432650\mathrm{e}+002) - 8.791324203084603\mathrm{e}+000\mathrm{i},\\ (3.270304131496312\mathrm{e}+002) - 8.453482697096458\mathrm{e}+000\mathrm{i},\\ (3.020996757260479\mathrm{e}+002) - 6.232073185455284\mathrm{e}+000\mathrm{i},\\ (2.889224705331136\mathrm{e}+002) - 4.937579404570077\mathrm{e}+000\mathrm{i},\\ (2.753247618357106\mathrm{e}+002) - 3.513785421406298\mathrm{e}+000\mathrm{i},\\ (2.599814606290563\mathrm{e}+002) - 1.799344961942028\mathrm{e}+000\mathrm{i},\\ (2.449407272303421\mathrm{e}+002) - 7.146831574203648\mathrm{e}-003\mathrm{i},\\ (2.322782654921158\mathrm{e}+002) + 1.587748875652816\mathrm{e}+000\mathrm{i},\\ (2.213884644417550\mathrm{e}+002) + 3.022652971105967\mathrm{e}+000\mathrm{i},\\ (2.168038059608033\mathrm{e}+002) + 3.644341792781596\mathrm{e}+000\mathrm{i}\big]$$

Then from (8.15), $\overline{\overline{F}}(j\omega_0)$ was determined as

$$\overline{\overline{F}}(j\omega_0) = \big[3.355850061999765\mathrm{e}+002 \ +9.147787717329777\mathrm{e}+000\mathrm{i},\\ -0.09260545518186 - 0.00733079515829\mathrm{i}\\ 7.802545290190465\mathrm{e}-005 \ +4.196941358069068\mathrm{e}-006\mathrm{i}\\ -8.171412395831490\mathrm{e}-008 \ -3.472552369765044\mathrm{e}-009\mathrm{i}\\ 7.983194136013857\mathrm{e}-011 \ +2.975659825236403\mathrm{e}-012\mathrm{i}\\ -6.014819558373321\mathrm{e}-014 \ -2.095287675780629\mathrm{e}-015\mathrm{i}\\ 3.139462445085954\mathrm{e}-017 \ +1.055716258995395\mathrm{e}-018\mathrm{i}\\ -1.065920417366710\mathrm{e}-020 \ -3.515136904764629\mathrm{e}-022\mathrm{i}\\ 2.214834610655676\mathrm{e}-024 \ +7.220197982843919\mathrm{e}-026\mathrm{i}\\ -2.536564081104798\mathrm{e}-028 \ -8.209302192093296\mathrm{e}-030\mathrm{i}\\ 1.219975622824295\mathrm{e}-032 \ +3.929425356306088\mathrm{e}-034\mathrm{i}\big].$$

Consequently, the parametric relationship for the OFRF of system (8.12a,b) subject to the input $u(t) = 100\sin(8.1t)$ at frequency $\omega_0 = 8.1$ was obtained as

$$Y(j\omega_0) = \begin{bmatrix}1 & c_1 & c_1^2 & c_1^3 & \cdots & c_1^{10}\end{bmatrix} \cdot \begin{bmatrix}\overline{\overline{F}}_1(j\omega_0) & \overline{\overline{F}}_2(j\omega_0) & \cdots & \overline{\overline{F}}_{11}(j\omega_0)\end{bmatrix}^T \tag{8.16}$$

For each order component of the OFRF, it can be obtained from Theorem 8.1 and Proposition 8.1 that for $n = 1,2,3,\ldots$

$$Y_{2n-1}(j\omega_0) = c_1^n \cdot \overline{\overline{F}}_n(j\omega_0) \text{ and } Y_{2n}(j\omega_0) = 0 \text{ and} \tag{8.17}$$

From (8.16), the effect of the nonlinear parameter c_1 on the system output frequency response at frequency ω_0 can readily be analysed. Figure 8.3 shows a comparison of the magnitudes of the output spectrum evaluated by (8.16) and their real values under different values of the nonlinear parameter c_1. Note that the error between the computed values and the real values is very small. Furthermore, the frequency domain analysis and design of system (8.12a,b) to achieve a

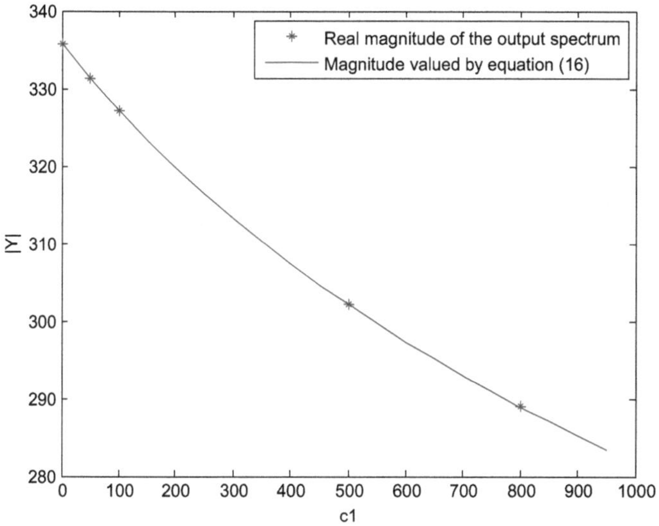

Fig. 8.3 Relationship between the output spectrum and nonlinear parameter c_1 (Jing et al. 2009d)

desired output response $y(t)$ can now be conducted from (8.16). Given a desired output spectrum Y^* at frequency ω_0, the nonlinear parameter c_1 can be optimized using (8.16) such that the difference $|Y(j\omega_0) - Y^*|$ can be made as small as possible.

8.5 Conclusions

This chapter shows that, the analytical parametric relationship described by the OFRF with a polynomial structure in terms of any model parameters of interest for Volterra systems given by a NDE model can be determined explicitly up to any high order by using a simple Least Square method with some simulation or experimental data, and every specific component of the OFRF can also be determined effectively. Moreover, it should be noted that the main result established in Theorem 8.1 is not only applicable for the OFRF based method, but also has significance for the determination of any analytical parametric relationship for this kind of system polynomial functions by using numerical methods.

To fully understand the results of this chapter, the readers can refer to Jing et al. (2008e, 2009d), Jing and Lang (2009a), Chen et al. (2013), and also other corresponding chapters.

8.6 Proofs

A. Proof of Proposition 8.1

When the input function is $u(t) = F_d \sin(\Omega t)$, it can be obtained from (3.3) in Chap. 3 that

$\omega_{k_l} = k_l \Omega$, $k_l = \pm 1$ and $F(\omega_{k_l}) = -jk_l F_d$, for $l = 1, \cdots, n$

From the results in Sect. 3.3 of Jing et al. (2006), it can be obtained in this case

$$\overline{\overline{F}}_i(j\Omega) = \frac{1}{2^i} \sum_{\omega_{k_1}+\cdots+\omega_{k_i}=\Omega} f_i(j\omega_{k_1}, \cdots, j\omega_{k_i}) \cdot F(\omega_{k_1}) \cdots F(\omega_{k_i})$$

Note that when $i = 2n + 1$, the condition $\omega_{k_1} + \cdots + \omega_{k_{2n+1}} = \Omega$ means that there are n frequencies $\omega_{k_l} = -\omega$ and $n + 1$ frequencies $\omega_{k_l} = \omega$. Thus, when $i = 2n$, $\omega_{k_1} + \cdots + \omega_{k_{2n}} \neq \Omega$ for any cases. This shows that $f_{2n}(j\omega_{k_1}, \cdots, j\omega_{k_{2n}}) = 0$, which further yields that $\overline{\overline{F}}_{2n}(j\Omega) = 0$ for $n = 1, 2, 3, \ldots$. Therefore, the parametric characteristics of the OFRF in this case can be written as (8.7). This completes the proof. □

B. Proof of Lemma 8.2

(1) Consider a parameter $c_{p,q}(\cdot)$. If $p + q = n$, it is from Lemma 8.1 that $c_{p,q}(\cdot)$ is an element of $CE(H_n(j\omega_1, \cdots, j\omega_n))$. If $p + q > n$, this parameter cannot appear in $H_n(j\omega_1, \cdots, j\omega_n)$, since the conditions in Lemma 8.1, e.g., $p + q + \sum_{i=1}^{k} (p_i + q_i) = n + k$, cannot be satisfied. If $p + q < n$, then there must exist a $k \geq 0$ such that $p + q + \sum_{i=1}^{k} (p_i + q_i) = n + k$. That is, $c_{p,q}(\cdot)$ must appear in a monomial which is an element of $CE(H_n(j\omega_1, \cdots, j\omega_n))$.

(2) For $(c_{p,q}(\cdot))^{k+1}$, it is from Lemma 8.1 that $p + q + \sum_{i=1}^{k} (p + q) = n + k$, which yields $k = \frac{n-p-q}{p+q-1}$. This completes the proof. □

C. Proof of Lemma 8.3

This can be proved by contradiction. Suppose there is a parameter monomial $c_{p,q}(\cdot) c_{p_1,q_1}(\cdot) c_{p_2,q_2}(\cdot) \cdots c_{p_k,q_k}(\cdot)$ which is not only an element of $CE(H_n(j\omega_1, \cdots, j\omega_n))$ but also an element of $CE(H_m(j\omega_1, \cdots, j\omega_m))$, where $m \neq n$. Then from Lemma 8.1, it can be derived that $p + q + \sum_{i=1}^{k} (p_i + q_i) = n + k$ and $p + q + \sum_{i=1}^{k} (p_i + q_i) = m + 1 + k$. Thus $n + k = m + k$, i.e., $n = m$. This is a contradiction. The Lemma is proved. □

D. Proof of Lemma 8.4

Since $\zeta|_{(c_1(i),c_2(i),\ldots,c_m(i))}$ for $i=1$ to n is a base of $\mathfrak{R}^n$, then for any $\zeta \in \mathfrak{R}^n$ there exist a series of real numbers $\alpha_1 \cdots \alpha_n$, such that

$$\zeta = \alpha_1 \zeta|_{(c_1(1),c_2(1),\ldots,c_m(1))} + \cdots + \alpha_n \zeta|_{(c_1(n),c_2(n),\ldots,c_m(n))}$$

which yields

$$\zeta \cdot \varphi^T = \alpha_1 \zeta|_{(c_1(1),c_2(1),\ldots,c_m(1))} \cdot \varphi^T + \cdots + \alpha_n \zeta|_{(c_1(n),c_2(n),\ldots,c_m(n))} \cdot \varphi^T = 0$$

(p2) is proved. (p1) is equivalent to (p2). For any point in the parameter space S_C, there is a corresponding vector $\zeta \in \mathfrak{R}^n$, thus it follows from (p2) that $\zeta \cdot \varphi^T = 0$. (p3) is proved. This completes the proof. □

E. Proof of Lemma 8.5

To proceed with the proof of this lemma, two special cases are studied first.

Case 1. Consider two different monomials $c_1^{r_1} c_2^{r_2} \cdots c_m^{r_m}$ and $c_1^{l_1} c_2^{l_2} \cdots c_m^{l_m}$, where c_1, $c_2, \ldots, c_m$ are parameters in C, and $r_1, r_2, \ldots, r_m$, $l_1, l_2, \ldots, l_m$ are non-negative integers. There exists at least one $1 \le i \le m$ for the two monomials satisfying $r_i \ne l_i$. Without speciality, suppose $r_1 \ne l_1$. Then suppose for any points $(c_1, c_2, \ldots, c_m)$ satisfying $c_i \ne 0$, there is a nonzero constant β, such that $c_1^{l_1-r_1} c_2^{l_2-r_2} \cdots c_m^{l_m-r_m} = \beta$. Letting $\bar{c}_1 = {}^{c_1}\big/_{\beta^{\frac{1}{l_1-r}}}$ gives $\bar{c}_1^{l_1-r_1} c_2^{l_2-r_2} \cdots c_m^{l_m-r_m} = 1$ for any points $(c_1, c_2, \ldots, c_m)$ satisfying $c_i \ne 0$. This further yields $(l_1 - r_1)\lg\bar{c}_1 + (l_2 - r_2)\lg c_2 + \cdots + (l_m - r_m)\lg c_m = 0$ for any points $(c_1, c_2, \ldots, c_m)$ satisfying $c_i \ne 0$. This shows that $r_i = l_i$, which results in a contradiction. Therefore, $c_1^{l_1-r_1} c_2^{l_2-r_2} \cdots c_m^{l_m-r_m}$ cannot be a nonzero constant. For any two points $(c_1(1), c_2(1), \ldots, c_m(1))$ and $(c_1(2), c_2(2), \ldots, c_m(2))$ such that

$$c_1^{l_1-r_1}(1) c_2^{l_2-r_2}(1) \cdots c_m^{l_m-r_m}(1) \ne c_1^{l_1-r_1}(2) c_2^{l_2-r_2}(2) \cdots c_m^{l_m-r_m}(2)$$

it can be obtained by equivalent row transforms that

$$\begin{bmatrix} c_1^{r_1}(1)c_2^{r_2}(1)\cdots c_m^{r_m}(1)) & c_1^{l_1}(1)c_2^{l_2}(1)\cdots c_m^{l_m}(1) \\ c_1^{r_1}(2)c_2^{r_2}(2)\cdots c_m^{r_m}(2) & c_1^{l_1}(2)c_2^{l_2}(2)\cdots c_m^{l_m}(2) \end{bmatrix} \Rightarrow \begin{bmatrix} 1 & c_1^{l_1-r_1}(1)c_2^{l_2-r_2}(1)\cdots c_m^{l_m-r_m}(1) \\ 1 & c_1^{l_1-r_1}(2)c_2^{l_2-r_2}(2)\cdots c_m^{l_m-r_m}(2) \end{bmatrix}$$
$$\Rightarrow \begin{bmatrix} 1 & c_1^{l_1-r_1}(1)c_2^{l_2-r_2}(1)\cdots c_m^{l_m-r_m}(1) \\ 0 & c_1^{l_1-r_1}(2)c_2^{l_2-r_2}(2)\cdots c_m^{l_m-r_m}(2) - c_1^{l_1-r_1}(1)c_2^{l_2-r_2}(1)\cdots c_m^{l_m-r_m}(1) \end{bmatrix}$$

It is obvious that the matrix is non-singular.

Case 2. Consider the same two different monomials $c_1^{r_1} c_2^{r_2} \cdots c_m^{r_m}$ and $c_1^{l_1} c_2^{l_2} \cdots c_m^{l_m}$ as Case 1. Let $a = c_1^{r_1} c_2^{r_2} \cdots c_m^{r_m} + a_1$, and $b = c_1^{l_1} c_2^{l_2} \cdots c_m^{l_m} + b_1$, where a_1 and b_1 are two constant real numbers, and suppose $a_1 \ne 0$ without speciality. Suppose for any points $(c_1, c_2, \ldots, c_m)$, there is a nonzero constant β, such that $b/a \equiv \beta$, which gives

$$c_1^{l_1}c_2^{l_2}\cdots c_m^{l_m}+b_1\equiv\beta c_1^{r_1}c_2^{r_2}\cdots c_m^{r_m}+\beta a_1$$

If $b_1=\beta a_1$, then it is the case 1. Consider the case $b_1\neq\beta a_1$. There must be some points (c_1, c_2, ..., c_m) such that $c_1^{r_1}c_2^{r_2}\cdots c_m^{r_m}=\frac{b_1-\beta a_1}{\beta}$. Thus for these points, $c_1^{l_1}c_2^{l_2}\cdots c_m^{l_m}\equiv 0$. This results in a contradiction with $c_1^{r_1}c_2^{r_2}\cdots c_m^{r_m}=\frac{b_1-\beta a_1}{\beta}$. Therefore, b/a cannot be a nonzero constant. For any two points ($c_1(1)$, $c_2(1)$, ..., $c_m(1)$) and ($c_1(2)$, $c_2(2)$, ..., $c_m(2)$) such that

$$b(1)\big/a(1)\neq b(2)\big/a(2)$$

it can be obtained by equivalent row transformation that

$$\begin{bmatrix} c_1^{r_1}(1)c_2^{r_2}(1)\cdots c_m^{r_m}(1))+a_1 & c_1^{l_1}(1)c_2^{l_2}(1)\cdots c_m^{l_m}(1)+b_1 \\ c_1^{r_1}(2)c_2^{r_2}(2)\cdots c_m^{r_m}(2)+a_1 & c_1^{l_1}(2)c_2^{l_2}(2)\cdots c_m^{l_m}(2)+b_1 \end{bmatrix}\Rightarrow\begin{bmatrix} 1 & b(1)/a(1) \\ 1 & b(2)/a(2) \end{bmatrix}$$
$$\Rightarrow\begin{bmatrix} 1 & b(1)/a(1) \\ 0 & b(2)/a(2)-b(1)/a(1) \end{bmatrix}$$

It is obvious that the matrix is non-singular.

Now consider the proof of the lemma. As mentioned, it is from Remark 8.1 and Lemma 8.3 that there are no repetitive elements in ψ. That is, each element in ψ is a non-repetitive monomial of some nonlinear parameters in C. Choose different points $C(i)$ in S_C for $i=1$ to σ_2, then produce a matrix row by row. For the first two rows, it is Case 1 if only considering the first two columns. Thus by equivalent row transformation, the first two rows can be transformed into an upper triangle form as Case 1, i.e., the entries in the first two columns and below the diagonal line are zero, while the diagonal entries in the first two rows are nonzero. For the next two rows, it is Case 2 if only considering the next two columns. Then by equivalent row transformation, the next two rows can also be transformed into an upper triangle form as Case 2, i.e., the entries in the first four columns and below the diagonal line are zero, while the diagonal entries in the first four rows are nonzero. Proceed this process forward until the last two or one rows. Therefore, the matrix can be equivalently transformed into an upper triangle form with nonzero diagonal entries, which is obviously non-singular. This shows that, there exist a series of points $C(i)$ in S_C for $i=1$ to σ_2 such that each rows of the generated matrix as mentioned above, i.e., $\psi|_{C(i\ )}$ for $i=\mathrm{t}$ to σ_2 are independent. This completes the proof. □

F. Proof of Theorem 8.1

When there is no truncation error, from Lemma 8.5, there exist a series of $C(1)$, $C(2)...C(\sigma_2)$ such that

$\psi|_{C(i)}$ for $i=1$ to σ_2 is a base of $\mathfrak{R}^{\sigma_2}$

and additionally from (8.6a) and (8.11a), it can be obtained that for each $\psi|_{C(i)}$

$$\psi\big|_{C(i)} \cdot \left(\widetilde{\Phi}(j\omega) - \Phi(j\omega)\right)^T = 0$$

Then from Lemma 8.4, for all the points in S_C

$\psi \cdot \left(\widetilde{\Phi}(j\omega) - \Phi(j\omega)\right)^T = 0$ and $\widetilde{\Phi}(j\omega) = \Phi(j\omega)$.

In case that S_C includes all the nonlinear parameters of NDE model (8.1), from Lemma 8.3, $CE(H_n(j\omega_1,\cdots,j\omega_n)) \wedge CE(H_m(j\omega_1,\cdots,j\omega_m)) = 0$ for $m \neq n$, then

$$\widetilde{Y}_n(j\omega) = CE(H_n(\cdot)) \cdot \widetilde{F}_n(j\omega)^T$$

Consider the truncation error $o(N+1) \neq 0$. In this case,

$$Y(j\omega) = \psi \cdot \Phi(j\omega)^T + o(N+1)$$

Therefore,

$$\begin{aligned}
\Phi(j\omega)^T &= \left[\psi\big|_{C(1)}{}^T\ \psi\big|_{C(2)}{}^T \cdots \psi\big|_{C(\sigma_2)}{}^T\right]^{-T} \\
&\quad \cdot \left[(Y(j\omega) - o(N+1))\big|_{C(1)}\ (Y(j\omega) - o(N+1))\big|_{C(2)} \cdots (Y(j\omega) - o(N+1))\big|_{C(\sigma_2)}\right]^T \\
&= \Psi^{-1}\big|_{C(1\cdots\sigma_2)} \cdot (Y - o_{N+1})\big|_{C(1\cdots\sigma_2)} = \Psi^{-1}\big|_{C(1\cdots\sigma_2)} \cdot Y\big|_{C(1\cdots\sigma_2)} - \Psi^{-1}\big|_{C(1\cdots\sigma_2)} \cdot o_{N+1}\big|_{C(1\cdots\sigma_2)} \\
&= \widetilde{\Phi}(j\omega)^T - \Psi^{-1}\big|_{C(1\cdots\sigma_2)} \cdot o_{N+1}\big|_{C(1\cdots\sigma_2)}
\end{aligned}$$

where $Y\big|_{C(1\cdots\sigma_2)} = \left[Y(j\omega)\big|_{C(1)}\ Y(j\omega)\big|_{C(2)} \cdots Y(j\omega)\big|_{C(\sigma_2)}\right]^T$. This leads to (8.11d). Note that $\widetilde{Y}(j\omega) = \psi \cdot \widetilde{\Phi}(j\omega)^T$. Therefore,

$$\widetilde{Y}(j\omega) - Y(j\omega) = \psi \cdot \left(\widetilde{\Phi}(j\omega) - \Phi(j\omega)\right)^T - o(N+1)$$

This, together with (8.11d), leads to (8.11e). The proof is completed. □

Chapter 9
Nonlinear Characteristic Output Spectrum

9.1 Introduction

Nonlinear analysis takes an important role in system analysis and design in practice. Several methods are available in the literature to this aim including perturbation method, averaging method and harmonic balance method etc (Judd 1998; Mees 1981; Gilmore and Steer 1991). As shown in the previous chapters, nonlinear analysis can be conducted in the frequency domain systematically based on the Volterra series theory.

However, the nonlinear analysis based on the GFRFs usually involves complicated computation cost especially for the orders higher than 3. Importantly, the traditional recursive algorithms for the GFRFs are easy to implement but actually complicate the relationship between the GFRFs and model parameters. In Chaps. 4–6, it is shown that both the GFRF and the output spectrum can be formulated into a polynomial in terms of nonlinear parameters of system model. Especially, through a parametric characteristic analysis and a mapping function (see Jing et al. 2008b or Chap. 11), the GFRFs and output spectrum can all be expressed into a straightforward polynomial function with respect to any nonlinear parameters of interest. Thus, for a nonlinear system described by a NDE or NARX model, these results could provide a significant insight or powerful approach into the nonlinear influence on system output frequency response (which will be discussed further in Chaps. 11 and 12). However, quantitative analysis of the nonlinear dynamics and its effect on system dynamic response still encounters problems due to computation complexity. Although a numerical method could be adopted for an estimation of the output frequency response function (OFRF) (Chaps. 6–8), biased or even wrong estimates may happen since the truncation order of the underlying Volterra series expansion for the nonlinear system under study is difficult to know in advance (it may also be varying with different input magnitudes). This may affect the effectiveness and reliability of the OFRF-based analysis.

X. Jing, Z. Lang, *Frequency Domain Analysis and Design of Nonlinear Systems based on Volterra Series Expansion*, Understanding Complex Systems,
DOI 10.1007/978-3-319-12391-2_9

In this chapter, a systematic frequency domain method for nonlinear analysis, design and estimation of nonlinear systems is established based on the discussions in the previous chapters. This method allows accurate determination of the linear and nonlinear components in system output spectrum of a given nonlinear system described by NDE, NARX or NBO (nonlinear block-oriented) models, with some simulation or experiment data. These output spectrum components can then be used for system identification or nonlinear analysis for different purposes such as fault detection etc. Noticeably, the OFRF in Chaps. 7 and 8 is expressed into a much improved polynomial function, referred to here as nonlinear characteristic output spectrum (nCOS) function, which is an explicit expression for the relationship between nonlinear output spectrum and system characteristic parameters of interest including nonlinear parameters, frequency variable, and input excitation magnitude (not just nonlinear parameters as that in Chaps. 7 and 8) with a more generic parametric structure. With the accurate determination of system output spectrum components in the previous step, the nCOS function can therefore be accurately determined up to any high orders, with less simulation trials and computation cost compared with a pure simulation based study or traditional theoretical computation (Yue et al. 2005; Jing et al. 2008e). These results can provide a significant approach for qualitative and quantitative analysis and design of nonlinear dynamics in the frequency domain. The study on a nonlinear vehicle suspension system is given to illustrate the results.

9.2 Nonlinear Characteristic Output Spectrum (nCOS) and the Problem

Nonlinear systems can usually be identified or modeled into a parametric model such as NDE, NARX or NBO models in practice. The nonlinear output spectrum of those nonlinear systems is not only a complex-valued function of frequency variables but also a function of model parameters and input magnitude of interest (which are all referred to as characteristic parameters in this study). An explicit relationship between system output spectrum and characteristic parameters would be of great significance for system analysis and design. Consider the NDE system (2.11) again. It is shown in Chap. 6 (and also Chap. 11) that nonlinear output spectrum of NDE models can be written into an explicit polynomial function of system characteristic parameters as

$$Y(j\omega) = \sum_{n=1}^{N} \chi_n \cdot \varphi_n(j\omega)^T. \tag{9.1}$$

where χ_n denotes the nth-order characteristic parameter vector composed of nonlinear parameters and $\varphi_n(j\omega)$ its correlative complex-valued function of the nth-order output spectrum, both of which can be (analytically) determined with the method in Chaps. 5, 6 and 11. Any nonlinear parameters of interest in analysis and design will be included in χ_n, given by

$$\chi_n = C_{n,0} \oplus \bigoplus_{\substack{k=0 \\ \sum_{i=0}^{k}(p_i+q_i)=n+k \\ 1 \le p_0 \le n-k,\ 2 \le p_i+q_i \le n-k}}^{n-2} \left(\bigotimes_{i=0}^{k} C_{p_i,q_i} \right) \tag{9.2}$$

where $C_{p,q} = [c_{p,q}(0,\cdots,0), c_{p,q}(0,\cdots,1), \cdots, c_{p,q}(\underbrace{K,\cdots,K}_{p+q=m})]$, $\oplus$ and $\oplus$ are two operators defined for symbolic manipulation (see the details in Chap. 4). For example, for two vectors C_1 and C_2 consisting of some symbolic variables, $C_1 \oplus C_2$ is a vector including all the elements in C_1 and C_2 without repetition, and $C_1 \oplus C_2$ is a vector including all the elements produced by the Kronecker product without repetition. If some model parameters will not be considered in the analysis and design, they can be set to 1 by default in (9.2). If there is an element 1 in χ_n, it means that there will be a pure frequency-dependent term in the polynomial (9.1). Similar results hold for the NARX model.

Obviously, (9.1) is an analytical and straightforward function of system characteristic parameters, which could considerably facilitate the analysis and design of nonlinear systems in the frequency domain. To emphasize the parametric relationship between the nonlinear output spectrum (nOS) and the characteristic parameters (including nonlinear parameters, excitation magnitude and frequency), (9.1) is referred to here as nonlinear characteristic output spectrum (nCOS) function and the nth-order component as the nth-order nCOS. In order to conduct a nonlinear analysis based on the nCOS function in (9.1), both χ_n and $\varphi_n(j\omega)$ must be determined up to a sufficiently high order. Since nonlinear systems can always be identified into a NARX, NDE or NBO model with experiment data in practice (Worden and Tomlinson 2001; Jing 2011; Ahn and Anh 2010; Wei and Billings 2008), this chapter suppose that a nonlinear model of the system of interest is already known. Therefore χ_n is known from (9.2) with the nonlinear model, and thus only $\varphi_n(j\omega)$ is yet to be determined. Although $\varphi_n(j\omega)$ can be computed analytically with the method in Chap. 11, the computation cost is usually high and it is even worse for high orders (>5). Therefore, the objective is to develop an effective method such that $\varphi_n(j\omega)$ can be determined accurately and directly with only some simulation data.

From (9.1), a simple least square algorithm could be applied in order to compute $\varphi_n(j\omega)$ under different parameter excitations as discussed in Chaps. 6 and 7. However, the difficulty is that the maximum truncation order N is not known and it is also varying with different input magnitudes. A larger input magnitude or a larger range of a specific model parameter of interest would result in a larger truncation order N (for an accurate Volterra series expansion). An inappropriate guessed truncation order N will lead to large error in the computation of $\varphi_n(j\omega)$. A sufficiently larger N could be attempted to avoid this problem (if realistic), but will easily result in singularity of the matrix inverse in the least square (LS) algorithm. If $\varphi_n(j\omega)$ is not computed correctly, the simple least square algorithm can only guarantee accurate fitting at the training points, and the generalization of (9.1) could be very worse.

To solve the problems mentioned above, new algorithms will be developed so that $\varphi_n(j\omega)$ can be determined accurately with only some simulation data and without knowing the best truncation order N^*.

9.3 Accurate Determination of the nCOS Function

For determination of the nCOS function (9.1), the first step of the proposed method is to compute the nth-order output spectrum (for any n) based on numerical or experimental data, and then to determine the nth-order nCOS function.

9.3.1 Computation of the nth-Order Output Spectrum

The nonlinear output spectrum (nOS) in (2.4) can be rewritten by considering the truncation error $\sigma_{[N]}(j\omega)$ and input function $\rho R(j\omega)$ as

$$\begin{aligned} Y(j\omega)_\rho &= \sum_{n=1}^{\infty} \rho^n Y_n(j\omega) = \rho Y_1(j\omega) + \rho^2 Y_2(j\omega) + \rho^3 Y_3(j\omega) + \cdots \\ &= \sum_{n=1}^{N} \rho^n Y_n(j\omega) + \sigma_{[N,\rho]}(j\omega) \end{aligned} \tag{9.3}$$

where $\sigma_{[N,\rho]}(j\omega)$ represents the truncation error, including all the remaining higher order output spectrum components in Volterra series expansion; ρ is a constant which is used to represent different magnitude of the input.

To determine $Y_n(j\omega)$ for $n \in \aleph = \{1,2,3,\ldots N\}$, a multi-level excitation method can be adopted as shown in Chap. 7. The system can be excited by the same input $R\,(j\omega)$ of different magnitudes $\rho_0, \rho_1, \rho_2, \ldots, \rho_{N-1}$, and there will be a series of output obtained accordingly, which are denoted by $Y(j\omega)_{\rho 0}, Y(j\omega)_{\rho 1}, Y(j\omega)_{\rho 2}, \ldots, Y(j\omega)_{\rho N-1}$. Through a LS method, it gives

$$\begin{bmatrix} \hat{Y}_1(j\omega) \\ \hat{Y}_2(j\omega) \\ \vdots \\ \hat{Y}_N(j\omega) \end{bmatrix} = \begin{bmatrix} \rho_0 & \rho_0^2 & \cdots & \rho_0^N \\ \rho_1 & \rho_1^2 & \cdots & \rho_1^N \\ \vdots & \vdots & \ddots & \vdots \\ \rho_{N-1} & \rho_{N-1}^2 & \cdots & \rho_{N-1}^N \end{bmatrix}^{-1} \begin{bmatrix} Y(j\omega)_{\rho_0} \\ Y(j\omega)_{\rho_1} \\ \vdots \\ Y(j\omega)_{\rho_{N-1}} \end{bmatrix} \tag{9.4}$$

The square matrix above is nonsingular if $\rho_0 \neq \rho_1 \neq \ldots \neq \rho_{N-1}$. Note that the computation of $Y_n(j\omega)$ for $n \in \aleph = \{1,2,3,\ldots N\}$ is basically accurate if the truncation error $\sigma_{[N,\rho]}(j\omega)$ is trivial. Otherwise, the computation for $Y_n(j\omega)$ could be significantly biased, and the LS in (9.4) only results in a good fitting at the training points. To overcome this problem, an alternative method can be employed by choosing the excitation magnitudes $\rho_0, \rho_1, \ldots, \rho_{N-1}$ so that the accurate computation through (9.4) can be achieved for any given N. The following results are derived for this purpose.

Proposition 9.1

$$\Delta Y_1(j\omega) = (-1)^{N-1}\rho_0\rho_1\cdots\rho_{N-1}\cdot\sum_{k=0}^{\infty}(\rho_0+\rho_1+\cdots+\rho_{N-1})^{[k]}Y_{N+k+1}(j\omega) \quad (9.5a)$$

$$\Delta Y_N(j\omega) = \sum_{k=1}^{\infty}(\rho_0+\rho_1+\cdots+\rho_{N-1})^{[k]}Y_{N+k}(j\omega) \quad (9.5b)$$

where for N(>1) nonzero distinct real numbers and for any non-negative integer r,

$$(\mathrm{x}_1+\mathrm{x}_2+\cdots+\mathrm{x}_\mathrm{N})^{[\mathrm{r}]} = \sum_{\substack{l_1+l_2+\cdots+l_N=r \\ l_1,l_2,\ldots,l_N\in\{0,1,2,\ldots,\mathrm{r}\}}} \mathrm{x}_1^{l_1}\mathrm{x}_2^{l_2}\cdots\mathrm{x}_\mathrm{N}^{l_\mathrm{N}} \quad (9.5c)$$

Proof See the proof in Sect. 9.6A.

Proposition 9.1 provides a straightforward insight into the computation error incurred by excitation magnitudes and truncation order. Given a truncation order N, different values of the excitation magnitudes could bring very different computation error. Proposition 9.1 demonstrates an effective method for designing $\rho_0\rho_1\cdots\rho_{N-1}$ so that the computation errors (i.e., $\Delta Y_1(j\omega)$, $\Delta Y_N(j\omega)$) can be mitigated for any N.

In (9.5a,b), the computation of $(\rho_0+\rho_1+\cdots+\rho_{N-1})^{[k]}$ is involved. Lemma 9.1 gives a recursive method for this.

Lemma 9.1

$$(\rho_0+\rho_1+\cdots+\rho_{N-1})^{[k]} = \sum_{r=0}^{N-1}\rho_r(\rho_r+\cdots+\rho_{N-1})^{[k-1]}$$

Proof By a mathematical induction, it is easy to have this conclusion.

Based on Proposition 9.1, different excitation methods can be used to minimize the computation error. The following results can be obtained by applying Proposition 9.1.

Corollary 9.1 Case 1: N=3,$\rho_1=-\rho,\rho_2=-\rho/\alpha,\rho_0=-\rho_1-\rho_2, 0<\rho<1, \alpha>1$.

$$\Delta Y_1(j\omega)=\frac{\alpha+1}{\alpha^2}\rho^3\cdot\left\{Y_4(j\omega)+\frac{\alpha^2+\alpha+1}{\alpha^2}\rho^2Y_6(j\omega)+\frac{\alpha+1}{\alpha^2}\rho^3Y_7(j\omega)+\cdots\right\} \quad (9.6a)$$

$$\begin{aligned}\Delta Y_3(j\omega) &= \frac{\alpha^2+\alpha+1}{\alpha^2}\rho^2Y_5(j\omega)+\frac{\alpha+1}{\alpha^2}\rho^3Y_6(j\omega)\\ &+\frac{\alpha^4+2\alpha^3+3\alpha^2+2\alpha+1}{\alpha^4}\rho^4Y_7(j\omega)\cdots\end{aligned} \quad (9.6b)$$

Case 2: N=5,$\rho_0=\rho,\rho_1=-\rho,\rho_2=-\rho/\beta,\rho_3=-\rho_2/\alpha,\rho_4=-\rho_2-\rho_3,0<\rho<1, \beta>1, \alpha>1$.

$$\begin{aligned}\Delta Y_1(j\omega) &= -\frac{1}{\alpha\beta}\left(\frac{1}{\alpha}+\frac{1}{\beta}\right)\rho^5\{Y_6(j\omega)\\ &+\left(1+\frac{1}{\alpha^2}+\frac{1}{\beta^2}+\frac{1}{\alpha\beta}\right)\rho^2Y_8(j\omega)+\frac{1}{\alpha\beta}\left(\frac{1}{\alpha}+\frac{1}{\beta}\right)\rho^3Y_9(j\omega)+\cdots\}\end{aligned} \quad (9.6c)$$

$$\Delta Y_5(j\omega)=\left(1+\frac{1}{\alpha^2}+\frac{1}{\beta^2}+\frac{1}{\alpha\beta}\right)\rho^2Y_7(j\omega)+\frac{1}{\alpha\beta}\left(\frac{1}{\alpha}+\frac{1}{\beta}\right)\rho^3Y_8(j\omega)+\cdots \quad (9.6d)$$

Proof See the proof in Sect. 9.6B.

Corollary 9.1 indicates that properly choosing the excitation magnitudes as specified would produce accurate computation of $Y_1(j\omega)$ and $Y_3(j\omega)$ or $Y_5(j\omega)$, respectively, although the truncation order is chosen as N=3 or 5. The result in Proposition 9.1 actually provides many choices for this purpose. For example, in Case 2 of Corollary 9.1, if $\rho=0.001$, $\beta=2$, and $\alpha=10$, then (9.6c,d) can be written as

$$\Delta Y_1(j\omega)=-0.03\cdot 0.001^5\{Y_6(j\omega)+1.31\cdot 0.001^2Y_8(j\omega)+0.03\cdot 0.001^3Y_9(j\omega)+\cdots\}$$

$$\Delta Y_5(j\omega)=1.31\cdot 0.001^2Y_7(j\omega)+0.03\cdot 0.001^3Y_8(j\omega)+\cdots$$

From the equations above it can be seen that higher order output spectra (>5) would have very limited effect on the determination of lower order output spectra. The computation error in $Y_5(j\omega)$ incurred by $Y_7(j\omega)$ would be $1.31\cdot 0.001^2Y_7(j\omega)$. Only when the magnitude of $Y_7(j\omega)$ is larger than 10^6, the influence could be significant. Similar conclusions hold for the other higher orders. Moreover, it can be checked by a mathematical induction that the other orders of output spectra between 1 and 5 can also be computed accurately if $Y_5(j\omega)$ can be computed sufficiently accurately. For different order N, similar results can be obtained.

Corollary 9.1 leads to the following estimation algorithm for the nth-order output spectrum (referred to as nth-order Output Spectrum Estimation (nth-OSE) algorithm).

Step 1. Choose N=3 or 5 etc, and properly set ρ, α and β, satisfying $0<\rho<1$, $\beta>1$, $\alpha>1$, e.g., $\rho=0.001$, $\alpha=10$ and $\beta=2$.

Step 2. Use (9.4) to find the estimates for $Y_n(j\omega)$, i.e., $\hat{Y}_n(j\omega)$, for n=1,2,. . .,N. Note that the computation error for n=1 or N is given by (9.6a,b) or (9.6c,d).

Step 3. Note that (9.3) can be written as $Y(j\omega)_\rho/\rho - Y_1(j\omega) = \rho Y_2(j\omega) + \rho^2 Y_3(j\omega) + \cdots$ Therefore, replacing $Y(j\omega)_{\rho_i}$ by $Y(j\omega)_{\rho_i}/\rho_i - \hat{Y}_1(j\omega)$ in (9.4) and re-applying (9.4) with a larger ρ (usually 5–10 times than the previous one) lead to the estimation of $Y_n(j\omega)$, i.e., $\hat{Y}_n(j\omega)$ for n=2,. . .,N+1. The estimation error is still given by (9.6a–d) accordingly by replacing $Y_i(j\omega)$ with $Y_{i+1}(j\omega)$.

Step 4. Similarly, (9.3) can be written as $(Y(j\omega)_\rho - \rho Y_1(j\omega) - \rho^2 Y_2(j\omega))/\rho^2 = \rho Y_3(j\omega) + \rho^2 Y_4(j\omega) + \cdots$ Therefore, replacing $Y(j\omega)_{\rho_i}$ by $(Y(j\omega)_\rho - \rho Y_1(j\omega) - \rho^2 Y_2(j\omega))/\rho^2$ in (9.4) and re-applying (9.4) with a larger ρ (usually 5–10 times than the previous one) leads to the estimation of $Y_n(j\omega)$, i.e., $\hat{Y}_n(j\omega)$ for $n=3,\ldots,$N+2.

Step 5. Follow a similar process as Step 3 and Step 4 until a sufficiently high order N^*. The truncation order N^* is not known but can be determined by evaluating the magnitude of $\hat{Y}_{N^*}(j\omega)$ according to a predefined threshold (e.g., $\left|\hat{Y}_{N^*}(j\omega)\right| < \varepsilon$).

With the nth-OSE algorithm, the nth-order output spectrum can be determined accurately (by properly choosing ρ (discussed later)) into a nonparametric form and an appropriate truncation order N^* can also be obtained. This estimation process does not necessarily need a system model but input–output data only from simulations or experiments.

9.3.2 Determination of the nth-Order nCOS Function

From (9.1), the nth-order nCOS function can be written as

$$Y_n(j\omega) = \chi_n \cdot \varphi_n(j\omega)^T \tag{9.7}$$

where χ_n is the nth-order characteristic parameter vector given by (9.2). If the system model under study is not known, several methods as mentioned before can be employed to identify a NDE, NARX or NBO model from experimental data. Here supposes that the NDE model of the system is known and similar results hold for other models. With the NDE model, χ_n is known explicitly for any characteristic parameters (Chaps. 5 and 6).

At any frequency ω, an identification method can be used to determine $\varphi_n(j\omega)$. Let τ denote the dimension of χ_n, and suppose there are m characteristic parameters in χ_n, i.e., $c_1, c_2, \ldots, c_m$. Denote $\overline{c} = (c_1, c_2, \ldots, c_m)$. Then,

Step 1: Compute the nth-order characteristic vector χ_n using (9.2) with respect to $\overline{c}$. If $\chi_n = 1$ implying that this order output spectrum has no relationship with $\overline{c}$, then let n→n+1 and repeat Step 1; otherwise, go to Step 2–5;

Step 2: Randomly generate τ distinct points $\overline{c} = (c_1, c_2, \ldots, c_m)$ so that $\left[\left[1\ \chi_{n|\overline{c}_1}\right]; \left[1\ \chi_{n|\overline{c}_2}\right]; \ldots; \left[1\ \chi_{n|\overline{c}_m}\right]\right]$ is nonsingular.

Step 3: At each point $\overline{c} = (c_1, c_2, \ldots, c_m)$, the system can be simulated subject to a specific input and applying the nth-OSE algorithm leads to determination of the nth-order output spectrum denoted by $\hat{Y}_n(j\omega)_{|\overline{c}}$, i.e.,

$$\hat{Y}_n(j\omega)_{|\overline{c}} = \hat{Y}_n(j\omega)_{|\overline{c}=0} + \chi_{n|\overline{c}} \cdot \phi_n(j\omega)^T$$

Step 4: It can be obtained that

$$\begin{aligned}\left[\hat{Y}_n(j\omega)_{|\overline{c}=0}, \hat{\varphi}_n(j\omega)\right]^T = & \left[\left[1\ \chi_{n|\overline{c}_1}\right]; \left[1\ \chi_{n|\overline{c}_2}\right]; \ldots; \left[1\ \chi_{n|\overline{c}_m}\right]\right]^{-1} \\ & \cdot \left[\hat{Y}_n(j\omega)_{|\overline{c}_1}, \hat{Y}_n(j\omega)_{|\overline{c}_2}, \ldots, \hat{Y}_n(j\omega)_{|\overline{c}_m}\right]^T\end{aligned} \quad (9.8)$$

Step 5: Thus, the nth-order nCOS function (9.7) is achieved, and finally the nCOS function can be obtained as

$$\hat{Y}(j\omega) = \rho \hat{Y}_1(j\omega) + \sum_{n=2,3,\ldots}^{\chi_{n|\overline{c}}=1} \rho^n \hat{Y}_n(j\omega) + \sum_{n=2,3,\ldots}^{\chi_{n|\overline{c}} \neq 1} \rho^n \left(\hat{Y}_n(j\omega)_{|\overline{c}=0} + \chi_{n_{|\overline{c}}} \hat{\varphi}_n(j\omega)^T\right)$$

The method above is referred to as nth-order nCOS estimation (nth-COSE) algorithm. Because the elements in χ_n is symbolically independent, there must exist τ distinct points $\overline{c} = (c_1, c_2, \ldots, c_m)$ such that the matrix $\left[\chi_{n|\overline{c}_1}; \chi_{n_{|\overline{c}_2}}; \ldots; \chi_{n_{|\overline{c}_m}}\right]$ is nonsingular (Chap. 8). If the estimation of $\hat{Y}_n(j\omega)_{|\overline{c}_i}$ is unbiased, the determination of $\hat{\varphi}_n(j\omega)$ in (9.8) will be unbiased. Compared with the result in Chap. 7, (a) the nCOS function here is determined as not only a polynomial function of model parameters but also an explicit function of the input magnitude ρ; (b) The best truncation order N* is not necessarily known in advance (this must be given in the previous result); (c) With the help of the nth-OSE algorithm, the complex-valued polynomial coefficients can be determined accurately, while the previous algorithm may result in biased (if not wrong) estimation; (d) The parameter-independent terms such as $\hat{Y}_n(j\omega)_{|\overline{c}=0}$ are explicitly considered in the nCOS function, which are incurred by those which have no relationship with model parameter $\overline{c}$; Thus the characteristic parametric structure of the nCOS function is more generic.

9.4 Example Studies

Applications of the results can be found in different areas. The nth-OSE algorithm can be used for estimation of nonlinear polynomial like (9.3). The accurate determination of each order of nonlinear output spectrum components can be used for nonlinear detection in non-destructive evaluation (Jing 2011; Chatterjee 2010; Lang and Peng 2008). The method can also be used to accurately estimate the linear part of a nonlinear system where the nonlinearity could be complicated. Importantly, the accurate determination of the nCOS function can be used for nonlinear analysis and design. These will be demonstrated with a nonlinear vehicle suspension system.

9.4.1 Identification of a Polynomial Function

Given a polynomial function

$$f(x) = 10x - 20x^2 + 300x^3 - 40x^4 + 500x^5 - 600x^6 + 700x^7 - 800x^8 + 900x^9$$

For convenience in understanding, the equation above can be regarded as a polynomial in (9.3) with $\rho = x$, $Y(j\omega)_\rho = f(x)$ and $Y_i(j\omega) = 10, -20, \ldots$ for $i = 1, 2, \ldots$. The coefficients can be estimated accurately with any guessed truncation order N (the real one is N=9) using the nth-OSE algorithm. For example, by Corollary 9.1, taking N=5, $\rho = 0.001$, $\alpha = 10$ and $\beta = 2$, the coefficients of the first 5 orders are given in Fig. 9.1a by applying Steps 1–2 of the nth-OSE algorithm. By applying Step 3 with $\rho = 0.01$, the estimation for the coefficients from the second to the sixth orders is shown in Fig. 9.1b. Both clearly show that the estimated coefficients accurately match the real values.

One issue with the nth-OSE algorithm is how to choose the excitation magnitude ρ for both α and β are given heuristically satisfying $\alpha > 1$ and $\beta > 1$. From Table 9.1, it can be seen that given $\alpha = 10$ and $\beta = 2$, different values for ρ result in a little difference in the estimation of lower order coefficients (especially for order 1), but could lead to very different estimates for higher order coefficients such as the coefficient of order 5. Table 9.1 shows that there is a large range for ρ to choose (e.g., $0 < \rho < 0.1$) to have an accurate estimation for each coefficient especially for lower order coefficients. It also clearly indicates that the estimation could be greatly biased without a proper value for ρ (e.g., $\rho = 0.5$).

To find the best excitation magnitude ρ, define the following excitation sensitivity function

$$S(\rho) = \left|\frac{\partial Y_\rho}{\partial \rho}\right| = \left|\frac{Y_\rho - Y_{\rho+\Delta}}{\Delta}\right|$$

where Y_ρ denotes the variable to be estimated under the excitation magnitude ρ, and Δ is a small positive number. By the definition, for each order of the coefficients

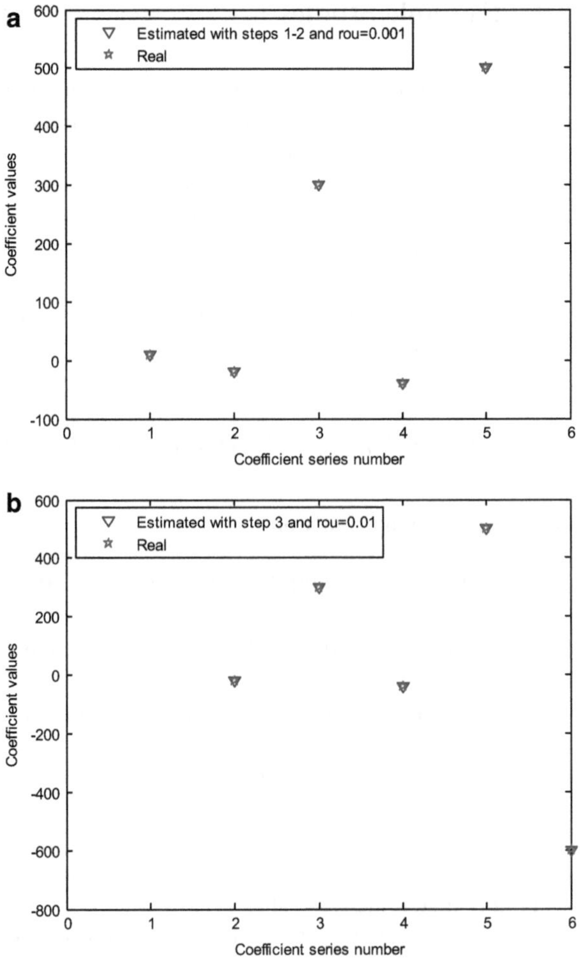

Fig. 9.1 Estimation with the nth-OSE algorithm with (**a**) steps 1–2, (**b**) step 3 (Jing 2014 ©IEEE)

there will be a V-shape or U-shape curve given by the sensitivity function (Fig. 9.2), and the best excitation magnitude should locate around the bottom of the curve which corresponds to the slightest estimation error for the corresponding order coefficient. In Fig. 9.2a, the effective excitation range for the orders 4 and 5 in Table 9.1 are highlighted, which are clearly around the bottom or turning corner of the V-shape sensitivity function; for the first 3 orders the effective excitation ranges cover the whole testing excitation range, having very small sensitivity values (below 1). Similar cases hold for different other values of α and β, for example $\alpha=5$, $\beta=5$ in Fig. 9.2b, indicating that ρ is actually the sensitive parameter to tune in the nth-OSE algorithm.

Table 9.1 Estimation with different excitation magnitude (the bold and italic numbers indicate the relatively accurate estimation)

ρ	Real				
	10	−20	30	−40	500
0.00001	***10.000***	***−20.000***	***300.0000***	***−36.0000***	−1,048,576.0000
0.00005	***10.000***	***−20.000***	***300.0000***	***−40.0156***	512.0000
0.0001	***10.000***	***−20.000***	***300.0000***	***−40.0156***	512.0000
0.0005	***10.000***	***−20.000***	***300.0000***	***−40.0002***	***500.0625***
0.001	***10.000***	***−20.000***	***300.0000***	***−40.0008***	***500.0000***
0.005	***10.000***	***−20.000***	***300.0000***	***−40.0192***	***500.0224***
0.01	***10.000***	***−20.000***	***300.0000***	***−40.0767***	***500.0894***
0.05	***10.000***	***−19.999***	***299.9977***	−41.9218	502.2419
0.1	***10.000***	***−19.983***	***299.9720***	−47.7635	509.0534
0.5	10.3669	−5.6135	281.7501	−297.5458	798.3785
The effective range for ρ	$0<\rho<0.1$	$0<\rho<0.1$	$0<\rho<0.1$	$0.00005<\rho<0.01$	$0.0005<\rho<0.01$

From the discussions above, considering Table 9.1 and Fig. 9.2 and based on some other testing results, it can be concluded for the nth-OSE algorithm that

(a) The parameters α and β can be chosen heuristically satisfying $\alpha > 1$ and $\beta > 1$. The excitation magnitude ρ is the key factor to be tuned properly.
(b) Testing can be done to generate plots of the excitation sensitivity function, in which the best excitation magnitude locates around the bottom or turning corner of the V-shape curve, corresponding to the estimation with smallest error (e.g., the orders 4–5 in Fig. 9.2).
(c) If the V-shape curve is very flat or very low in values at all testing range, this indicates that all the testing magnitudes are effective for the estimation of the corresponding order (e.g., the orders 1–3 in Fig. 9.2). Moreover, a flat curve around the bottom of the V-shape curve implies the estimation with the corresponding excitation magnitudes is consistently accurate.
(d) The best ρ^* obtained in this method implies that the obtained lower order polynomial estimated by the nth-OSE algorithm will be applicable to the excitation input $\rho R(j\omega)$ of magnitude $\rho \in [0, \overline{\rho}]$, where $\overline{\rho} = \rho^*$ when N=5 and $\overline{\rho} = \frac{1+\alpha}{\alpha}\rho^*$ when N=3 by Corollary 9.1.

This could be used as a useful guidance in selection of the excitation magnitude ρ in the nth-OSE algorithm (referred to as the ρ-selection method). The value of ρ in step 1 and 3 of the nth-OSE algorithm can both be obtained with this method.

9.4.2 Analysis of Nonlinear Suspension Systems

Consider the analysis of a nonlinear vehicle suspension system (Fig. 9.3) in this example. It is known that vehicle suspension systems usually have inherent nonlinearity, which brings difficulties in active or semi- active control. To certain

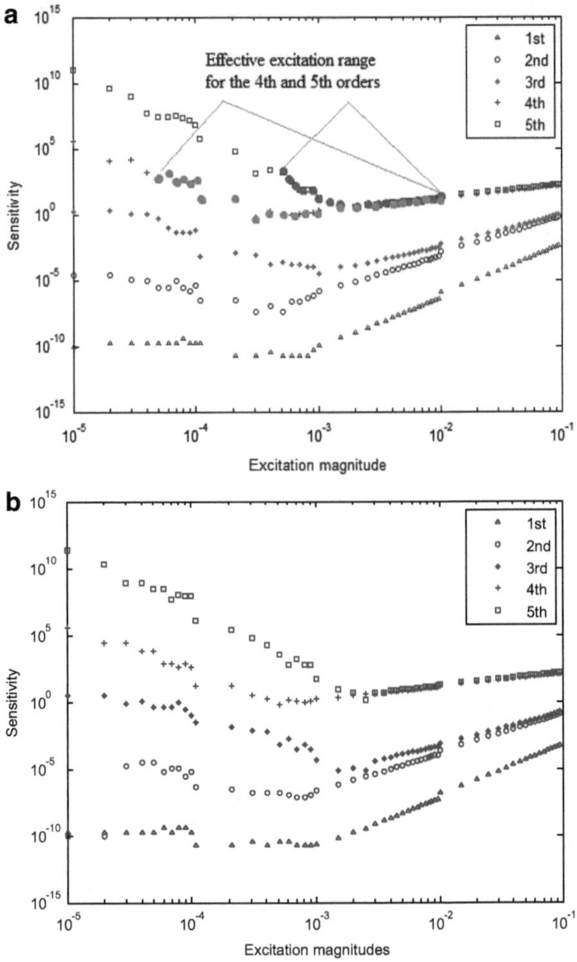

Fig. 9.2 Sensitivity to excitation in estimation of different order coefficients, (**a**) $\alpha=10$, $\beta=2$ (**b**) $\alpha=5$, $\beta=5$ (Jing 2014 © IEEE)

extent, accurate identification of the linear part and each nonlinear component would be very helpful in the analysis and design of a desired suspension system in practice. The nth-OSE algorithm could be very helpful to this objective. This will be demonstrated in Case I of this example. On the other hand, it is more and more noticed that nonlinear damping characteristics could produce superior performance in vibration suppression compared with linear damping (Chap. 12). To achieve adjustable damping characteristics, MR dampers are often used as an ideal shock absorber (Case et al. 2012; Zapateiro et al. 2012). However, the question is how to design a desired nonlinear damping characteristic for a given vehicle suspension system. The nonlinear COS function above can provide an alternative approach to this problem for the nonlinear analysis and design. This will be demonstrated in Case II of this example.

Fig. 9.3 A nonlinear vehicle suspension system (Jing 2014 © IEEE)

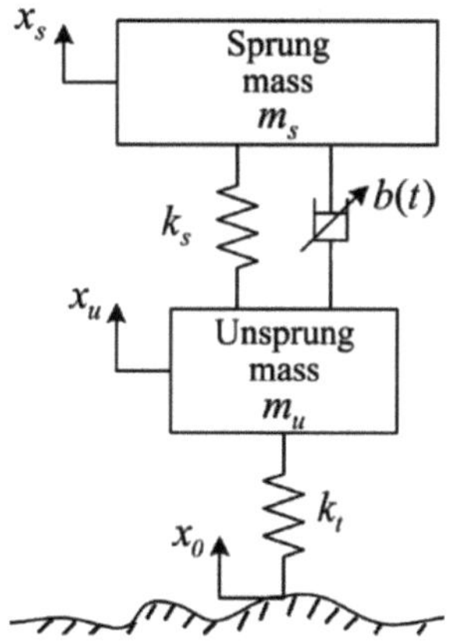

The one quarter vehicle suspension model is given as

$$m_s\ddot{x}_s = -f_s - b(t)(\dot{x}_s - \dot{x}_u) \tag{9.9a}$$

$$m_u\ddot{x}_u = f_s + b(t)(\dot{x}_s - \dot{x}_u) - k_t(x_u - x_0) \tag{9.9b}$$

where the spring force is a nonlinear function given by

$$f_s = k_{s1}(x_s - x_u) + k_{s2}(x_s - x_u)^2 + k_{s3}(x_s - x_u)^3 \tag{9.9c}$$

and b(t) denotes the nonlinear damping coefficient to be designed, which can generally be any nonlinear function. In this study, it is as an example written as

$$b(t) = b_0 + b_1 z^2 \tag{9.9d}$$

where $z = x_s - x_u$. Previous results have studied the cubic nonlinear damping $b_1\dot{z}^2$(Chap. 7, Jing et al. 2011). For the model parameters above, a model in Dixit and Buckner (2011) is adopted here as: $m_s = 240$ kg, $k_{s1} = 12{,}394$ N/m, $k_{s2} = -73{,}696$ N/m^2, $k_{s3} = 3{,}170{,}400$ N/m^3, $m_u = 25$ kg, $k_t = 160{,}000$ N/m, and $b_0 = 1{,}385.4$ Ns/m.

Case I: The estimation of the linear part and the nonlinear output spectrum without knowing the truncation order N and system model. In this part, it is supposed that the system model is not known and $b_1 = 10^6$ in (9.9d). It can be seen that the suspension system has strong nonlinearity both in stiffness and damping elements. The nth-OSE algorithm can be used to estimate the linear part of the system and therefore to estimate how the nonlinearity takes a role in the dynamic response using the method discussed in Jing (2011). To this aim, the input excitation is considered as a multi-tone function given by

$$u(t) = \rho \sum_{i=1}^{10} \sin((5i)t) \tag{9.10}$$

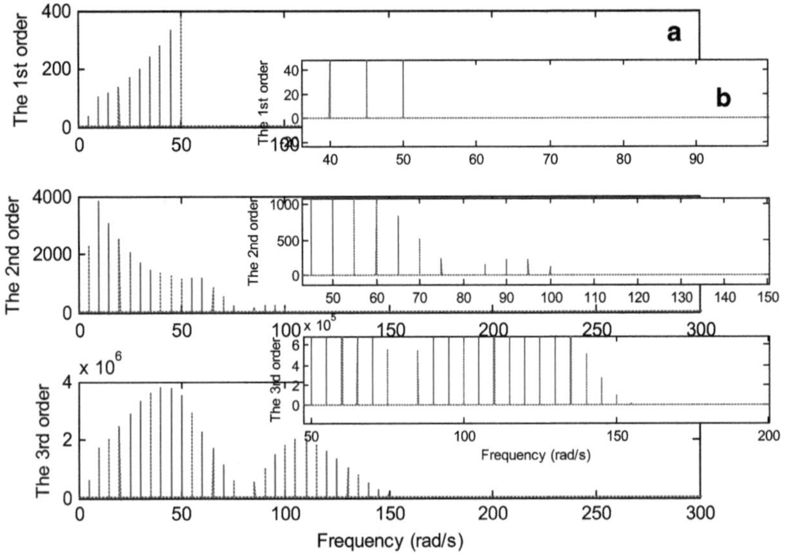

Fig. 9.4 The estimated output spectra up to order 3: (**a**) the estimated results, (**b**) the zoomed results of (**a**) (Jing 2014 © IEEE)

Note that the multi-tone signal above can actually include more frequency points and thus provide a sufficient excitation to the system at different frequencies of interest. The parameter ρ is to be determined in the estimation.

Usually the nonlinear output spectrum of system (9.9a–d) would have an infinite order of output spectrum components but could be truncated at an appropriate order N to approximate the original nonlinear dynamic response in practice. With the nth-OSE algorithm, the nth-order output spectrum of any given nonlinear systems could be estimated in a non-parametric form up to any high order without knowing the truncation order and system model. If only the estimation of the linear part of the system is interested, it would be much easier to choose a proper excitation magnitude ρ. To estimate accurately all the output spectrum components simultaneously, ρ must be chosen using the ρ-selection method above. For example, if taking $N=3$, the corresponding parameters can be chosen as $\rho=0.0005$ and $\alpha=10$ based on the ρ-selection method, and the estimated results for the nth-order output spectrum up to order 3 are given in Figs. 9.4 and 9.5.

In Fig. 9.4, it can be seen that the output frequencies in the first order are exactly the same as the input frequencies, those in the second order are doubled and in the third order tripled. This is completely consistent with the known results in Chap. 3 that is, the output frequencies in the nth-order output spectrum can only appear in the range [0, nb] given that the input frequencies are in [a, b]. To further validate the estimation results in Fig. 9.4, the real output spectrum of the linear part is given in Fig. 9.5. It is clearly shown that the estimated values of the first order output spectrum exactly match the real ones with overall root-mean-square error 8.5068e-004.

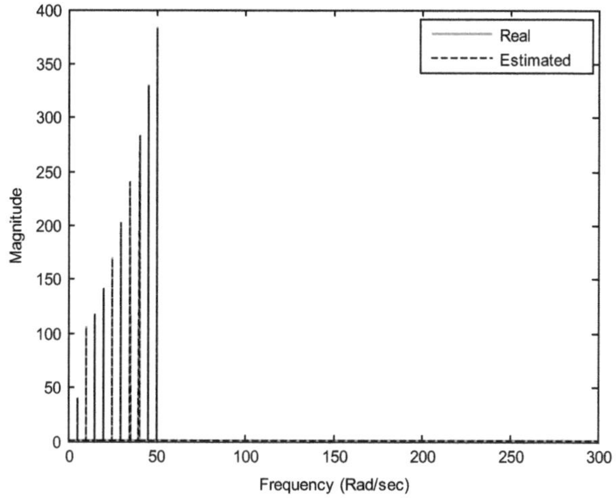

Fig. 9.5 The estimated and real first order output spectrum (Jing 2014 © IEEE)

Similarly, if taking N=5 and the other parameters by Corollary 9.1 and the ρ–selection method as ρ=0.001, α=10 and β=2, the estimation results of the output spectrum up to order 5 are shown in Fig. 9.6. Comparisons indicate that the estimated first three orders of output spectrum are exactly the same in Figs. 9.4 and 9.6, implying the consistence and validation of the nth-OSE algorithm in estimation of the nth-order output spectrum for the system. Figures 9.4 and 9.6 both show that the magnitude of the higher order output spectrum is becoming larger with the increase of order n. Recalling (9.3), this implies that when the input magnitude is becoming larger, the nonlinearity will take more dominant roles in system dynamic response and the vibration performance could become worse (see Fig. 9.7). Figure 9.7 clearly confirms that when the input magnitude is increased to ρ =0.1, the nonlinear response is dominant and vibration transmissibility becomes obviously worse. Therefore, a proper design of the linear or nonlinear damping coefficient would be very crucial in vibration control. This consequently incurs a need for a systematic method for the nonlinear analysis and design, which would be focused in the following section.

It should be noted that if the excitation magnitude ρ_1, ρ_2, ρ_3 are not properly chosen according to Corollary 9.1, the estimation results would be greatly biased (for paper length this is not illustrated here). With the accurately estimated first order output spectrum, many methods are available to estimate a linear parametric model (Levi 1959; Young 1985; Ibrahim 2008). This is not demonstrated here either. Moreover, it is interesting to mention that the nth-OSE algorithm developed here provides an effective tool to accurately estimate the linear and nonlinear components in the dynamical response of a system. This would also be of significance to non-destructive evaluation and similar topics can be referred to (Jing 2011; Chatterjee 2010; Lang and Peng 2008). It is noted that in application of the method to experimental data, noise corruption can never be avoided. Preliminary results indicate that the additive noise can basically have little effects on the

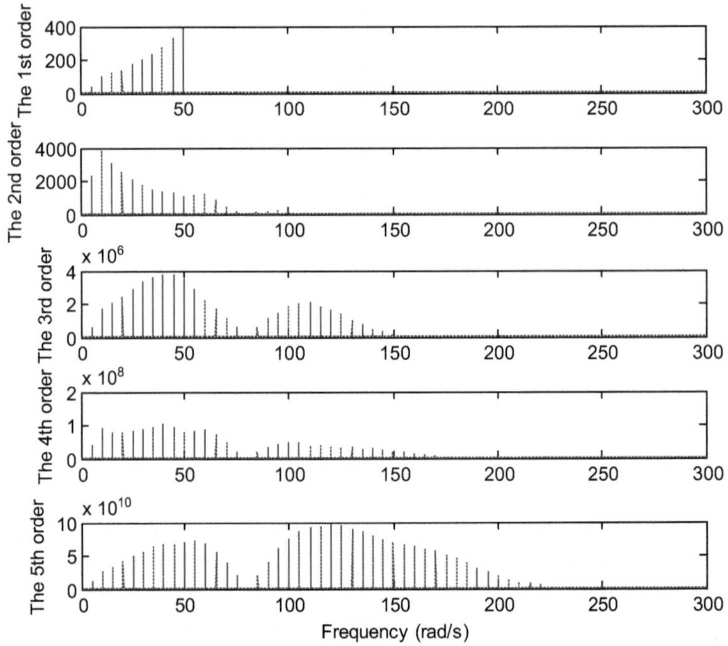

Fig. 9.6 The estimated output spectrum up to order 5 (Jing 2014 © IEEE)

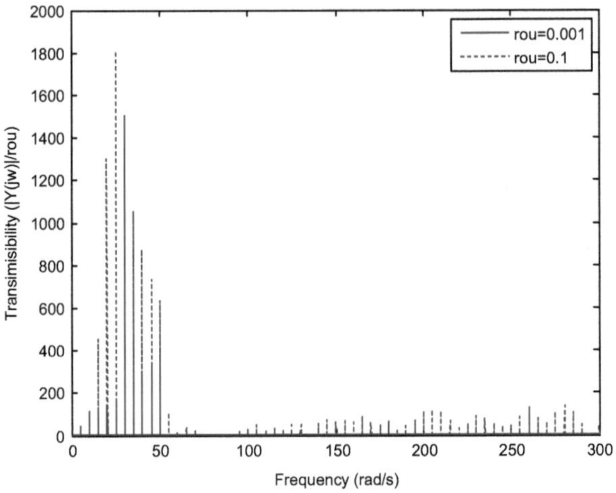

Fig. 9.7 Transmissibility $\left(\frac{|Y(j\omega)|}{\rho}\right)$ for input magnitude $\rho = 0.001$ and 0.1 (Jing 2014 © IEEE)

estimation with the nth-OSE algorithm after using some noise processing methods. This will be further investigated in future studies.

Case II: The estimation of the nCOS function with respect to the nonlinear damping. To explore the nonlinear benefit in vibration control, the nonlinear damping characteristic can be analyzed by regarding $\ddot{x}_s$ as the system output, i.e.,

$$y = \ddot{x}_{\text{tf=\}TTf90d833a}_{\text{I}}\text{\}s}} \tag{9.11}$$

and deriving the nonlinear COS function in (5) in terms of the nonlinear parameter b_1. To this aim, consider the input as

$$u(t) = \rho \sin(\Omega t) \tag{9.12}$$

where Ω could be any frequency of interest. Since only a cubic order nonlinear term $b_1 z^2 \dot{z}$ would be considered, it can be obtained using the parametric characteristic analysis (Chaps. 5 and 6) that

$$\begin{gathered}\chi_2 = 1;\ \chi_3 = b_1;\ \chi_4 = b_1;\ \chi_5 = \left[b_1,\ b_1^2\right]; \chi_6 = \left[b_1,\ b_1^2\right];\\ \chi_7 = \left[b_1,\ b_1^2,\ b_1^3\right];\ \chi_8 = \left[b_1,\ b_1^2,\ b_1^3\right];\ \cdots\end{gathered}$$

Then in this case the nonlinear COS function can be written as

$$\begin{aligned}Y(j\omega) = {} & Y_1(j\omega) + Y_2(j\omega) + Y_3(j\omega)_{|b_1=0}\\ & + b_1 \cdot \varphi_3(b_1; j\omega) + Y_4(j\omega)_{|b_1=0} + b_1 \cdot \varphi_4(b_1; j\omega)\\ & + Y_5(j\omega)_{|b_1=0} + b_1 \cdot \varphi_5(b_1; j\omega) + b_1^2 \cdot \varphi_5\left(b_1^2; j\omega\right) + \cdots\end{aligned} \tag{9.13}$$

Note that $Y_1(j\omega)$ and $Y_2(j\omega)$ are the first and second order output spectrum, $Y_3(j\omega)_{|b_1=0}$, $Y_4(j\omega)_{|b_1=0}$, and $Y_5(j\omega)_{|b_1=0}$ are all incurred by the second and third order nonlinearity in the stiffness, which have no relationship with the nonlinear damping term. By the parametric characteristic analysis, there will be a term in χ_5 given by the multiplication between b_1 and the coefficient of the third order nonlinearity in the stiffness. While the latter is a constant, it thus yields the term b_1 in χ_5. This results in the term $b_1\varphi_5(b_1;j\omega)$ in (9.13). By applying the nth-COSE algorithm, the estimation of (9.13) can be obtained as

$$\begin{aligned}Y(j\omega) = {} & \rho\hat{Y}_1(j\omega) + \rho^2\hat{Y}_2(j\omega) + \rho^3 \underbrace{\left[\hat{Y}_3(j\omega)_{|b_1=0} + b_1 \cdot \hat{\varphi}_3(b_1; j\omega)\right]}_{\hat{Y}_3(j\omega, b_1)}\\ & + \rho^4 \underbrace{\left[\hat{Y}_4(j\omega)_{|b_1=0} + b_1 \cdot \hat{\varphi}_4(b_1; j\omega)\right]}_{\hat{Y}_4(j\omega, b_1)}\\ & + \rho^5 \underbrace{\left[\hat{Y}_5(j\omega)_{|b_1=0} + b_1 \cdot \hat{\varphi}_5(b_1; j\omega) + b_1^2 \cdot \hat{\varphi}_5(b_1^2; j\omega)\right]}_{\hat{Y}_5(j\omega, b_1) + \cdots}\end{aligned} \tag{9.14}$$

Note that the nCOS function in (9.14) is an explicit function of the frequency ω, nonlinear parameter b_1 and excitation magnitude ρ. In (9.12), consider $\Omega = 18$ rad/s as the input frequency, which is around the resonant frequency of the system and also a sensitive frequency to human body. If in (9.14), take $\omega=\Omega$, the even order output spectrum at this frequency must be zero (Jing et al. 2006, 2010). Thus, (9.14) can be written as

$$\hat{Y}(j\Omega) = \rho\hat{Y}_1(j\Omega) + \rho^3 \underbrace{\left[\hat{Y}_3(j\Omega)_{|b_1=0} + b_1 \cdot \hat{\varphi}_3(b_1;j\Omega)\right]}_{\hat{Y}_3(j\Omega,b_1)} + \rho^5 \underbrace{\left[\hat{Y}_5(j\Omega)_{|b_1=0} + b_1 \cdot \hat{\varphi}_5(b_1;j\Omega) + b_1^2 \cdot \hat{\varphi}_5\left(b_1^2;j\Omega\right)\right]}_{\hat{Y}_5(j\Omega,b_1)} + \cdots \tag{9.15}$$

By the nth-COSE algorithm, to estimate (9.15), the following steps can be used (referred to as the nCOSE-procedure):

(a) Setting $b_1=10^6$ (or any other value), apply the nth-OSE algorithm to estimate $Y_n(j\omega)$ for $n=1,2,3,\ldots,5\ldots$;
(b) Take the estimated values of the output spectrum $Y_n(j\omega)$ at the frequency Ω, denoted by $\hat{Y}_1(j\Omega), \hat{Y}_3(j\Omega)$ and $\hat{Y}_5(j\Omega)$ ($\hat{Y}_2(j\Omega)$ and $\hat{Y}_4(j\Omega)$ must be zero);
(c) Take $b_1=10^5$ (or any other value) and $b_1=0$ (preferable), and repeat (a-b);
(d) Equation (9.15) can therefore be estimated using a least square method for each component.

Based on Steps (a–c), Table 9.2 can be obtained with the parameter setting in the nth-OSE algorithm as $\rho=0.001$, $\alpha=10$ and $\beta=2$ as suggested in the previous case study.

In Table 9.2, the estimates for $Y_1(j\Omega)$ under each parameter value are consistent since $Y_1(j\Omega)$ is the dynamic response of the linear part of the system, independent of system nonlinearity. This confirms again the effectiveness and reliability of the nth-OSE algorithm.

For the estimation of $\hat{Y}_3(j\Omega, b_1)$ in (9.15), Table 9.2 yields

$$\hat{Y}_3(j\Omega)_{|b_1=0} = 5.3988\text{e}+002 \ + 1.2252\text{e}+004\text{i}$$

Therefore,

$$\begin{aligned}\hat{\varphi}_3(b_1;j\Omega) &= \hat{Y}_3(j\Omega,\text{b}_1) - \hat{Y}_3(j\Omega)_{|b_1=0})/\text{b}_1 \\ &= \begin{cases}(2.8583\text{e}+003 \ - 1.2300\text{e}+002\text{i})10^{-5} \\ (2.8583e+004 \ - 1.2310e+003i)10^{-6}\end{cases}\end{aligned}$$

for $b_1=10^5$ and 10^6, respectively, which are almost the same. The averaged value can be used

Table 9.2 The estimated output spectrum at Ω rad/s for different values of b_1

b_1	$\hat{Y}_1(j\Omega)$	$\hat{Y}_3(j\Omega)$	$\hat{Y}_5(j\Omega)$
0	1.0876e+002+7.2310e+001i	5.3988e+002+1.2252e+004i	1.8958e+005+2.5757e+006i
10^5	1.0876e+002+7.2310e+001i	3.3982e+003+1.2129e+004i	4.3776e+005+2.4906e+006i
10^6	1.0876e+002+7.2310e+001i	2.9123e+004+1.1021e+004i	1.5670e+006+1.7068e+006i

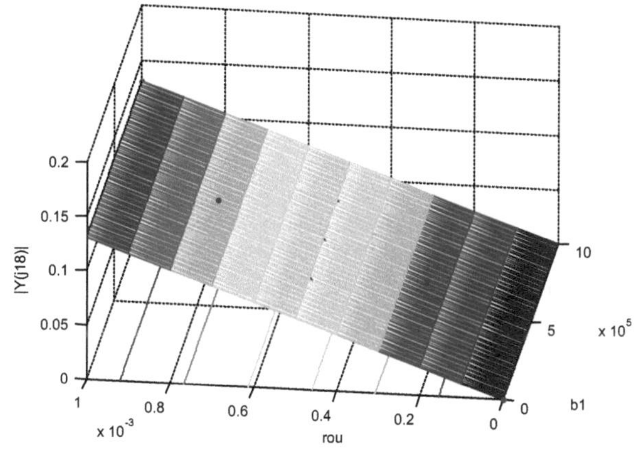

Fig. 9.8 The estimated nCOS function and its validation for ρ∈[0,0.001] (Jing 2014 © IEEE)

$$\hat{\varphi}_3(b_1;j\Omega) = (2.8583\text{e}+003 \; - 1.2305\text{e}+002\text{i})10^{-5}$$

Similarly, it can be obtained that

$$\begin{bmatrix} \hat{\varphi}_5(b_1;j\Omega) \\ \hat{\varphi}_5(b_1^2;j\Omega) \end{bmatrix} = \begin{bmatrix} 10^5 & 10^{10} \\ 10^6 & 10^{12} \end{bmatrix}^{-1} \begin{bmatrix} \hat{Y}_5(j\Omega,10^5) - \hat{Y}_5(j\Omega)_{|b_1=0} \\ \hat{Y}_5(j\Omega,10^6) - \hat{Y}_5(j\Omega)_{|b_1=0} \end{bmatrix} = \begin{bmatrix} 2.6045 \; - 0.8490\text{i} \\ -1.2271\text{e-}007 \; - 1.9889\text{e}-009\text{i} \end{bmatrix}$$

Therefore, the nCOS function (up to the fifth order) of the system with respect to input (9.12) and nonlinear damping parameter b_1 is obtained. To verify the nCOS function in output prediction, different excitation magnitudes and different nonlinear damping coefficient b_1 can be used for a validation, which is shown in Fig. 9.8 and Table 9.3, indicating clearly the exact prediction by using the estimated nCOS function (9.15) in the excitation range ρ ∈[0,0.001].

Importantly, it can also be checked that the prediction of the estimated nCOS (up to the fifth order) is still reliable even when the excitation magnitude is larger for example up to 0.05 (see Fig. 9.9 and Table 9.3), which is much larger than the excitation range [0, 0.001] used in the estimation. Similar conclusion can be drawn for a larger parameter value b_1. This demonstrates clearly the reliability and advantage using the nCOS function in system analysis. Moreover, comparing

Table 9.3 Prediction of the estimated nCOS function with (9.15) (Jing 2014 © IEEE)

ρ	b_1	Real magnitude	Prediction by (19)
0	0	0	0
$0.5*10^{-3}$	$5*10^5$	0.0653	0.0653
$0.5*10^{-3}$	$2.5*10^5$	0.0653	0.0653
$0.5*10^{-3}$	$7.5*10^5$	0.0653	0.0653
$0.25*10^{-3}$	$5*10^5$	0.0327	0.0327
$0.75*10^{-3}$	$5*10^5$	0.098	0.098
10^{-3}	10^6	0.1306	0.1306
0.01	$2*10^5$	1.3181	1.3181
0.01	$6*10^5$	1.3274	1.3274
0.03	$6*10^5$	4.5275	4.5504
0.05	$6*10^5$	9.5741	9.9726
0.05	10^6	10.655	11.3523

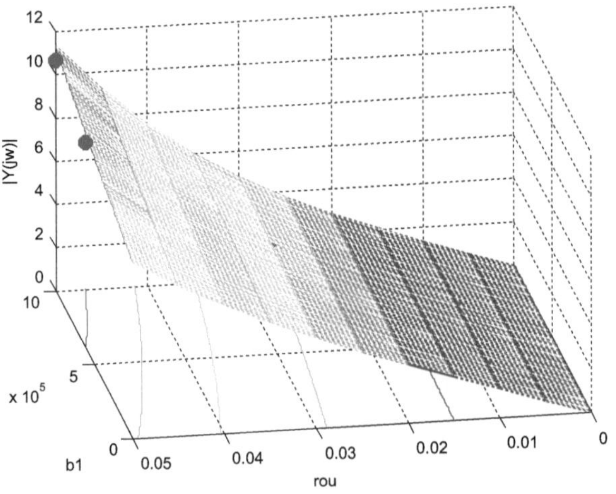

Fig. 9.9 The estimated nCOS function and its validation for $\rho \in [0, 0.05]$ (Jing 2014 © IEEE)

Fig. 9.9 with Fig. 9.8, it can also be seen that the nonlinear dynamics become more obviously when the input magnitude becomes larger. Using the nth-COSE algorithm, it is easy to estimate the nCOS function up to any high order. This therefore provides a useful tool for the analysis and design of nonlinear damping in vehicle suspension systems, and the parameter optimization would also be possible in terms of any nonlinear parameters of interest to achieve a desired output spectrum in vibration suppression. The advantages of the nCOS function based method proposed in this study for nonlinear analysis and design could be that it provides an explicit expression for the relationship between nonlinear output spectrum and system parameters of interest (including frequency, nonlinear parameters and excitation magnitude), and this relationship can be accurately determined with less simulation trials and computation cost compared with a pure simulation based study or traditional theoretical computations. Regarding the last point, for example, to obtain the nCOS in (9.15), the nCOSE-procedure above involves at

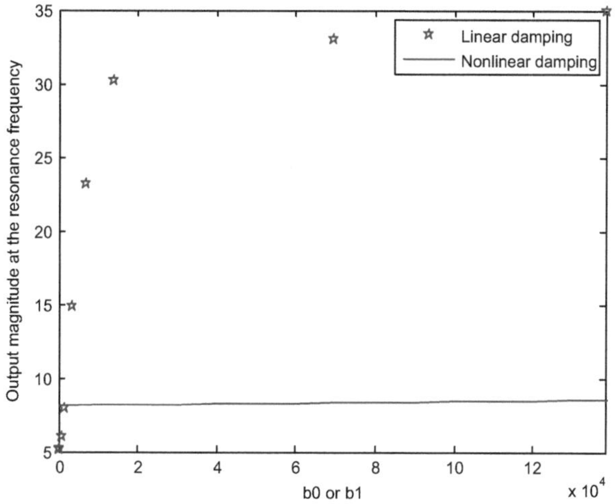

Fig. 9.10 Linear and nonlinear damping in suppression of resonance peak (Jing 2014 © IEEE)

most 15 simulation trials and 15 runs of the FFT algorithms. However, only to generate Fig. 9.9, there are totally 11*101 grid points for (ρ,b_1) (i.e., $\rho=$ 0:0.005:0.05 and $b_1=$0:10000:10^6), and thus it involves 1,111 times of simulation trials and 1,111 FFTs. If there are more nonlinear parameters involved, the nCOS based method would be more efficient and effective in practice.

Moreover, to further illustrate the nonlinear damping given in (9.9d), comparisons between the nonlinear damping (b_1) and linear damping (b_0) are conducted for the peak suppression of the vehicle suspension system. For the same input (9.12) with excitation magnitude $\rho=0.05$ and $\Omega=18$ rad/s, the linear damping is used with $b_0 \in [0, 1385.4*100]$ and $b_1=0$ firstly, and the nonlinear damping is then used with $b_0=1{,}385.4$ and $b_1 \in [0, 1385.4*100]$ secondly. The output magnitudes at Ω for both cases are shown in Fig. 9.10.

The results for the nonlinear damping can directly be obtained by using the nCOS function (9.15) developed above. When the nonlinear damping coefficient is increased, the vibration magnitude is slightly increased starting from 7.9 (which is the output magnitude when there is no nonlinear damping but only linear damping $b_0=1{,}385.4$); while the vibration magnitude is greatly increased with the increase of b_0 when there is no nonlinear damping. This implies an alternative approach to vibration control in practice since introducing nonlinear damping brings little effect on the damping performance of the original systems at the resonance frequency but could bring vibration suppression over a broad band of frequencies. For example, Fig. 9.10 implies that a small linear damping could be better in vibration control at steady state. But too small damping would definitely bring strong vibration considering the transient response. A proper nonlinear damping could be introduced, together with a small linear damping, to the system, which could suppress transient vibrations quickly and simultaneously keep a small linear damping performance in vibration transmissibility. This is confirmed by Fig. 9.11, where a small linear damping with $b_0=1{,}385.4*0.0005$ results in a very slow suppression speed in vibration control, while together with a nonlinear damping $b_1=10{,}000$ could it bring a much faster convergence speed and thus lead to better vibration

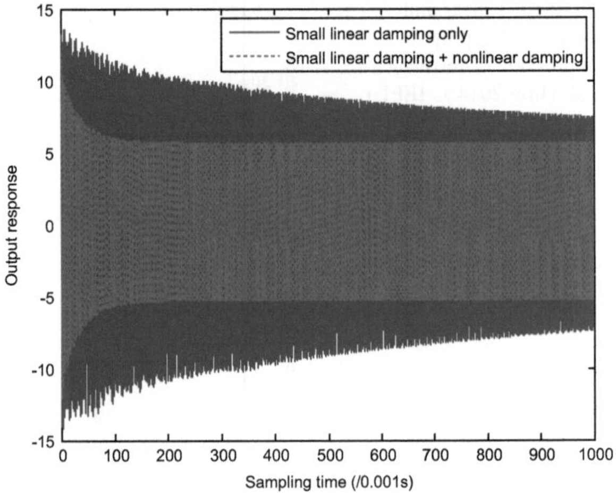

Fig. 9.11 Vibration suppression with linear and nonlinear damping (Jing 2014 © IEEE)

suppression. More discussions about nonlinear damping could be referred to Jing and Lang (2009b), Xiao, et al (2013b), Chapters 10 and 12.

9.5 Conclusions

A systematic frequency domain method for nonlinear analysis, design and estimation of nonlinear systems is discussed in this chapter using the concept of nonlinear characteristic output spectrum (nCOS) function, which is referred to here as the nCOS based method. This method allows accurate determination of the linear and nonlinear components in the nonlinear output spectrum of a given nonlinear system described by NDE, NARX or NBO models, provided with some simulation or experiment output data. These output spectrum components can then be used for system identification or nonlinear analysis for different purposes such as signal processing, vibration control and fault detection etc. Importantly, the nCOS function can therefore be developed, which is an explicit expression for the relationship between nonlinear output spectrum and system characteristic parameters of interest (including nonlinear parameters, frequency, and excitation magnitude). This relationship can be accurately obtained with less simulation trials and computation cost compared with a pure simulation based study or traditional theoretical computation, and could provide a significant and straightforward method for nonlinear analysis and design. The nCOS function can be regarded as a greatly-improved version of the OFRF function established in Chaps. 6–8.

9.6 Proofs

A. Proof of Proposition 9.1

The proof needs three more lemmas. An outline of the proof is given here and the detailed proof for Lemmas 9.2–9.4 can be referred to Jing (2012). The inverse matrix in (9.4) is denoted by A whose element is denoted by $A_{i,r}$. An explicit expression for $A_{i,r}$ is given in Lemma 9.2. Lemma 9.3 provide an analytical computation of the estimation error in terms of the truncation order N and the excitation magnitudes $\rho_1, \rho_2, \rho_3\ldots$ based on Lemma 9.2 and using (9.3), while Lemma 9.4 provides a method for the simplification of the estimation error in Lemma 9.3 for different excitation magnitudes $\rho_1, \rho_2, \rho_3\ldots$. With Lemma 9.3 and Lemma 9.4, the result of this proposition can be established.

Lemma 9.2

$$A_{i,r} = \frac{(-1)^{N-i} \sum\limits_{m=N-i;j_1\neq\cdots\neq j_m\in\aleph_0\backslash\{r-1\}\text{ with no repetition}} \rho_{j_1}\rho_{j_2}\cdots\rho_{j_m}}{\prod\limits_{k\in\aleph_0\backslash\{r-1\}}(\rho_{r-1}-\rho_k)},$$

where $1\leq i, r\leq N$, $\aleph_0=\{0,1,2,\cdots,N-1\}$, $\rho_{j_0}=1$.

Lemma 9.3 Given the truncation order N, the estimation error with (8) for the nth-order output spectrum, i.e., $\Delta Y_n(j\omega)$, is

$$\Delta Y_n(j\omega) = \sum_{j=1}^{N} A_{n,j}\cdot\rho_{j-1}^{N}\overline{\sigma}_{[N,\rho_{j-1}]}(j\omega) =$$

$$\sum_{j=1}^{N}\left(\frac{(-1)^{N-n}\sum\limits_{\substack{m=N-n;\\ j_1\neq\cdots\neq j_m\in\aleph_0\backslash\{j-1\}}}\rho_{j_1}\rho_{j_2}\cdots\rho_{j_m}}{\prod_{k\in\aleph_0\backslash\{j-1\}}(\rho_{j-1}-\rho_k)}\cdot\rho_{j-1}^{N}\overline{\sigma}_{[N,\rho_{j-1}]}(\cdot)\right)$$

where $A_{i,j} = \dfrac{(-1)^{N-i}\sum\limits_{\substack{m=N-i;\\ j_1\neq\cdots\neq j_m\in\aleph_0\backslash\{j-1\}}}\rho_{j_1}\rho_{j_2}\cdots\rho_{j_m}}{\prod_{k\in\aleph_0\backslash\{j-1\}}(\rho_{j-1}-\rho_k)}$, $\aleph_0=\{0,1,2,\cdots,N-1\}$

and $\rho_{j_0}=1$;

$$\overline{\sigma}_{[N,\rho]}(j\omega) = Y_{N+1}(j\omega)+\rho Y_{N+2}(j\omega)+\rho^2 Y_{N+3}(j\omega)+\cdots,\text{ and }\sigma_{[N,\rho]}(j\omega) = \rho^{N+1}\overline{\sigma}_{[N,\rho]}(j\omega)$$

.

Lemma 9.4 For $\forall\, x_1, x_2, \ldots, x_N\in R\backslash 0$ satisfying $x_1\neq x_2\neq\ldots\neq x_N$ and $N>1$, it holds that

$$\sum_{i=1}^{N}\frac{x_i^m}{\prod_{k=1,k\neq i}^{N}(x_i-x_k)} = \begin{cases}(x_1+x_2+\cdots+x_N)^{[m-N+1]} & m\geq N-1\\ 0 & 0\leq m<N-1\end{cases}$$

The following proof is given. From (9.3),

$$\rho_{r-1}^{N-1}\overline{\sigma}_{[N,\rho_{r-1}]}(j\omega)=\rho_{r-1}^{N-1}Y_{N+1}(j\omega)+\rho_{r-1}^{N}Y_{N+2}(j\omega)+\rho_{r-1}^{N+1}Y_{N+3}(j\omega)+\cdots \quad \text{(A1)}$$

Using Lemmas 9.3 and 9.4 and (A1),

$$\begin{aligned}\Delta Y_1(j\omega) &= (-1)^{N-1}\rho_0\rho_1\cdots\rho_{N-1}\sum_{r=1}^{N}\frac{\rho_{r-1}^{N-1}\overline{\sigma}_{[N,\rho_{r-1}]}(j\omega)}{\prod_{k\in\aleph_0\backslash\{r-1\}}(\rho_{r-1}-\rho_k)}\\ &= (-1)^{N-1}\rho_0\rho_1\cdots\rho_{N-1}\left(\sum_{r=1}^{N}\frac{\rho_{r-1}^{N-1}Y_{N+1}(j\omega)}{\prod_{k\in\aleph_0\backslash\{r-1\}}(\rho_{r-1}-\rho_k)}+\sum_{r=1}^{N}\frac{\rho_{r-1}^{N}Y_{N+2}(j\omega)}{\prod_{k\in\aleph_0\backslash\{r-1\}}(\rho_{r-1}-\rho_k)}+\cdots\right)\\ &= (-1)^{N-1}\rho_0\rho_1\cdots\rho_{N-1}\begin{pmatrix}(\rho_0+\rho_1+\cdots+\rho_{N-1})^{[0]}Y_{N+1}(j\omega)\\ +(\rho_0+\rho_1+\cdots+\rho_{N-1})^{[1]}Y_{N+2}(j\omega)\\ +(\rho_0+\rho_1+\cdots+\rho_{N-1})^{[2]}Y_{N+3}(j\omega)\cdots\end{pmatrix} \quad \text{(A2)}\end{aligned}$$

This gives (9.5a). Similarly, (9.5b) can be obtained.

B. Proof of Corollary 9.1

Considering N=3 firstly, the output spectrum is truncated at N=3, which can be written as

$$Y(j\omega) = Y_1(j\omega) + Y_2(j\omega) + Y_3(j\omega) + \sigma_{[3]}(j\omega)$$

where $\sigma_{[3]}(j\omega)=Y_4(j\omega)+Y_5(j\omega)+Y_6(j\omega)+\cdots$ is the truncation error. Using (9.4) to estimate the different orders of output spectra, the estimation errors are given by Proposition 1

$$\Delta Y_1(j\omega) = -\rho_0\rho_1\rho_2\sum_{k=0}^{\infty}(\rho_0+\rho_1+\rho_2)^{[k]}Y_{k+3}(j\omega)$$

$$\Delta Y_3(j\omega) = \sum_{k=1}^{\infty}(\rho_0+\rho_1+\rho_2)^{[k]}Y_{k+2}(j\omega)$$

If choosing $\rho_0+\rho_1+\rho_2=0$, $\rho_1=-\rho$, $\rho_2=-\rho/\alpha$, and using Lemma 9.1, computation of $(\rho_0+\rho_1+\rho_2)^{[k]}$ can be obtained as

$$(\rho_0+\rho_1+\rho_2)^{[1]} = \rho_0+\rho_1+\rho_2 = 0$$

$$(\rho_0+\rho_1+\rho_2)^{[2]} = \rho_0(\rho_0+\rho_1+\rho_2)^{[1]} + \rho_1(\rho_1+\rho_2)^{[1]}$$

$$+\rho_2^2 = (\rho_1+\rho_2)^2 - \rho_1\rho_2 = \frac{\alpha^2+\alpha+1}{\alpha^2}\rho^2$$

$$(\rho_0+\rho_1+\rho_2)^{[3]} = \rho_0\rho_1\rho_2 = \frac{\alpha+1}{\alpha^2}\rho^3$$

$$\begin{aligned}(\rho_0+\rho_1+\rho_2)^{[4]} &= \rho_0\rho_0\rho_1\rho_2 + \rho_1^4 + \rho_1^3\rho_2 + \rho_1^2\rho_2^2 + \rho_1\rho_2^3 + \rho_2^4 \\ &= \frac{\alpha^4+2\alpha^3+3\alpha^2+2\alpha+1}{\alpha^4}\rho^4\end{aligned}$$

......

Therefore, (9.6a,b) can be obtained. Following a similar process, (9.6c,d) can be derived.

Chapter 10
Using Nonlinearity for Output Vibration Suppression: An Application Study

10.1 Introduction

Suppression of periodic disturbances covers a wide range of engineering practices, involved in active control and isolation of vibrations in mechanical, vehicle and aerospace systems. Traditionally, an increase in damping can reduce the response at the resonance. However, this is often at the expense of degradation of isolation at high frequencies (Graham and McRuer 1961). Many methods have been proposed to deal with this problem, such as optimal control, H-infinity control, "skyhook" damper, repetitive learning control, and optimization etc (Graham and McRuer 1961; Housner et al. 1997; Karnopp 1995; Lee and Smith 2000). A much more comprehensive and up-to-date survey can refer to Hrovat (1997). Nonlinear feedback is an approach that has been noted recently by some researchers (Alleyne and Hedrick 1995; Chantranuwathanal and Peng 1999; Zhu et al. 2001). It is shown in Lee and Smith (2000) that, although it is not possible to use linear time-invariant controllers to robustly stabilize a controlled plant and to achieve asymptotic rejection of a periodic disturbance, the problem is solvable by using a nonlinear controller for a linear plant subjected to a triangular wave disturbance. Based on the Hamiltonian system theory, an optimal nonlinear feedback control strategy is proposed in Zhu et al. (2001) for randomly excited structural systems. It has also been reported many times that existing nonlinearities or deliberately introduced nonlinearities may bring benefits to control system design (Graham and McRuer 1961). Hence, the design of a nonlinear feedback controller to suppress periodic disturbances has great potential to achieve a considerably improved control performance. However, it should be noted that most of these existing methods mentioned above are based on state space and in the time domain, and some of those usually involve a complicated design procedure.

Based on the results discussed previous Chapters, the OFRF (output frequency response function) for nonlinear systems can be obtained explicitly, which reveals

X. Jing, Z. Lang, *Frequency Domain Analysis and Design of Nonlinear Systems based on Volterra Series Expansion*, Understanding Complex Systems,
DOI 10.1007/978-3-319-12391-2_10

an analytical relationship between system output spectrum and system model parameters for a wide class of nonlinear systems and provides an important basis for the analysis and design of output response behaviour of nonlinear systems in the frequency domain. For a linear controlled plant subject to periodic disturbances, if a nonlinear feedback is introduced to produce a nonlinear closed loop system, the relationship between the disturbance and the system output is nonlinear and can, under certain conditions, be described in the frequency domain by using the OFRF to explicitly relate the controller parameters to the system output frequency response. Therefore, by properly designing the controller parameters based on this explicit relationship, the effect of the periodic disturbance on the system output frequency response could be significantly suppressed. Motivated by this idea, a frequency domain approach to the analysis and design of nonlinear feedback for the exploitation of the potential advantage of nonlinearities is proposed in this study to suppress sinusoidal exogenous disturbances for a linear controlled plant.

This chapter is organized as follows. The problem formulation is given in Sect. 10.2, which is divided into several basic problems that can be addressed separately. Section 10.3 is concerned with some fundamental issues of the analysis and design of nonlinear feedback corresponding to different basic problems. Some theoretical results and techniques needed to solve these basic problems are established. Section 10.4 illustrates the implementation of the new approach by tackling a simple vibration system. Some proofs for the theoretical results are provided in Sect. 10.6 and a conclusion is given in Sect. 10.5.

10.2 Problem Formulation

Consider an SISO linear system described by the following differential equation:

$$\sum_{l=0}^{L} C_x(l) D^l x + b \cdot u + e \cdot \eta = 0 \tag{10.1}$$

$$y = \sum_{l=0}^{L-1} C_y(l) D^l x + d \cdot u \tag{10.2}$$

where, x, y, u, $\eta \in \mathbb{R}^1$ represent the system state, output, control input, and an exogenous disturbance input respectively; η stands for a known, external, bounded and periodical vibration, which can be described by a summation of multiple sinusoidal functions; L is a positive integer; D^l is an operator defined by $D^l x = d^l x / dt^l$. The model of system (10.1)–(10.2) can also be written in a state-space form:

$$\dot{\mathbf{X}} = \mathbf{AX} + \mathbf{B}u + \mathbf{E}\eta \tag{10.3}$$

$$y = \mathbf{CX} + du \tag{10.4}$$

where, $\mathbf{X}=[x, D^1x, \ldots, D^{L-1}x]^{\mathrm{T}} \in \Re^L$ is the system state variable, **A** and **C** are matrixes with appropriate dimensions, $\mathbf{B}=[0_{1\times(L-1)}, \mathrm{b}]^{\mathrm{T}}$, $\mathbf{E}=[0_{1\times(L-1)}, \mathrm{e}]^{\mathrm{T}}$. The problem to be addressed as case study of the theory and methods discussed in previous chapters is:

Given a frequency interval $I(\omega)$ and a desired magnitude level of the output frequency response Y^* over this frequency interval, find a nonlinear feedback control law

$$u = -\varphi\left(x, D^1x, \ldots, D^{L-1}x\right) \tag{10.5}$$

such that

$$\max_{\omega \in I(\omega)} \left(Y(j\omega)Y(-j\omega)\right) \le Y^* \tag{10.6a}$$

where the feedback control law $-\varphi(x, D^1x, \ldots, D^{L-1}x)$ is generally a nonlinear function of $x, D^1x, \ldots, D^{L-1}x$, with the linear state/output feedback as a special case; $Y(j\omega)$ is the spectrum of the system output.

For the purpose of implementation, the control objective (10.6a) is transformed to be

$$\max_{\substack{\omega_k \in I(\omega) \\ k=1,2,\ldots,\bar{k}}} \left(Y(j\omega_k)Y(-j\omega_k)\right) \le Y^* \tag{10.6b}$$

That is, evaluate the output spectrum at a series frequency point such that the maximum value is suppressed to a desired level. The control law (10.5) should therefore achieve the control objective defined by (10.6b). In the following, assume $I(\omega)=\omega_0$, that is only the output response at a specific frequency is considered. Let $Y = Y(j\omega)Y(-j\omega)|_{(\omega_0,u)}$, then $Y_0 = Y(j\omega)Y(-j\omega)|_{(\omega_0,0)}$ shows the magnitude of the system output frequency response at frequency ω_0 under zero control input. Obviously,

$$Y(j\omega)Y(-j\omega)|_{(\omega_0,u)} \le Y^* < Y_0 = Y(j\omega)Y(-j\omega)|_{(\omega_0,0)} \tag{10.7}$$

To obtain a nonlinear feedback controller, $\varphi(x, D^1x, \ldots, D^{L-1}x)$ is written in a polynomial form in terms of $x, D^1x, \ldots, D^{L-1}x$ as

$$\varphi\left(x, D^1x, \ldots, D^{L-1}x\right) = \sum_{p=1}^{M} \sum_{l_1 \cdots l_p=0}^{L-1} C_{p0}\left(l_1, \cdots, l_p\right) \prod_{i=1}^{p} D^{l_i}x \tag{10.8}$$

where M is a positive integer representing the maximum degree of nonlinearity in terms of $\mathrm{D}^{\mathrm{i}}x(t)$ $(i=0\ldots L-1)$; $\sum_{l_1 \cdots l_p=0}^{L-1}(\cdot) = \sum_{l_1=0}^{L-1} \cdots \sum_{l_p=0}^{L-1}(\cdot)$. The nonlinear function in

(10.8) includes a general class of possible linear and nonlinear functions of $D^i x$ ($i=0\ldots L-1$). Since $D^i x=e(i+1)^{\mathrm{T}}\mathbf{X}$, where $e(i+1)$ is an L-dimensional column vector whose $(i+1)$th element is 1 with all other terms zero, $\varphi(x, D^1x, \ldots, D^{L-1}x)$ can also be written as a function of $\mathbf{X}$, *i.e.*, $\varphi(\mathbf{X})$. As mentioned before, for the parameters $C_{p0}(.)$ ($p=1,\ldots,M$), when $p=1$ the parameters will be referred to as the linear parameters corresponding to the linear terms in (10.8), *e.g.*, $C_{1,0}(2)\frac{d^2x(t)}{dt^2}$. All other parameters in (10.8) will be referred to as nonlinear parameters corresponding to the nonlinear terms $\prod_{i=1}^{p} D^{l_i}x(t)$. p is the nonlinear degree of nonlinear parameter $C_{p0}(\cdot)$. Let

$$C(M,L) = \left(C_{p0}(l_1,\cdots,l_p) \middle| \begin{array}{l} p = 1\cdots M \\ l_i = 0\cdots L-1 \\ i = 1\cdots p \end{array} \right) \tag{10.9}$$

which includes all the parameters in (10.8). Substituting (10.8) into (10.1) and (10.2) yields the closed loop system as

$$\sum_{p=1}^{M}\sum_{l_1\cdots l_p=0}^{L} \overline{C}_{p0}(l_1,\cdots,l_p)\prod_{i=1}^{p} D^{l_i}x + e\cdot\eta = 0 \tag{10.10a}$$

$$\sum_{p=1}^{M}\sum_{l_1\cdots l_p=0}^{L-1} \tilde{C}_{p0}(l_1,\cdots,l_p)\prod_{i=1}^{p} D^{l_i}x = y \tag{10.10b}$$

where,

$$\overline{C}_{10}(l_1) = C_x(l_1) - bC_{10}(l_1),\ \tilde{C}_{10}(l_1) = C_y(l_1) - dC_{10}(l_1)$$
$$\overline{C}_{p0}(l_1,\cdots,l_p) = -bC_{p0}(l_1,\cdots,l_p),\ \tilde{C}_{p0}(l_1,\cdots,l_p) = -dC_{p0}(l_1,\cdots,l_p),$$

for $p=2\cdots M$, $l_i=0\cdots L$, and $i=1\cdots p$. Equation (10.10a,b) is a nonlinear differential equation model, whose generalized frequency response function can be obtained by using the results in Chap. 2. According to the results in Chen and Billings (1989), the model can represent a wide class of nonlinear systems. This implies that the nonlinear control law (10.8) can be used for many control purposes of interests. The task for the nonlinear feedback controller design is to determine M and a range for the controller parameters in (10.9) to make the closed loop system (10.10a,b) bounded stable around its zero equilibrium, and then to determine the specific values for the controller parameters from the OFRF which defines the relationship between the closed loop system output spectrum and controller parameters to achieve the required steady state performance (10.7).

There are generally four fundamental issues to be addressed for the nonlinear feedback design problem as follows:

(a) Determination of the analytical relationship between the system output spectrum and the nonlinear controller parameters.
(b) Determination of an appropriate structure for the nonlinear feedback controller. Only nonlinear terms which are useful for the control purpose are needed in the controller to achieve the design objective.
(c) Derivation of a range for the values of the control parameters over which the stability of the closed loop nonlinear system is guaranteed.
(d) Development of an effective numerical method for the practical implementation of the feedback controller design.

Section 10.3 is to investigate these fundamental issues. A simulation study will be presented thereafter to illustrate these results.

10.3 Fundamental Results for the Analysis and Design of the Nonlinear Feedback Control

10.3.1 Output Frequency Response Function

In this section, the output frequency response of the closed loop nonlinear system (10.10a,b) is derived. The relationship between the system output spectrum and the controller parameters are investigated to produce some useful results for the nonlinear feedback analysis and design.

A. Output Spectrum of the Closed Loop System

As discussed before, any time invariant, causal, nonlinear system with fading memory can be approximated by a finite Volterra series. With the BIBO stability condition for the controller parameters which will be studied in Sect. 10.3.3, the relationship between the output $y(t)$ and the input $\eta(t)$ of system (10.10a,b) can be approximated by a Volterra functional series up to a finite order N as described by (2.1), i.e.,

$$y(t)=\sum_{n=1}^{N}y_n(t),\ \ y_n=\int_{-\infty}^{\infty}\cdots\int_{-\infty}^{\infty}h_n(\tau_1,\cdots,\tau_n)\prod_{i=1}^{n}\eta(t-\tau_i)d\tau_i \tag{10.11}$$

where $h_n(\tau_1,\cdots,\tau_n)$ is the nth order Volterra kernel of system (10.10a,b) corresponding to the input–output relationship from $\eta(t)$ to $y(t)$. When the input in (10.11) is a multi-tone function in (3.2), i.e.,

$$\eta(t)=\sum_{i=1}^{K}|F_i|\cos(\omega_i t+\angle F_i) \tag{10.12}$$

the system output spectrum can be obtained as given in (3.3), i.e.,

$$Y(j\omega) = \sum_{n=1}^{N} \frac{1}{2^n} \sum_{\omega_{k_1}+\cdots+\omega_{k_n}=\omega} H_n(j\omega_{k_1}, \cdots, j\omega_{k_n}) F(\omega_{k_1}) \cdots F(\omega_{k_n}) \tag{10.13}$$

where,

$$F(\omega) = \begin{cases} |F_i| e^{j\angle F_i} & \text{if } \omega \in \{\omega_k, k = \pm 1, \cdots, \pm K\} \\ 0 & \text{else} \end{cases} \tag{10.14}$$

$$H_n(j\omega_{k_1}, \cdots, j\omega_{k_n}) = \int_{-\infty}^{\infty} \cdots \int_{-\infty}^{\infty} h_n(\tau_1, \cdots, \tau_n) e^{-j(\omega_1\tau_1+\cdots+\omega_n\tau_n)} d\tau_1 \cdots d\tau_n \tag{10.15}$$

Equation (10.15) is the nth-order generalised frequency response function (GFRF) of system (10.10a,b) for the relationship between $\eta(t)$ and y(t), which can be obtained by directly following the results in Sect. 2.3.

Proposition 10.1 The GFRFs $H_n(j\omega_{k_1}, \cdots, j\omega_{k_n})$ from the disturbance $\eta(t)$ to the output y(t) of nonlinear system (10.10a,b) can be determined as

$$H_n(j\omega_1, \cdots, j\omega_n) = \sum_{p=1}^{n} \sum_{l_1 \cdots l_p=0}^{L-1} \widetilde{C}_{p0}(l_1 \cdots l_p) H_{np}^1(j\omega_1, \cdots, j\omega_n) \tag{10.16a}$$

$$H_{np}^1(j\omega_1, \cdots, j\omega_n) = \sum_{i=1}^{n-p+1} H_i^1(j\omega_1, \cdots, j\omega_i) H_{n-i,p-1}^1(j\omega_{i+1}, \cdots, j\omega_n)(j\omega_1 + \cdots + j\omega_i)^{l_p} \tag{10.16b}$$

$$\begin{aligned} H_{n1}^1(j\omega_1, \cdots, j\omega_n) &= H_n^1(j\omega_1, \cdots, j\omega_n)(j\omega_1 + \cdots + j\omega_n)^{l_1}, \quad H_1^1(j\omega_1) \\ &= e \Big/ \sum_{l_1=0}^{L} \overline{C}_{10}(l_1)(j\omega_1)^{l_1} \end{aligned} \tag{10.16c}$$

$$H_n^1(j\omega_1, \cdots, j\omega_n) = -\frac{1}{e} H_1^1(j\omega_1 + \cdots + j\omega_n) \left(\sum_{p=2}^{n} \sum_{l_1 \cdots l_p=0}^{L-1} \overline{C}_{p0}(l_1 \cdots l_p) H_{np}^1(j\omega_1, \cdots, j\omega_n) - e\delta(n-1) \right) \tag{10.16d}$$

and $\delta(n) = \begin{cases} 1 & n = 0 \\ 0 & \text{otherwise} \end{cases}$. □

From Proposition 10.1, the GFRFs can be computed recursively from the time domain model (10.10a,b), and the output spectrum of system (10.10a,b) can be obtained analytically from (10.13) and (10.16a–d), which are an explicit function of the parameters in the control law (10.8). Therefore, the design of controller (10.8) can be studied in the frequency domain. In order to obtain an analytical relationship between the system output spectrum and model parameters from these recursive computations the OFRF of system (10.10a,b) is expressed as a polynomial function of the nonlinear controller parameters in (10.9) according to Chaps. 6 and 7, *i.e.*,

$$Y(j\omega) = P_0(j\omega) + a_1 P_1(j\omega) + a_2 P_2(j\omega) + \cdots \tag{10.17a}$$

where $P_0(j\omega)$ is the linear part of the system output frequency response, $P_i(j\omega)$ $(i \geq 1)$ represents the effects of higher order nonlinearities, and a_i $(i=1,2,\cdots)$ are functions of the nonlinear controller parameters which can be determined by following Chaps. 5 and 6. Moreover, for a nonlinear controller parameter c in (10.9), there exists a series of functions of frequency ω $\{\overline{P}_i(j\omega),\ \ \mathrm{i} = 0, 1, 2, 3, \ldots\}$ such that

$$Y(j\omega) = \overline{P}_0(j\omega) + c\overline{P}_1(j\omega) + c^2\overline{P}_2(j\omega) + \cdots \tag{10.17b}$$

Equation (10.17b) explicitly shows the relationship between the system output spectrum and the nonlinear controller parameters, and therefore enables the OFRF to be determined by using a simple numerical method which will be discussed in Sect. 10.3.4. Obviously, this considerably facilitates the analysis and design of the nonlinear feedback controller in the frequency domain. In order to reveal the contribution of the nonlinear controller parameters of different degrees to the output spectrum more clearly and thus shed light on the issue of the structure determination for the control law (10.8), some useful results regarding the parametric characteristic of the OFRF are discussed in the following section.

B. Parametric Characteristic Analysis of the Output Spectrum

The parametric characteristic analysis of the system output spectrum is to investigate the polynomial structure of OFRF (10.17a), and to reveal how the frequency response functions in (10.13) and (10.16a–d) depend on the nonlinear controller parameters (*i.e.*, $C_{p0}(.)$ for $p > 1$) in (10.9).

Following the results in Sect. 6.4, the parametric characteristics of the GFRF $H_n^1(j\omega_1,\cdots,j\omega_n)$ from $u(t)$ to $y(t)$ can be obtained as for n > 1

$$\begin{aligned}
\mathrm{CE}\left(H_n^1(j\omega_1,\cdots,j\omega_n)\right) &= \mathop{\oplus}_{p=2}^{n}\left(C_{p,0} \otimes \mathrm{CE}\left(H_{n,p}^1(j\omega_1,\cdots,j\omega_n)\right)\right) \\
&= \mathop{\oplus}_{p=2}^{n}\left(C_{p,0} \otimes \mathrm{CE}\left(H_{n-p+1}^1(j\omega_1,\cdots,j\omega_n)\right)\right) \\
&= C_{n0} \oplus \mathop{\oplus}_{p=2}^{\left[{}^{n}+1/_{2}\right]}\left(C_{p0} \otimes \mathrm{CE}\left(H_{n-p+1}^1(\cdot)\right)\right)
\end{aligned} \tag{10.18}$$

For n = 1, $\mathrm{CE}(H_1^1(j\omega_1)) = 1$. Here, $[n/2]$ means to take the integer part of [.]. From the invariant property of the CE operator, it follows for the nonlinear controller parameters in (10.9) that

$$\mathrm{CE}\left(\overline{C}_{p0}(l_1,\cdots,l_p)\right) = C_{p0}(l_1,\cdots,l_{p+q}),\ \ \mathrm{CE}\left(\widetilde{C}_{p0}(l_1,\cdots,l_p)\right) = C_{p0}(l_1,\cdots,l_p)$$

Applying CE operator to (10.16a) for the nonlinear parameters in (10.9),

$$\begin{aligned}
&\mathrm{CE}(H_n(j\omega_1,\cdots,j\omega_n)) = \mathrm{CE}\left(\sum_{l_1=0}^{L}\widetilde{C}_{1,0}(l_1)H^1_{n,1}(j\omega_1,\cdots,j\omega_n) + \sum_{p=2}^{n}\sum_{l_1\cdots l_p=0}^{L}\widetilde{C}_{p0}(l_1\cdots l_p)H^1_{np}(j\omega_1,\cdots,j\omega_n)\right)\\
&= \mathrm{CE}\left(\sum_{l_1=0}^{L}(C_y(l_1) - C_{10}(l_1))H^1_{n1}(j\omega_1,\cdots,j\omega_n) + \sum_{p=2}^{n}\sum_{l_1\cdots l_p=0}^{L}(-d)C_{p0}(l_1\cdots l_p)H^1_{np}(j\omega_1,\cdots,j\omega_n)\right)\\
&= \begin{cases} 1 & n=1\\ \overset{n}{\underset{p=2}{\oplus}}\left(C_{p0}\otimes CE\left(H^1_{np}(j\omega_1,\cdots,j\omega_n)\right)\right) & n>1 \end{cases}
\end{aligned} \tag{10.19}$$

Therefore, with respect to the nonlinear parameters in (10.9), the parametric characteristics of the GFRFs $H_n(j\omega_1,\cdots,j\omega_n)$ from $\eta(t)$ to y(t) is the same as those of the GFRFs $H^1_n(j\omega_1,\cdots,j\omega_n)$ from u(t) to y(t), i.e.,

$$\mathrm{CE}\left(H^2_n(\cdot)\right) = \mathrm{CE}\left(H^1_n(\cdot)\right) \quad \text{for n} > 0 \tag{10.20}$$

That is, the effect of the nonlinear parameters in (10.9) on the GFRFs $H_n(j\omega_1,\cdots,j\omega_n)$ is the same as that on the GFRFs $H^1_n(j\omega_1,\cdots,j\omega_n)$. Equations (10.18)–(10.20) reveal how the GFRFs depend on the nonlinear controller parameters in (10.9). Based on these results, the parametric characteristic of the OFRF can be obtained as

$$\begin{aligned}
&\mathrm{CE}(Y(j\omega)) = \mathrm{CE}\left(\sum_{n=1}^{N}\frac{1}{2^n}\sum_{\omega_{k_1}+\cdots+\omega_{k_n}=\omega} H^2_n(j\omega_{k_1},\cdots,j\omega_{k_n})F(\omega_{k_1})\cdots F(\omega_{k_n})\right)\\
&= \mathrm{CE}\left(\sum_{n=1}^{N}\sum_{\omega_{k_1}+\cdots+\omega_{k_n}=\omega} H^2_n(j\omega_{k_1},\cdots,j\omega_{k_n})\right) = \mathrm{CE}\left(\sum_{n=1}^{N}H^2_n(j\omega_{k_1},\cdots,j\omega_{k_n})\right)\\
&= \mathrm{CE}(H^2_1(\cdot))\oplus\mathrm{CE}(H^2_2(\cdot))\oplus\cdots\oplus\mathrm{CE}(H^2_N(\cdot)) = \mathrm{CE}(H^1_1(\cdot))\oplus\mathrm{CE}(H^1_2(\cdot))\oplus\cdots\oplus\mathrm{CE}(H^1_N(\cdot))
\end{aligned} \tag{10.21a}$$

Therefore, according to the results in Chap. 6, there exist a complex valued function vector $\widetilde{F}_n(j\omega)$ with appropriate dimension such that

$$Y(j\omega) = \left(\overset{N}{\underset{n=1}{\oplus}}\mathrm{CE}\left(H^1_n(j\omega_1,\cdots,j\omega_n)\right)\right)\cdot\widetilde{F}_n(j\omega) \tag{10.21b}$$

This is the detailed polynomial function of (10.17a). Equation (10.21b) provides an analytical and straightforward expression for the relationship between system output spectrum and the controller parameters. Now the coefficients of the polynomial function (10.17a) can be determined as

$$\begin{aligned}[a_1 \ \ a_2 \ \ a_3 \ \ \cdots \ \ a_K] &= \mathrm{CE}(Y(j\omega)) \\ &= \mathrm{CE}\big(H_1^1(\cdot)\big) \oplus \mathrm{CE}\big(H_2^1(\cdot)\big) \oplus \cdots \oplus \mathrm{CE}\big(H_N^1(\cdot)\big) \end{aligned} \quad (10.21c)$$

where K is the dimension of the vector $\mathrm{CE}(H_1^1(\cdot)) \oplus \mathrm{CE}(H_2^1(\cdot)) \oplus \cdots \oplus \mathrm{CE}(H_N^1(\cdot))$.

In order to better understand these parametric characteristics, the following results are given, which is a special case of Proposition 5.1.

Proposition 10.2 The elements in $\mathrm{CE}(H_n^1(j\omega_1,\cdots,j\omega_n))$ include and only include all the parameter monomials (consisting of the nonlinear parameters in (10.9)) in $\mathbf{C}_{p0} \otimes \mathbf{C}_{r_1 0} \otimes \mathbf{C}_{r_2 0} \otimes \cdots \otimes \mathbf{C}_{r_k 0}$ for $0 \le k \le n-2$, satisfying $p + \sum_{i=1}^{k} r_i = n + k$, $2 \le r_i \le n-1$, and $2 \le p \le n$. □

Proposition 10.2 shows whether and how a nonlinear parameter in (10.9) is included in $\mathrm{CE}(H_n^1(j\omega_1,\cdots,j\omega_n))$. Different parameters may form one monomial acting as an element in $\mathrm{CE}(H_n^1(j\omega_1,\cdots,j\omega_n))$, and thus have a coupled effect on $H_n^1(j\omega_1,\cdots,j\omega_n)$. If a nonlinear parameter appears in $\mathrm{CE}(H_n^1(j\omega_1,\cdots,j\omega_n))$, this implies that it has an effect on $H_n^1(j\omega_1,\cdots,j\omega_n)$ and thus on $Y(j\omega)$. If this nonlinear parameter is an independent element in $\mathrm{CE}(H_n^1(j\omega_1,\cdots,j\omega_n))$, then it has an independent effect on $Y(j\omega)$. Furthermore, if a parameter frequently appears in $\mathrm{CE}(H_n^1(j\omega_1,\cdots,j\omega_n))$ with different monomial degrees, this may implies that this parameter has more strong effect on $H_n^1(j\omega_1,\cdots,j\omega_n)$ and thus $Y(j\omega)$. For this reason, the parametric characteristic analysis of $H_n^1(j\omega_1,\cdots,j\omega_n)$ can shed light on the effect of different nonlinear parameters on $H_n^1(j\omega_1,\cdots,j\omega_n)$ and thus $Y(j\omega)$.

From Proposition 10.2 (also referring to Property 5.3 for the general case), the term $(\mathbf{C}_{n0})^i$ should be included in the GFRF $H_m(.)$, where m is computed as $m+k=m+i-1=ni$. Hence, $m=ni-i+1=1+(n-1)i$. It can be seen that, when n is smaller, $\mathbf{C}_{n,0}$ will contribute independently to more GFRFs whose orders are $(n-1)i+1$ for $i=1,2,3,\ldots$; and if n is larger, $\mathbf{C}_{n,0}$ can only affect the GFRFs of orders higher than n. It is known that for a Volterra system, the system nonlinear dynamics could be dominated by low order GFRFs (Boyd and Chua 1985). This implies that the nonlinear terms with coefficient $\mathbf{C}_{\mathrm{n},0}$ of smaller nonlinear degree, e.g., 2 and 3, may play greater roles than other pure output nonlinear terms. This property is significant for the selection of possible nonlinear terms in the feedback design. Moreover, it can be verified from Proposition 10.2 that, If the second and third degree nonlinear control parameters are all zero, *i.e.*, $\mathbf{C}_{20}=0$ and $\mathbf{C}_{30}=0$, then $\mathrm{H}_2(.)=0$, and $\mathrm{H}_3(.)=0$. However, even if $\mathbf{C}_{n0}=0$ (for $n>3$), the nth order GFRF $H_n(.)$ is not zero, providing there are nonzero terms in $\mathbf{C}_{20}$ or $\mathbf{C}_{30}$. This further demonstrates that the nonlinear controller parameters in $\mathbf{C}_{20}$ and $\mathbf{C}_{30}$ have a more important role in the determination of the GFRFs than other nonlinear parameters, and thus has a more important effect on the output spectrum. These imply that a lower degree nonlinear feedback may be sufficient for some control problems. These provide a guidance for the selection of the candidate terms in (10.9).

10.3.2 The Structure of the Nonlinear Feedback Controller

The determination of the structure for the nonlinear feedback controller (10.8) is an important task to be tackled. Firstly, as discussed in Sect. 10.3.1(B), the structure parameter M in (10.8) should be chosen as small as possible since lower degree of nonlinear terms have greater contributions to the output spectrum. It can be increased gradually until the control objective is achieved. Secondly, after M is determined, whether a term in $\mathbf{C}_{p0}$ is effective or not should be checked. An effective controller must satisfy the inequality (10.7). Thus for the effectiveness of a specific nonlinear controller parameter c, this requirement can be written as

$$\frac{\partial |Y(j\omega_0)|}{\partial c} < 0 \text{ for some c} \tag{10.22}$$

Consider the specific nonlinear controller parameter c in $\mathbf{C}_{p0}$ and let all the other nonlinear controller parameters be zero or assumed to be a constant. Then only the nonlinear coefficient c^i appears in CE($H^1_{1+(p-1)i}(.)$) according to Proposition 10.2. Therefore, only the GFRFs for the orders $1+(p-1)i$ (for i=1,2,3,...) need to be computed to obtain the system output spectrum in (10.13). According to (10.21a–c), the output spectrum can be written as

$$Y(j\omega; c) = \overline{P}_0(j\omega) + c\overline{P}_1(j\omega) + c^2\overline{P}_2(j\omega) + \cdots \tag{10.23}$$

It can easily be shown that if $\mathrm{Re}\big(\overline{P}_0(j\omega)\cdot\overline{P}_1(-j\omega)\big) < 0$ then there must exist $\varepsilon >$ 0 such that $\frac{\partial |Y(j\omega)|}{\partial c} < 0$ for $0 < c < \varepsilon$ or $-\varepsilon < c < 0$, where Re($\cdot$) is to take the real part of (.). This can be used to find the nonlinear terms which are effective. Only the effective nonlinear terms in $C(M)$ is considered. By this way, the structure of the nonlinear function (10.8) can be determined. It should be noted that, in this process the output spectrum needs to be analytically computed up to at most the third order by using (10.12)–(10.16a–d). The structure of the control law (10.8) can also be determined by simply including all the possible nonlinear terms of degree up to M. Once the output spectrum is determined by the numerical method in Sect. 10.3.4, the values of the coefficients of these nonlinear terms can be optimized for the control objective (10.7) in the stability region developed in the following section. If the objective (10.7) cannot be achieved after M is enough large, this may implies that the objective (10.7) cannot be achieved by the controller (10.8) and a better possible solution can be used for this case.

10.3.3 Stability of the Closed-Loop System

As mentioned above, the stability of a nonlinear system should be guaranteed such that the nonlinear system can be approximated by a locally convergent Volterra series. Therefore, a range for the nonlinear controller parameters which can ensure the stability of the closed loop system (10.10a,b) can be determined. For simplicity, (10.10a,b) can also be written in a state space form as

$$\dot{\mathbf{X}} = \mathbf{A}\mathbf{X} - \mathbf{B}\varphi(\mathbf{X}) + \mathbf{E}\eta := f(\mathbf{X}) + \mathbf{E}\eta \tag{10.24a}$$

$$y = \mathbf{C}\mathbf{X} - \mathbf{D}\varphi(\mathbf{X}) := h(\mathbf{X}) \tag{10.24b}$$

A, B, C, D, E are appropriate matrices which are the same as the matrices in (10.3) and (10.4). Note that the exogenous disturbance in (10.24a,b) is a periodic bounded signal, and the objective in a vibration control is often to suppress the output vibration below a desired level, a concept of asymptotic stability to a ball is adopted in this section. This concept implies that the magnitude of the output for a system is asymptotically controlled to a satisfactory predefined level. Based on this concept, a general result is then derived to ensure the stability of the closed loop nonlinear system (10.24a,b), which can be regarded as an application of some existing theories in Isidori (1999).

A Ball $B_\rho(\mathbf{X})$ is defined as: $B_\rho(\mathbf{X})=\{\mathbf{X}|\|\mathbf{X}\|\leq\rho,\ \rho>0\}$. A K-function $\gamma(s)$ is an increasing function of s, and a KL-function $\beta(s,t)$ is an increasing function of s, but a decreasing function of t. For detailed definitions of K/KL-functions can refer to Isidori (1999).

Asymptotic Stability to a Ball Given an initial state $\mathbf{X}_0\in\Re^n$ and disturbance input η of a nonlinear system, if there exists a KL-function β such that the solution $\mathbf{X}(t,\mathbf{X}_0,\eta)$ (for $t\geq 0$) of the system satisfies $\|\mathbf{X}(t,\mathbf{X}_0,\eta)\|\leq\beta(\|\mathbf{X}_0\|,t)+\rho$, $\forall t>0$, then the system is said to be asymptotically stable to a ball $B_\rho(\mathbf{X})$, where ρ is an upper bound function of η, *i.e.*, there exist a K-function γ such that $\rho=\gamma(\|\eta\|_\infty)$.

Assumption 10.1 There exists a K-function o such that the output function $h(\mathbf{X})$ of the nonlinear system (10.24a,b) satisfies $\|h(\mathbf{X})\|\leq o(\|\mathbf{X}\|)$.

Proposition 10.3 If assumption 10.1 holds, then the following statements are equivalent:

(a) There exist a smooth function $V:\Re^L\to\Re_{\geq 0}$ and K_∞-functions β_1,β_2 and K-functions α, γ such that

$$\beta_1(\|\mathbf{X}\|) \leq V(\mathbf{X}) \leq \beta_2(\|\mathbf{X}\|) \text{ and } \frac{\partial V(\mathbf{X})}{\partial \mathbf{X}}\{f(\mathbf{X}) + \mathbf{E}\eta\} \leq -\alpha(\|\mathbf{X}\|) + \gamma\left(\|\eta\|_\infty\right) \tag{10.25}$$

(b) System (10.24a,b) is asymptotically stable to the ball $B_\rho(\mathbf{X})$ with $\rho=\beta_1(2\cdot\beta_2^{-1}\cdot\alpha^{-1}\cdot\gamma(\|\eta\|_\infty))$, and the output of system (10.24a,b) is asymptotically stable to the ball $B_{o(2\rho)}(y)$. □

Proof See the proof in Sect. 10.6. □

Note that Proposition 10.3 can guarantee the asymptotical stability to a ball of system (10.24a,b) when subject to bounded disturbance, and asymptotical stability to zero when the disturbance tends to zero. This is just the property of fading memory which is required for the existence of a convergent Volterra series approximation for the system input–output relationship (Boyd and Chua 1985). Although it is not easy to derive a general stability condition for the general controller (10.5), there are always various methods (Ogota 1996) to choose a proper Lyapunov function based on Proposition 10.3 to derive a stability condition for a specific controller.

10.3.4 A Numerical Method for the Nonlinear Feedback Controller Design

The nonlinear controller parameters can be determined by solving (10.17b) to satisfy the performance (10.6b) or (10.7) under the stability condition. However, it can be seen that the analytical derivation of the output spectrum of system (10.10a,b) involves complicated symbolic computation for orders higher than 5. To circumvent this problem, as discussed in Sect. 10.3.1(A), the numerical method discussed in Chaps. 7 and 8 can be used since the detailed polynomial structure of the OFRF can be determined by using the method in Sect. 10.3.1, which is summarized as follows:

(1) The system output frequency response function can be expressed as $Y(j\omega)Y(-j\omega) = |Y(j\omega)|^2 = \mathbf{C}\cdot\widetilde{\mathbf{P}}(j\omega)$ according to (10.21a–c) with a finite polynomial degree, where $\widetilde{\mathbf{P}}(j\omega)$ is a complex valued function vector,

$$\mathbf{C} = [1 \;\; c_1 \;\; c_2 \;\; c_3 \;\; \cdots \;\; c_{K!}] = \left(\mathrm{CE}\left(H_1^1(\cdot)\right)\oplus \mathrm{CE}\left(H_2^1(\cdot)\right)\oplus\cdots\oplus \mathrm{CE}\left(H_N^1(\cdot)\right)\right)$$
$$\otimes\left(\mathrm{CE}\left(H_1^1(\cdot)\right)\oplus \mathrm{CE}\left(H_2^1(\cdot)\right)\oplus\cdots\oplus \mathrm{CE}\left(H_N^1(\cdot)\right)\right)$$

(2) Collect the system time domain steady output $y_i(t)$ under different values of the controller parameters $\mathbf{C}_i = [1\; c_{1i}, c_{2i}, \ldots c_{(K!)i}]$ for $i=1,2,3,\ldots N_i$;
(3) Evaluate the FFT for $y_i(t)$ to obtain $Y_i(j\omega)$, then obtain the magnitude $|Y_i(j\omega_0)|^2$ at frequency ω_0, and finally form a vector $\mathbf{YY} = \left[|Y_1(j\omega_0)|^2, \cdots, |Y_{N_i}(j\omega_0)|^2\right]^T$
(4) Obtain the following equation,

$$\begin{bmatrix} 1, & c_{11}, & c_{12}, \cdots, c_{1,K!} \\ 1, & c_{21}, & c_{22}, \cdots, c_{2,K!} \\ \cdots, \cdots, & \cdots, & \cdots, \quad \cdots \\ 1, c_{N_i 1}, c_{N_i 2}, \cdots, c_{N_i,K!} \end{bmatrix} \cdot \begin{bmatrix} \widetilde{P}_0 \\ \widetilde{P}_1 \\ \vdots \\ \widetilde{P}_{K!} \end{bmatrix} = \begin{bmatrix} |Y_1(j\omega_0)|^2 \\ |Y_2(j\omega_0)|^2 \\ \vdots \\ |Y_{N_i}(j\omega_0)|^2 \end{bmatrix} \quad i.e., \; \boldsymbol{\psi}_{\mathbf{C}} \cdot \widetilde{\mathbf{P}}(j\omega_0) = \mathbf{YY}$$

(5) Evaluate the function $\widetilde{P}(j\omega_0)$ by using Least Squares,

$$\widetilde{P}(j\omega_0) = \left(\psi_C{}^T \cdot \psi_C\right)^{-1} \cdot \psi_C{}^T \cdot YY$$

(6) Finally, the nonlinear controller parameters $\mathbf{C}^*$ for given Y^* at a specific frequency ω_0 can be determined according to

$$Y^* == \mathbf{C}^* \cdot \widetilde{\mathbf{P}}(j\omega_0)$$

The numerical method above is very effective for the implementation of the design of the proposed nonlinear controller parameters, which will be verified by a simulation study in Sect. 10.5.

Although there are some time domain methods which can address the nonlinear control problems based on Lyapunov stability theory such as the back-stepping technique and feedback linearization (Isidori 1999) *etc*, few results are available for the design and analysis of a nonlinear feedback controller in the frequency domain to achieve a desired frequency domain performance. Based on the analytical relationship between system output spectrum and controller parameters defined by the OFRF, the analysis and design of a nonlinear feedback controller can be conducted in the frequency domain. For a summary, a general procedure for this new method is given as follows.

(A) Derivation of the output spectrum for the closed loop system given M and L.
Given M and L in (10.8), the general output spectrum with respect to the control law (10.8) for the closed loop system (10.10a,b) can be obtained according to (10.13) and (10.16a–d). This will be used for the validation of the effectiveness of nonlinear terms in the next step. L is the maximum derivative order which is dependent of the system model, and M is the maximum nonlinearity order which can be given as 2 or 3 at this stage.

(B) Determination of the structure of the nonlinear feedback function in (10.8).
This is to determine the value of M and choose the effective nonlinear controller parameters $C_{p0}(.)$ (p=2,3,...,M). Based on the analysis of the parametric characteristics in Sect. 10.3.1B, the nonlinear controller parameters included in $\mathbf{C}_{20}$ and $\mathbf{C}_{30}$ take a dominant role in the determination of GFRFs and output spectrum. Hence, M can be chosen as 2 or 3 at the beginning, and increased later if needed. The effectiveness of each nonlinear parameter can be checked by $\Re\left(\overline{P}_0(j\omega) \cdot \overline{P}_1(-j\omega)\right) < 0$, where $\overline{P}_1(-j\omega)$ can be computed from Step(A) by letting the other nonlinear parameters to be zero and $\overline{P}_0(j\omega)$ is the linear part of the output spectrum in this case. If the parameter is not effective, it can be discarded.

(C) Derivation of the region for the nonlinear feedback parameters in $C_{p0}(.)$ for p=2,3,...,M.
This is to ensure the stability of the nonlinear closed loop system (10.10a,b), which can be conducted by applying Proposition 10.3 to derive a stability condition for the closed loop system in terms of the nonlinear controller parameters. Although how to develop a systematic method for this purpose for a general nonlinear system is still an open problem, this can be easily done for some special or simple cases.

(D) Determination of the OFRF by using the numerical method and the optimal values for the nonlinear parameters

This is to derive a detailed polynomial expression for the output spectrum according to (10.21a–c) for the maximum nonlinearity order M larger than 3, and use the numerical method provided above to determine the desired value for each nonlinear controller parameter within the stability region to achieve the control objective (10.6b) or (10.7).

10.4 Simulation Study

Consider a simple case of the model in (10.1) and (10.2), which can be written as

$$\begin{cases} M\ddot{x} = -Kx - a_1\dot{x} + (\eta + u) \\ y = Kx + a_1\dot{x} - u \end{cases}$$

This is the model of a vibration isolation system studied in Daley et al. (2006) (Fig. 10.1), where $y(t)$ is the force transmitted from the disturbance $\eta(t)$ to the ground, K and a_1 are the spring and a damping characteristic parameters respectively.

Following the procedure in Sect. 10.3, a nonlinear feedback active controller $u(t)$ is designed and analysed for the suppression of the force transmitted to the ground. It will be shown that a simple nonlinear feedback can bring much better improvement for the system performance, compared with a linear feedback control. According to the general procedure above, the output spectrum under control law (10.8) for the closed loop system should first symbolically be determined. But for this simple example, it can be left to the next step.

10.4.1 *Determination of the Structure of the Nonlinear Feedback Controller*

Considering the nonlinear feedback in (10.8), for this simple system, M is directly chosen to be 3, and all the other nonlinear controller parameters are chosen to be

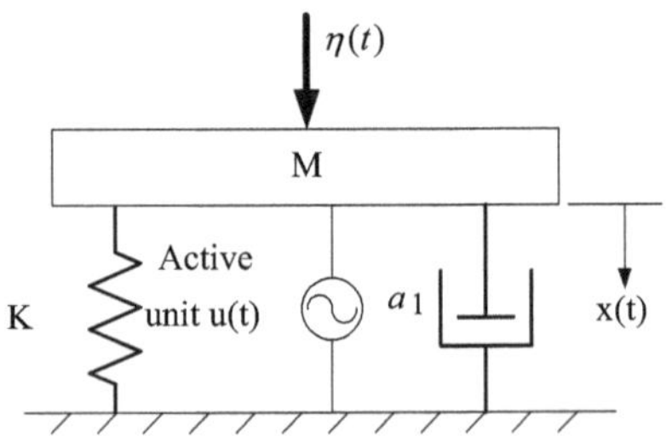

Fig. 10.1 A vibration isolation system

zero except $C_{30}(111)=a_3$ which represents a nonlinear damping and will be shown to be effective in the later analysis. If $C_{30}(111)=a_3$ is not effective, more other nonlinear terms can be chosen.

The nonlinear feedback control law now is

$$u = -a_3\dot{x}^3$$

and the closed loop system is therefore

$$\begin{cases} M\ddot{x} = -Kx - a_1\dot{x} - a_3\dot{x}^3 + \eta & \text{(a)} \\ y = Kx + a_1\dot{x} + a_3\dot{x}^3 & \text{(b)} \end{cases} \tag{10.26}$$

Note that system (10.26) is a very simple case of system (10.10a,b), that is, L=2, $\overline{C}_{10}(2) = M$, $\overline{C}_{10}(1) = a_1$, $\overline{C}_{10}(0) = K$, $\overline{C}_{30}(111) = a_3$, $\overline{C}_{01}(0) = -1$ and $\widetilde{C}_{10}(1) = a_1$, $\widetilde{C}_{10}(0) = K$, $\widetilde{C}_{30}(111) = a_3$; All other parameters in model (10.10a, b) are zero. Moreover, assume the disturbance input is $\eta(t)=F_d\sin(8.1t)$ (8.1 is the concerned working frequency of the system), which is a single tone function and a simple case of (10.12). Now the task for the nonlinear feedback controller design is to determine a_3 such that system (10.26) satisfies the control objective (10.7).

To verify the effectiveness of this nonlinear control, the output spectrum should be computed up to the third order as discussed in Step(B). Note that only $C_{30}(111)= a_3$ and other nonlinear parameters C_{p0} for $p>2$ are all zero. According to (10.18)–(10.20), the following parametric characteristics of the GFRFs can be obtained

$$\begin{aligned} \mathrm{CE}\left(H_2^1(\cdot)\right) &= \mathbf{C}_{20} \oplus \sum_{p=2}^{\left[{}^{2+1}/_{2}\right]} \mathbf{C}_{p0} \otimes \mathrm{CE}\left(H_{2-p+1}^1(\cdot)\right) = \mathbf{C}_{20} = 0, \ \ \mathrm{CE}\left(H_3^1(\cdot)\right) \\ &= \mathbf{C}_{30} \oplus \sum_{p=2}^{\left[{}^{3+1}/_{2}\right]} \mathbf{C}_{p0} \otimes \mathrm{CE}\left(H_{3-p+1}^1(\cdot)\right) = \mathbf{C}_{30} = a_3 \\ \mathrm{CE}\left(H_4^1(\cdot)\right) &= \mathbf{C}_{40} \oplus \sum_{p=2}^{\left[{}^{4+1}/_{2}\right]} \mathbf{C}_{p0} \otimes \mathrm{CE}\left(H_{4-p+1}^1(\cdot)\right) = 0, \ \ \mathrm{CE}\left(H_5^1(\cdot)\right) \\ &= \mathbf{C}_{50} \oplus \sum_{p=2}^{\left[{}^{5+1}/_{2}\right]} \mathbf{C}_{p0} \otimes \mathrm{CE}\left(H_{5-p+1}^1(\cdot)\right) = \mathbf{C}_{30} \otimes \mathrm{CE}\left(H_3^1(\cdot)\right) = a_3^2, \end{aligned}$$

It is easy to check from Propositions 10.2 that

$$CE\left(H_{2n+1}^1(\cdot)\right) = a_3^n \quad \text{for } n > 0 \text{ and all other } \mathrm{CE}\left(H_i^1(\cdot)\right) = 0 \tag{10.27}$$

This shows that only $H^1_{2n+1}(\cdot)$ for $n>0$ are nonzero and all others are zero. Therefore, the output spectrum can be computed from (10.13) and (10.16a–d) with only odd order GFRFs as

$$
\begin{aligned}
Y(j\omega) &= \sum_{n=1}^{N} \frac{1}{2^{2n+1}} \sum_{\omega_{k_1}+\cdots+\omega_{k_{2n+1}}=\omega} H^2_{2n+1}\left(j\omega_{k_1},\cdots,j\omega_{k_{2n+1}}\right)F\left(\omega_{k_1}\right)\cdots F\left(\omega_{k_{2n+1}}\right)\\
&= \frac{1}{2}H^2_1(j\omega)F(\omega) + \frac{a_3}{8}\sum_{\omega_{k_1}+\cdots+\omega_{k_3}=\omega} G^2_3(j\omega_{k_1},j\omega_{k_2},j\omega_{k_3})F(\omega_{k_1})F(\omega_{k_2})F(\omega_{k_3})\\
&+\frac{a_3^2}{32}\sum_{\omega_{k_1}+\cdots+\omega_{k_5}=\omega} G^2_5(j\omega_{k_1},\cdots,j\omega_{k_5})F(\omega_{k_1})\cdots F(\omega_{k_5}) + \cdots\\
&:= \overline{P}_0(j\omega) + a_3\overline{P}_1(j\omega) + a_3^2\overline{P}_2(j\omega) + \cdots
\end{aligned}
\tag{10.28a}
$$

where

$$
\begin{aligned}
&\overline{P}_0(j\omega) = \frac{1}{2}H^2_1(j\omega)F(\omega) = \frac{-j(a_1(j\omega)+K)F_d}{2M(j\omega)^2+2a_1(j\omega)^1+2K},\quad \overline{P}_1(j\omega) = -\frac{3}{8}MF_d^3\omega^5\left|H^1_1(j\omega)\right|^2\left[H^1_1(j\omega)\right]^2\\
&\overline{P}_2(j\omega) = -\frac{3j}{32}MF_d^5\left|j\omega H^1_1(j\omega)\right|^4\left[j\omega H^1_1(j\omega)\right]^2(j\omega)\cdot\left(j3\omega H^1_1(j3\omega) - j3\omega H^1_1(-j\omega) + j6\omega H^1_1(j\omega)\right),
\end{aligned}
\tag{10.28b}
$$

Note that carrying out the computation above, the analytical relationship between the output spectrum and nonlinear parameter a_3 can be obtained explicitly for up to any high orders. It can be checked that $\mathrm{Re}\left(\overline{P}_0(j\omega_0)\cdot\overline{P}_1(-j\omega_0)\right) = 0.5\left(\overline{P}_0(j\omega_0)\overline{P}_1(-j\omega_0) + \overline{P}_0(-j\omega_0)\overline{P}_1(j\omega_0)\right) = -31.132 < 0$ when $a_3>0, \omega_0=8.1$ rad/s and other system parameters as given in the simulation studies. Hence, the nonlinear control parameter a_3 is effective. If there are other nonlinear controller parameters, the same method can be used to check the effectiveness as discussed in Step(B). Only the effective nonlinear terms are used in the controller.

10.4.2 Derivation of the Stability Region for the Parameter $\mathbf{a_3}$

According to Proposition 10.3, the following result can be obtained.

Proposition 10.4 Consider the closed loop system (10.26), and assume the exogenous disturbance input satisfies $\|\eta(t)\| \le F_d$. The system is asymptotically stable to a ball $B_{F_d\sqrt{\lambda_{\min}(Q)^{-1}}\varepsilon}(\mathbf{X})$, if $a_3>0$ and additionally there exist $\mathbf{P}=\mathbf{P}^{\mathrm{T}}>0$, $\beta>0$ and $\varepsilon>0$ such that

$$\mathbf{Q}=\begin{bmatrix} -\mathbf{A}^{\mathrm{T}}\mathbf{P}-\mathbf{P}\mathbf{A}-\varepsilon^{-1}\mathbf{P}\mathbf{E}\mathbf{E}^{\mathrm{T}}\mathbf{P} & -\beta\mathbf{A}^{T}\mathbf{C}^{T}+\mathbf{P}\mathbf{B}-\beta\mathbf{P}\mathbf{E}\mathbf{E}^{T}\mathbf{C}^{T} \\ * & +2\beta\mathbf{C}\mathbf{B}-\varepsilon^{-1}\beta^{2}\mathbf{C}\mathbf{E}\mathbf{E}^{T}\mathbf{C}^{T} \end{bmatrix}>0$$

Moreover, the closed loop system (10.26) without a disturbance input is global asymptotically stable if the above inequality holds with $\mathbf{E}=0$. Here, $\mathbf{A}=\begin{bmatrix} 0 & 1 \\ -K/M & -a_1/M \end{bmatrix}$, $\mathbf{B}=[0,1/M]^{T}$, $\mathbf{C}=[0,1]$, $\mathbf{E}=[0,1/M]^{T}$.

Proof See the proof in Sect. 10.6. □

It is noted that the inequality in Proposition 10.4 has no relation with a_3 and is determined by the linear part of system (10.26) which can be checked by using the LMI technique by Boyd et al. (1994). This implies that the value of a_3 has no effect on the stability of the system if the inequality is satisfied. Hence, the nonlinear controller parameter a_3 is now only restricted to the region $[0, \infty)$, provided that the linear system satisfies the inequality condition.

10.4.3 Derivation of the OFRF and Determination of the Desired Value of the Nonlinear Parameter $\mathbf{a}_3$

By using (10.27), the parametric characteristics of the output spectrum of nonlinear system (10.26) can be obtained as

$$\begin{aligned} \mathrm{CE}(Y(j\omega)) &= \mathrm{CE}\big(H_1^1(\cdot)\big)\oplus\mathrm{CE}\big(H_2^1(\cdot)\big)\oplus\cdots\oplus\mathrm{CE}\big(H_N^1(\cdot)\big) \\ &= \begin{bmatrix} 1 & a_3 & a_3^2 & \cdots & a_3^Z \end{bmatrix} \end{aligned}$$

where $Z=\lfloor {}^{N-1}/_{2}\rfloor$. Therefore, the system output spectrum can be written as a polynomial expression as

$$Y(j\omega)=\overline{P}_0(j\omega)+a_3\overline{P}_1(j\omega)+a_3^2\overline{P}_2(j\omega)+\cdots+a_3^Z\overline{P}_Z(j\omega)$$

Hence,

$$\begin{aligned} Y(j\omega)Y(-j\omega)=|Y(j\omega)|^2 &= \left|\overline{P}_0(j\omega)\right|^2+a_3\big(\overline{P}_0(j\omega)\overline{P}_1(-j\omega)+\overline{P}_0(-j\omega)\overline{P}_1(j\omega)\big) \\ &\quad +a_3^2\Big(\left|\overline{P}_1(j\omega)\right|^2+\overline{P}_0(j\omega)\overline{P}_2(-j\omega)+\overline{P}_0(-j\omega)\overline{P}_2(j\omega)\Big)+\cdots \end{aligned} \tag{10.28c}$$

Clearly, $|Y(j\omega)|^2$ is also a polynomial function of a_3. Given the magnitude of a desired output frequency response Y^* at any frequency ω_0, a_3 can be solved from (10.28c) provided that $|Y(j\omega)|$ can be approximated by a polynomial expression of a finite order. In order to determine a desired value for a_3 to achieve the control objective (10.7), the numerical method proposed in Sect. 10.3.4 is used. Since

(10.28c) is a polynomial function of a_3, $|Y(j\omega)|^2$ can be directly approximated by a polynomial function of a_3 as follows:

$$Y(j\omega)Y(-j\omega) = |Y(j\omega)|^2$$
$$\approx a_3^{2Z}\widetilde{P}_{2Z} + \cdots a_3^n\widetilde{P}_n + a_3^{n-1}\widetilde{P}_{n-1} + \cdots + a_3\widetilde{P}_1 + \left|\overline{P}_0(j\omega)\right|^2 \quad (10.29a)$$

where $|Y(j\omega)|^2$ can be obtained via evaluating the FFT of the system output response from the system simulations or experimental data. Given 2Z different values of a_3, *i.e.*, a_{31}, a_{32}, …, $a_{3,2Z}$, (10.29a) can be further written as (for each values of a_3)

$$\left|Y(j\omega)_i\right|^2 \approx a_{3i}^{2Z}\widetilde{P}_{2Z} + \cdots a_{3i}^n\widetilde{P}_n + a_{3i}^{n-1}\widetilde{P}_{n-1} + \cdots + a_{3i}\widetilde{P}_1 + \left|\overline{P}_0(j\omega)\right|^2$$

for i=1,2,. . .,2Z, *i.e.*,

$$\begin{bmatrix} a_{31} & a_{31}^2 & a_{31}^3 & \cdots & a_{31}^{2Z} \\ a_{32} & a_{32}^2 & a_{32}^3 & \cdots & a_{32}^{2Z} \\ & & & \ddots & \vdots \\ a_{3,2Z} & a_{3,2Z}^2 & a_{3,2Z}^3 & \cdots & a_{3,2Z}^{2Z} \end{bmatrix} \cdot \begin{bmatrix} \widetilde{P}_1 \\ \widetilde{P}_2 \\ \vdots \\ \widetilde{P}_{2Z} \end{bmatrix} = \begin{bmatrix} \left|Y(j\omega)_1\right|^2 - \left|\overline{P}_0(j\omega)\right|^2 \\ \left|Y(j\omega)_2\right|^2 - \left|\overline{P}_0(j\omega)\right|^2 \\ \vdots \\ \left|Y(j\omega)_{2Z}\right|^2 - \left|\overline{P}_0(j\omega)\right|^2 \end{bmatrix}$$

Then $\widetilde{P}_1, \widetilde{P}_2, \cdots, \widetilde{P}_{2Z}$ are obtained as

$$\begin{bmatrix} \widetilde{P}_1 \\ \widetilde{P}_2 \\ \vdots \\ \widetilde{P}_{2Z} \end{bmatrix} = \begin{bmatrix} a_{31} & a_{31}^2 & a_{31}^3 & \cdots & a_{31}^{2Z} \\ a_{32} & a_{32}^2 & a_{32}^3 & \cdots & a_{32}^{2Z} \\ & & & \ddots & \vdots \\ a_{3,2Z} & a_{3,2Z}^2 & a_{3,2Z}^3 & \cdots & a_{3,2Z}^{2Z} \end{bmatrix}^{-1} \cdot \begin{bmatrix} \left|Y(j\omega)_1\right|^2 - \left|\overline{P}_0(j\omega)\right|^2 \\ \left|Y(j\omega)_2\right|^2 - \left|\overline{P}_0(j\omega)\right|^2 \\ \vdots \\ \left|Y(j\omega)_{2Z}\right|^2 - \left|\overline{P}_0(j\omega)\right|^2 \end{bmatrix} \quad (10.29b)$$

Consequently, (10.29a) is obtained. By using this method, a polynomial expression of $|Y(j\omega)|^2$ in any order can be achieved. Given a desired output frequency response Y^* at a frequency ω_0, a_3 can be solved from (10.29a) to implement the design. Note that roots of (10.29a) are multiple. According to Proposition 10.4, the solution a_3 should be a nonnegative real number.

10.4.4 Simulation Results

In the simulation study, the parameters of system (10.26) are: K=16,000 N/m, a_1=296 N S/m, M=240 kg. The resonant frequency of the system is ω_0=8.1 rad/s. In order to show the effectiveness and advantage of the nonlinear feedback controller $u=-a_3\dot{x}^3$, a linear controller $u=-a_2\dot{x}$ will be used for a comparison.

Table 10.1 Comparison between simulation and theoretical results (Jing et al. 2008a)

Simulation results from (10.29a,b)		Theoretical results from (10.28a–c)	
$\lvert\overline{P}_0(j\omega)\rvert^2$	1.1270e+05	$\lvert\overline{P}_0(j\omega)\rvert^2$	1.1257e+05
$\widetilde{P}_1$	−58.9652	$\overline{P}_0(j\omega)\overline{P}_1(-j\omega)$ $+\overline{P}_0(-j\omega)\overline{P}_1(j\omega)$	−62.2641
$\widetilde{P}_2$	0.0423	$\lvert\overline{P}_1(j\omega)\rvert^2+\overline{P}_0(j\omega)\overline{P}_2(-j\omega)$ $+\overline{P}_0(-j\omega)\overline{P}_2(j\omega)$	0.0615
$\widetilde{P}_3$	−2.3762e−005	—	—
$\widetilde{P}_4$	10.1382e−009	—	—
$\widetilde{P}_5$	−2.3593e−012	—	—
...	...	...	...

Firstly, let $F_d=100$ N. We need to obtain the polynomial function (10.29a). In order to have a larger working region of a_3, let $Z=6$ in (10.29a), and $a_3=500$, 1,000, 2,000, 4,000, 6,000, 8,000, 10,000, 12,000, 14,000, 16,000, 18,000, 20,000. Under these different values of a_3, the output frequency response of the system was obtained and the corresponding output spectrum was determined via FFT operations. Then $\widetilde{P}_n(j\omega)$ for n=1...12 were obtained according to (10.29b), which are summarized partly in Table 10.1. For comparisons, the corresponding theoretical results were also computed from (10.28a–c) and are given partly in Table 10.1. From Table 10.1, it can be seen that there is a good match between the numerical analysis results and the theoretical computations although there are some errors. This result shows that the theoretical computation results are basically consistent with the results from the simulation analyses. It can also be seen from the numerical analysis results in Table 10.1 that (10.29a) is in fact an alternative series in this case.

Figure 10.2 shows the results of the system output spectrum under different values of the nonlinear control parameter a_3 and provides a comparison between theoretical computations using polynomial expression (10.28c) up to the 3rd order and the numerical results using the polynomial expression (10.29a) up to the 12th order. This result demonstrates the analytical relationship between the nonlinear control parameter and the system output spectrum, and shows that the theoretical results have a good match with the numerical results when a_3 is small since only up to the third order GFRF are used in the theoretical computations. Hence, with an increase of a_3, the numerical method has to be used in order to give correct results. Moreover, it should be noted that the magnitude of the system output spectrum decreases with the increase of a_3. This verifies that the nonlinear control parameter a_3 is effective for the control problem.

Without a control input, the system output frequency spectrum is as shown in Fig. 10.3b, where $Y(j\omega)|_{\omega_0} = 335.71$. Note that the output response spectrum shown in the figures is 2|Y| not |Y|, which is also applied on the plot of the output spectrum using the theoretical computation. This is because 2|Y| represents the physical magnitude of the system output at the frequency ω_0. If the desired output

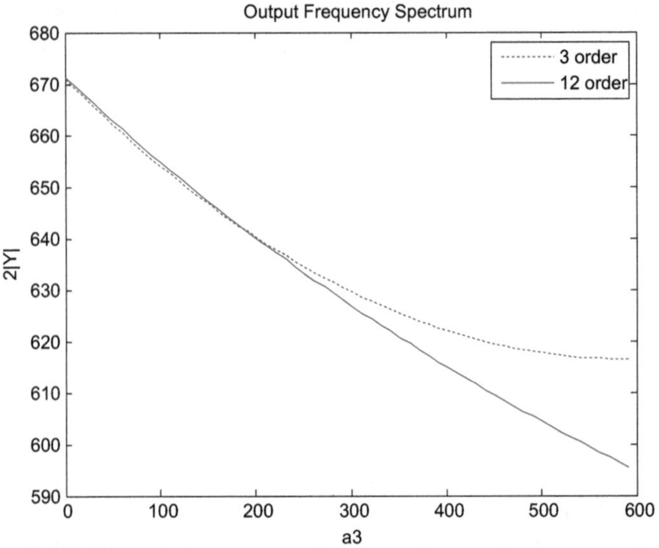

Fig. 10.2 Analytical relationship between the system output spectrum and the control parameter a_3 (Jing et al. 2008a)

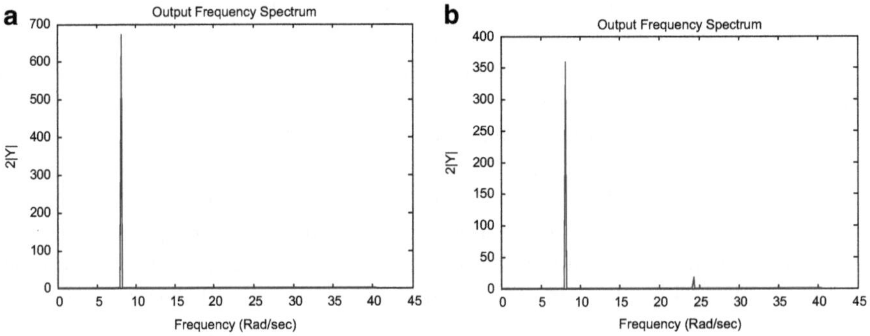

Fig. 10.3 Output spectrum (**a**) without a feedback control, (**b**) with the designed nonlinear feedback (Jing et al. 2008a)

frequency spectrum is set to be $Y^*=180$, then the calculation according to (10.29a, b) and Proposition 10.4 yields $a_3=118{,}610$. The output frequency spectrum under the nonlinear feedback control is shown in Fig. 10.3a, where $Y(j\omega)|_{\omega_0} = 180.08$, and hence the result matches the desired result quite well. The system outputs in the time domain without and under the nonlinear feedback control are given in Fig. 10.4. It can be seen that the system steady state performance is considerably improved when the nonlinear controller is used.

In order to further demonstrate the advantage of the nonlinear feedback control, consider a linear controller $u=-275\dot{x}$. Under this linear control, the system output

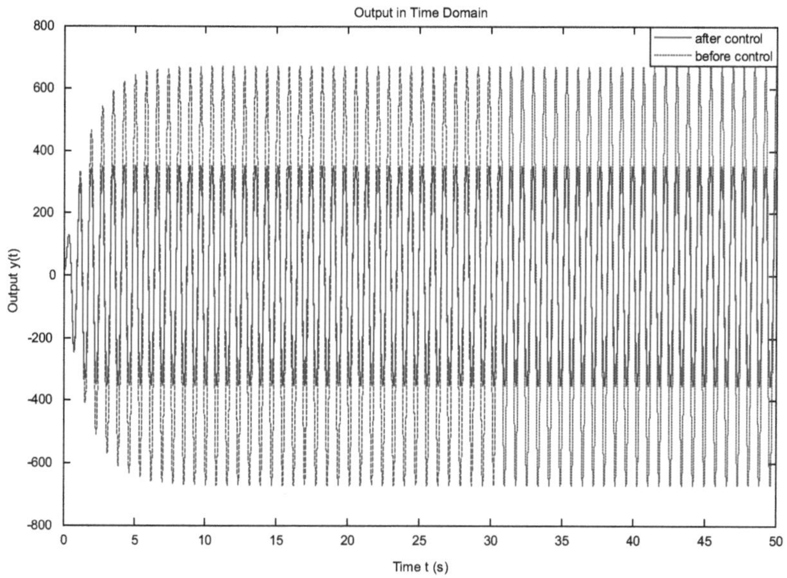

Fig. 10.4 System output in time domain: before and after control (Jing et al. 2008a)

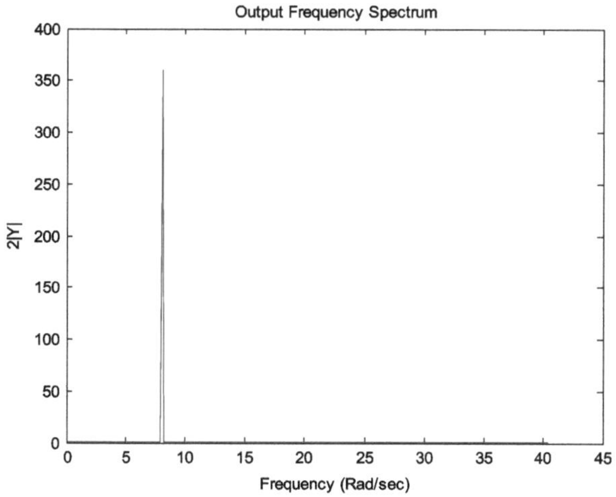

Fig. 10.5 Output spectrum with the linear feedback control (Jing et al. 2008a)

frequency response as shown in Fig. 10.5 is similar to that achieved under the nonlinear controller. However, when F_d is increased to 200 N, the output frequency response is quite different under the two controllers. The nonlinear feedback controller results in a much smaller magnitude of output frequency response at frequency ω_0, referring to Fig. 10.6. Figure 10.7 shows the results of the system outputs in the time domain under the two different control inputs, indicating the

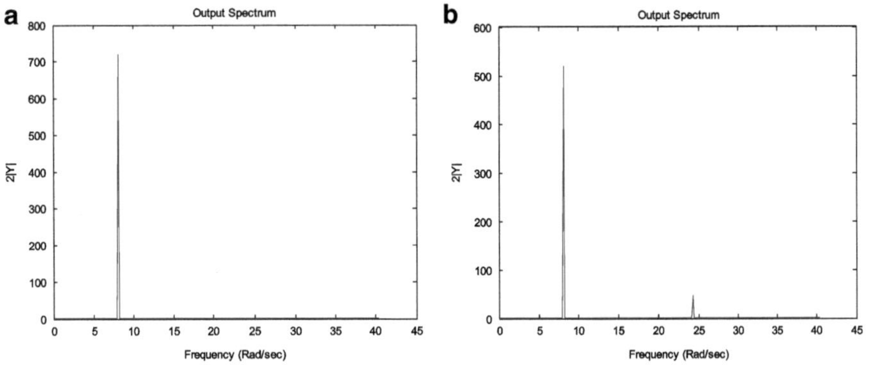

Fig. 10.6 Output spectrum (**a**) with the linear feedback control and (**b**) with the designed nonlinear feedback control, when F_d is increased to F_d=200 (a_2=275, a_3=11,869) (Jing et al. 2008a)

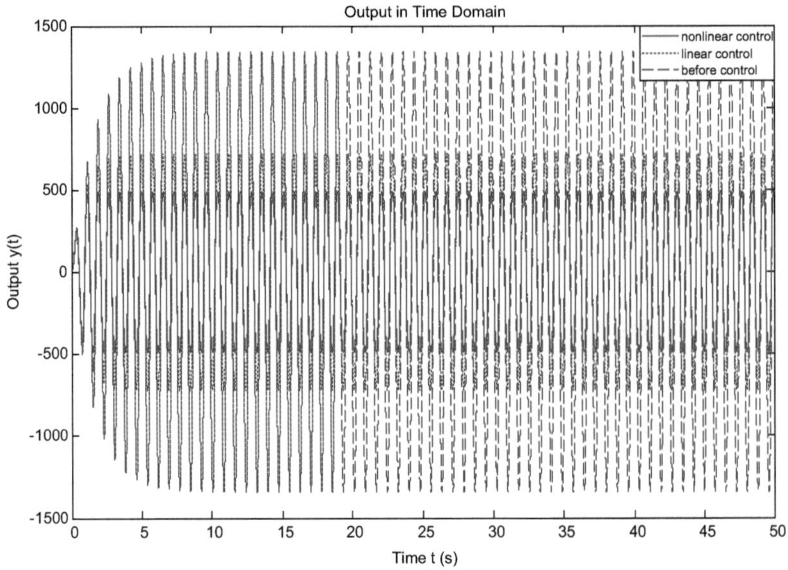

Fig. 10.7 The system outputs in time domain under different control inputs (F_d=200) (Jing et al. 2008a)

nonlinear controller has a much better result than the linear controller. When the input frequency ω_0 is increased to be 15 rad/s, the same conclusions can be reached for the two controllers, referring to Fig. 10.8. When the input frequency is decreased to be 5 rad/s, the output spectrums under the two controllers are similar (see Fig. 10.9). On the other hand, although increase of the liner damping can also achieve better output performance at the driving frequency, this will degrade the

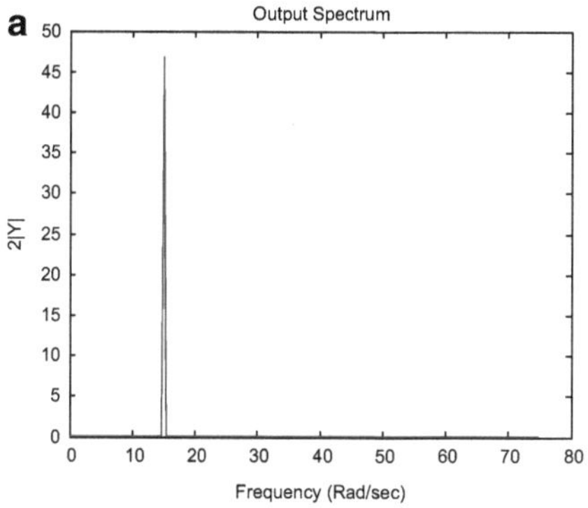

Fig. 10.8 Output spectrum (**a**) with the linear feedback control and (**b**) with the designed nonlinear feedback control, when $\omega_0 = 15$ rad/s, $F_d = 100$, $a_2 = 275$, $a_3 = 11{,}869$ (Jing et al. 2008a)

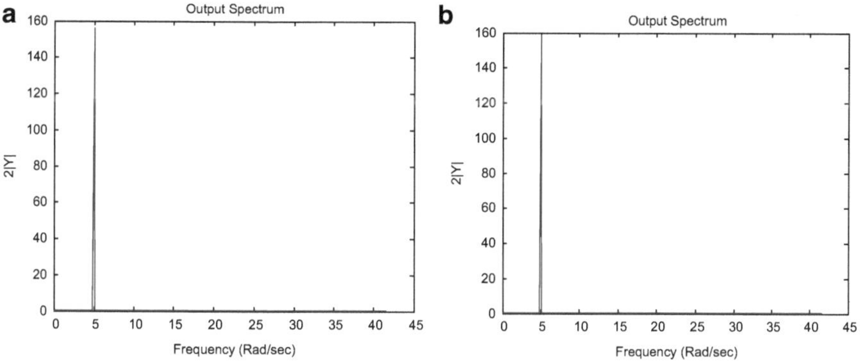

Fig. 10.9 Output spectrum (**a**) with the linear feedback control and (**b**) with the designed nonlinear feedback control when $\omega_0 = 5$ rad/s, $F_d = 100$, $a_2 = 275$, $a_3 = 11{,}869$ (Jing et al. 2008a)

output performance at high frequencies as known in literature (Fig. 10.10). However, the nonlinear damping has no obviously such a limitation (Fig. 10.11).

The results demonstrate that a cubic nonlinear damping as introduced by a simple nonlinear feedback control can achieve better performance than a linear damping control for vibration suppression both in low and high frequencies. The frequency domain method proposed in this study provides an effective approach to the analysis and design of the nonlinear feedback control. Although only a simple case with only one nonlinear term is studied in this simulation, much more complicated cases with multiple nonlinear parameters can also be analysed and designed by following a similar method. It should be noted that there may be some other methods in the literature which can be used to realize the same control purpose of this study, however, the advantage of this method is that it can directly relate the nonlinear controller parameters to system output frequency response and

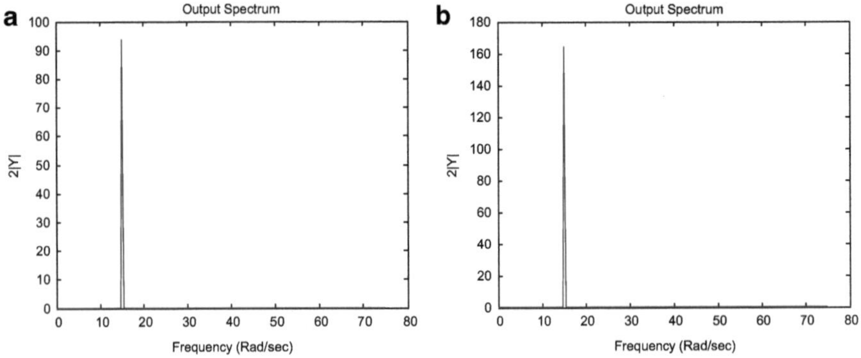

Fig. 10.10 Output spectrum with the linear feedback control when (**a**) a_2=275 and (**b**) a_2=2,750 (ω_0=15 rad/s, F_d=200) (Jing et al. 2008a)

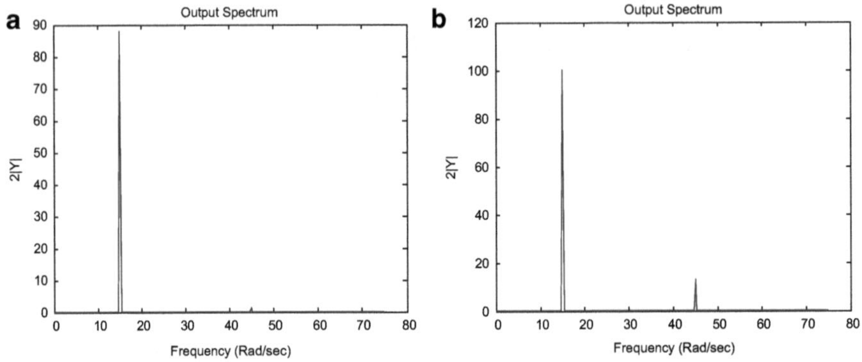

Fig. 10.11 Output spectrum with the nonlinear feedback control when (**a**) a_3=11,869 and (**b**) a_3=118,690 (ω_0=15 rad/s, F_d=200) (Here, a_3 is just arbitrarily increased to see the control effect) (Jing et al. 2008a)

therefore the nonlinear controller or structural parameters can be analysed and designed in the frequency domain, which is a more understandable way in engineering practice. Furthermore, the designed controller, for instance the nonlinear damping designed in the example study above, may also be realized by a passive unite, and the analysis by using this method can be performed directly for a physical characteristics of a structural unite in a system. This will have great significance in practical applications.

10.5 Conclusions

A frequency domain approach to the analysis and design of nonlinear feedback controller for suppressing periodic disturbances is studied and some preliminary results in this subject are provided. Although there are already some time-domain methods, which can address nonlinear control problems based on Lyapunov stability theory, few results are available for analysis and design of a nonlinear feedback controller in the frequency domain to achieve a desired frequency domain performance. Based on the analytical relationship between system output spectrum and controller parameters defined by the OFRF, this chapter demonstrates a systematic frequency domain approach to exploiting the potential advantage of nonlinearities to achieve a desired output frequency domain performance for the analysis and design of vibration systems. Compared with other existing methods for the same purposes, the method in this chapter can directly relate the nonlinear parameters of interest to the system output frequency response and the designed controller may also be realized by a passive unite in practice. Although the results in this paper are developed for the problem of periodic disturbance suppression for SISO linear plants, the idea can be extended to a more general case (*i.e.*, nonlinear controlled plants) and to address more complicated control problems.

Exploring nonlinear benefits for vibration control is an interesting and hot topic in the literature. More results in this topic can also be referred to Xiao et al. (2013a), Jing et al. (2009c, 2010, 2011), Jing and Lang (2009b) and Liu et al. (2015).

10.6 Proofs

A. Proof of Proposition 10.3
To prove Proposition 10.3, the following Lemmas are needed.

Lemma 10.3 Consider two positive, scalar and continuous process in time t, $x(t)$ and $y(t)$ satisfying $y(t) \le \alpha(x(t))$ (for $t \ge 0$), where α is a K-function. If $x(t)$ is asymptotically stable to a ball $B_\rho(x)$, then $y(t)$ is asymptotically stable to a ball $B_{\alpha(2\rho)}(y)$.

Proof There exists a KL-function β, such that function x(t) (for $t \ge 0$) satisfies $x(t) \le \beta(x(0),t)+\rho$, $\forall t>0$. Therefore,

$$
\begin{aligned}
y(t) \le \alpha(x(t)) &= \alpha(\beta(x(0),t)+\rho) \le \alpha(\max(2\beta(x(0),t),2\rho)) \\
&= \max(\alpha(2\beta(x(0),t)),\alpha(2\rho)) \le \alpha(2\beta(x(0),t)) + \alpha(2\rho)
\end{aligned}
$$

Note that $\alpha(2\beta(x(0),t))$ is still a KL-function of $x(0)$ and t, thus the lemma is concluded. □

From Lemma 10.3, if there exists a K-function o such that the output function $h(\mathbf{X})$ of a nonlinear system satisfies $\|h(\mathbf{X})\| \le o(\|\mathbf{X}\|)$, then the system output is asymptotically stable to a ball if the system is asymptotically stable to a ball.

Lemma 10.4 Consider a scalar differential inequality $\dot{y}(t) \le -\alpha(y(t)) + \gamma$, where α is a K-function and γ is a constant and $y(t)$ satisfies Lipschitz condition. Then there exists KL-function β such that

$$y(t) \le \beta\left(\left|y(t_0) - \alpha^{-1}(\gamma)\right|, t\right) + \alpha^{-1}(\gamma).$$

Proof Consider the differential equation $\dot{y}(t) = -\alpha(y(t))$. From Lemma 10.1.2 in Isidori (1999) it is known that, there is a KL-function β such that $y(t) = \beta(y(t_0), t)$. Similarly, considering the differential equation $\dot{y}(t) = -\alpha(y(t)) + \gamma$, then $y(t) = sign(y(t_0) - \alpha^{-1}(\gamma)) \cdot \beta(|y(t_0) - \alpha^{-1}(\gamma)|, t) + \alpha^{-1}(\gamma)$. Thus from the comparison principle and the differential inequality $\dot{y}(t) \le -\alpha(y(t)) + \gamma$, the lemma follows. □

Then to prove Proposition 10.3, it follows from (10.25) that

$$\dot{V}(\mathbf{X}(t)) \le -\alpha(\|\mathbf{X}\|) + \gamma\left(\|\eta\|_\infty\right) \tag{A1}$$

Noting $V(\mathbf{X}) \le \beta_2(\|\mathbf{X}\|)$, we have $\|\mathbf{X}\| \ge \beta_2^{-1}(V(\mathbf{X}))$. Substituting this inequality into (A1), we have

$$\dot{V}(\mathbf{X}(t)) \le -\alpha\left(\beta_2^{-1}(V(\mathbf{X}))\right) + \gamma\left(\|\eta\|_\infty\right)$$

From lemma 10.4, it follows that, there exist a KL-function β, such that

$$V(\mathbf{X}(t)) \le \beta(V_0, t) + \beta_2^{-1} \cdot \alpha^{-1} \cdot \gamma\left(\|\eta\|_\infty\right) \tag{A2}$$

where, $V_0 = |V(\mathbf{X}(t_0)) - \beta_2^{-1} \cdot \alpha^{-1} \cdot \gamma(\|\eta\|_\infty)|$. From (A2), $V(\mathbf{X}(t))$ is asymptotically stable to the ball $B_{\beta_2^{-1} \cdot \alpha^{-1} \cdot \gamma\left(\|\eta\|_\infty\right)}(V)$. Noting $\beta_1(\|\mathbf{X}\|) \le V(\mathbf{X})$, we have $\|\mathbf{X}\| \le \beta_1(V(\mathbf{X}))$. From lemma 10.3, $\mathbf{X}(t)$ is asymptotically stable to the ball $B_\rho(\mathbf{X})$. Furthermore, since assumption 10.1 holds, from lemma 10.3, $y(t)$ is asymptotically stable to the ball $B_{o(2\rho)}(y)$. This completes the proof of sufficiency. The proof of the necessity of the proposition can follow a similar method as demonstrated in the appendix of Hu et al. (2005). The proof completes. □

B. Proof of Proposition 10.4

The state-space equation of system (10.26a) can be written as $\dot{\mathbf{X}} = \mathbf{AX} - \mathbf{B}\phi + \mathbf{E}\eta$, where, $\mathbf{X} = [x, \dot{x}]^T$, $\phi = a_3\sigma^3$, $\sigma = \mathbf{CX}$. Choose a Lyapunov candidate as:

$$V = \mathbf{X}^T\mathbf{P}\mathbf{X} + \frac{\alpha}{2}\sigma^4 \tag{A3}$$

where, $\alpha > 0$. Equation (A3) further gives

$$\dot{V} = \mathbf{X}^T\mathbf{P}\dot{\mathbf{X}} + \dot{\mathbf{X}}^T\mathbf{P}\mathbf{X} + 2\alpha\sigma^3\mathbf{C}\dot{\mathbf{X}} = \mathbf{X}^T(\mathbf{A}^T\mathbf{P}+\mathbf{P}\mathbf{A})\mathbf{X} - 2\mathbf{X}^T\mathbf{P}\mathbf{B}\phi + 2\mathbf{X}^T\mathbf{P}\mathbf{E}\eta + \tfrac{2\alpha}{a_3}\phi\mathbf{C}(\mathbf{A}\mathbf{X} - \mathbf{B}\phi + \mathbf{E}\eta)$$

$$= \mathbf{X}^T(\mathbf{A}^T\mathbf{P}+\mathbf{P}\mathbf{A})\mathbf{X} - 2\mathbf{X}^T\mathbf{P}\mathbf{B}\phi + \frac{2\alpha}{a_3}\phi\mathbf{C}\mathbf{A}\mathbf{X} - \frac{2\alpha}{a_3}\phi\mathbf{C}\mathbf{B}\phi + 2\mathbf{X}^T\mathbf{P}\mathbf{E}\eta + \frac{2\alpha}{a_3}\phi\mathbf{C}\mathbf{E}\eta \quad \text{(A4)}$$

Let $\mathbf{Z} = \begin{bmatrix} \mathbf{X} \\ \phi \end{bmatrix}$, $\mathbf{T} = \begin{bmatrix} \mathbf{PE} \\ \frac{\alpha}{a_3}\mathbf{CE} \end{bmatrix}$, and $\beta = \alpha/a_3$ then (A4) gives

$$\dot{V} = \mathbf{Z}^T\begin{bmatrix} \mathbf{A}^T\mathbf{P}+\mathbf{P}\mathbf{A} & \beta\mathbf{A}^T\mathbf{C}^T - \mathbf{P}\mathbf{B} \\ * & -2\beta\mathbf{C}\mathbf{B} \end{bmatrix}\mathbf{Z} + \mathbf{Z}^T\mathbf{T}\eta \le \mathbf{Z}^T\begin{bmatrix} \mathbf{A}^T\mathbf{P}+\mathbf{P}\mathbf{A} & \beta\mathbf{A}^T\mathbf{C}^T - \mathbf{P}\mathbf{B} \\ * & -2\beta\mathbf{C}\mathbf{B} \end{bmatrix}\mathbf{Z} + \varepsilon^{-1}\mathbf{Z}^T\mathbf{T}\mathbf{T}^T\mathbf{Z} + \varepsilon\eta^T\eta$$

$$= \mathbf{Z}^T\left(\begin{bmatrix} \mathbf{A}^T\mathbf{P}+\mathbf{P}\mathbf{A} & \beta\mathbf{A}^T\mathbf{C}^T - \mathbf{P}\mathbf{B} \\ * & -2\beta\mathbf{C}\mathbf{B} \end{bmatrix} + \varepsilon^{-1}\mathbf{T}\mathbf{T}^T\right)\mathbf{Z} + \varepsilon\eta^2 = -\mathbf{Z}^T\mathbf{Q}\mathbf{Z} + \varepsilon\eta^2$$

Note that, in the inequality above, the following inequality is used

$$2\mathbf{Z}^T\mathbf{T}\eta \le \varepsilon^{-1}\mathbf{Z}^T\mathbf{T}\mathbf{T}^T\mathbf{Z} + \varepsilon\eta^T\eta, \text{ for any } \varepsilon > 0.$$

If $\mathbf{Q}=\mathbf{Q}^T>0$, then $\mathbf{Z}^T\mathbf{Q}\mathbf{Z} \ge \lambda_{\min}(\mathbf{Q})\|\mathbf{X}\|^2$ is a K-function of $\|\mathbf{X}\|$. Hence, according to Proposition 10.3, the system is asymptotically stable to a ball $B_\rho(\mathbf{X})$ with $\rho = \sqrt{\lambda_{\min}(\mathbf{Q})^{-1}\varepsilon \sup\left(\|\eta\|^2\right)} = F_d\sqrt{\lambda_{\min}(\mathbf{Q})^{-1}\varepsilon}$. Additionally, when there is no exogenous disturbance input, and if $\mathbf{Q}=\mathbf{Q}^T>0$ holds with $\mathbf{E}=0$, then it is obvious that the system without a disturbance input is globally asymptotically stable. This completes the proof. □

Chapter 11
Mapping from Parametric Characteristics to the GFRFs and Output Spectrum

11.1 Introduction

Frequency domain methods for nonlinear systems have been studied for many years (Taylor 1999; Solomou et al. 2002; Pavlov et al. 2007). The frequency domain theory for Volterra systems was initiated by the concept of the GFRF (George 1959). Thereafter, many significant results relating to the estimation and computation of the GFRFs and analysis of output frequency response for practical nonlinear systems have been developed (Bendat 1990; Billings and Lang 1996; Chua and Ng 1979; and also see Chaps. 2–10).

To compute the GFRFs of nonlinear systems, Bedrosian and Rice (1971) introduced the "harmonic probing" method. By applying the probing method (Rugh 1981), algorithms to compute the GFRFs for nonlinear Volterra systems described by the NDE, NARX and NBO models were derived, which enable the nth-order GFRF to be recursively obtained in terms of the coefficients of the governing nonlinear model (Chap. 2). Based on the GFRFs, frequency response characteristics of nonlinear systems can then be investigated as shown before. These results are important extensions of well known frequency domain methods of linear systems such as transfer function or Bode diagram, and provide a systematic and effective method for analysis of nonlinear systems in the frequency domain.

However, it can be seen that existing recursive algorithms for computations of the GFRFs and system output spectrum can not explicitly and simply reveal the analytical relationship between system time domain model parameters and system frequency response functions in a clear and straightforward way. Therefore, many problems remain unsolved, related to how the frequency response functions are influenced by the parameters of the underlying system, and how they are connected to complex non-linear behaviours, etc. In order to solve these problems, the parametric characteristics of the GFRFs were studied in Chaps. 4–5, which

X. Jing, Z. Lang, *Frequency Domain Analysis and Design of Nonlinear Systems based on Volterra Series Expansion*, Understanding Complex Systems,
DOI 10.1007/978-3-319-12391-2_11

effectively build up a mapping from the GFRF to its parametric characteristic and provide an explicit expression for the analytical relationship between the GFRFs and system time-domain model parameters. The significance of the parametric characteristic analysis of the GFRFs is that it can reveal what model parameters contribute to and how these parameters affect system frequency response functions including the GFRFs and output frequency response function (see the detailed results and discussions in the previous chapters). This provides an effective approach to the analysis of the frequency domain characteristics of nonlinear systems in terms of system time domain model parameters.

The study in this chapter is based on the results in Chap. 5. It is shown in Chaps. 5–6 that the *n*th-order GFRF and output spectrum of a nonlinear Volterra system can both be written as an explicit and straightforward polynomial function in terms of nonlinear model parameters, and this polynomial function is characterized by its parametric characteristics with its coefficients being complex valued functions of frequencies and dependent on the system linear dynamics and input. Note that, the parametric characteristics can be analytically determined by the results in Chap. 5. The objective of this chapter is to analytically determine the complex valued functions related to the parametric characteristics. An inverse mapping function from the parametric characteristics of the GFRFs to the GFRFs is therefore studied. By using this new mapping function, the *n*th-order GFRF can directly be recovered from its parametric characteristic as an *n*-degree polynomial function of the first order GFRF, revealing an explicit analytical relationship between higher order GFRFs and system linear frequency response function. Compared with the existing recursive algorithm for the computation of the GFRFs, the new mapping function enables the *n*th-order GFRF to be explicitly expressed in a more straightforward and meaningful way. Note from previous results that higher order GFRFs are recursively dependent on lower order GFRFs. This recursive relationship often complicates the qualitative analysis and understanding of system frequency characteristics. The new results can effectively overcome this problem, and unveil the system's linear and nonlinear factors included in the *n*th-order GFRF more clearly. This provides a useful insight into the frequency domain analysis and design of nonlinear systems based on the GFRFs, and can be regarded as an important application of the parametric characteristic theory established in the previous chapters.

11.1.1 Some Notations for This Chapter

Some notations are listed here especially for readers' convenience in understanding the discussions of this Chapter, although some of these notations have already appeared in previous chapters and will also be used in the following chapters.

$c_{p,q}(k_1,\cdots,k_{p+q})$	A model parameter in the NDE model, k_i is the order of the derivative, p represents the order of the involved output nonlinearity, q is the order of the involved input nonlinearity, and $p+q$ is the nonlinear degree of the parameter.
$H_n(j\omega_1,\cdots,j\omega_n)$	The nth-order GFRF
$C_{p,q} = \left[c_{p,q}(0,\cdots,0), c_{p,q}(0,\cdots,1), \cdots, c_{p,q}(\underbrace{K,\cdots,K}_{p+q=m})\right]$	A parameter vector consisting of all the nonlinear parameters of the form $c_{p,q}(k_1,\cdots,k_{p+q})$
$CE(.)$	The coefficient extraction operator (Chap. 4)
$CE(H_n(j\omega_1,\cdots,j\omega_n))$	The parametric characteristics of the nth-order GFRF
$f_n(j\omega_1,\cdots,j\omega_n)$	The correlative function of $CE(H_n(j\omega_1,\cdots,j\omega_n))$
$\otimes$	The reduced Kronecker product defined in the CE operator
$\oplus$	The reduced vectorized summation defined in the CE operator
$c_{p_0,q_0}(\cdot)c_{p_1,q_1}(\cdot)\cdots c_{p_k,q_k}(\cdot)$	A monomial consisting of nonlinear parameters
$s_{x_1}s_{x_2}\cdots s_{x_p}$	A p-partition of a monomial $c_{p_0,q_0}(\cdot)c_{p_1,q_1}(\cdot)\cdots c_{p_k,q_k}(\cdot)$
s_{x_i}	A monomial of x_i parameters of $\{c_{p_0,q_0}(\cdot),\cdots,c_{p_k,q_k}(\cdot)\}$ of the involved monomial, $0\le x_i\le k$, and $s_0=1$
$\varphi_n: S_C(n)\to S_f(n)$	A new mapping function from the parametric characteristics to the correlative functions, $S_C(n)$ is the set of all the monomials in the parametric characteristics and $S_f(n)$ is the set of all the involved correlative functions in the nth order GFRF.
$n(s_x(\bar{s}))$	The order of the GFRF where the monomial $s_x(\bar{s})$ is generated
$\bar{\lambda}_n(\omega_1,\cdots,\omega_n)$	The maximum eigenvalue of the frequency characteristic matrix Θ_n of the nth-order GFRF

11.2 The *n*th-Order GFRF and Its Parametric Characteristic

In this chapter, consider Volterra-type nonlinear systems described by the NDE model in (2.11). Similar results can be extended to the NARX model (2.10). For convenience, some basic results are restated in this section as follows.

Using the definitions in (2.25), i.e.,

$$H_{0,0}(\cdot)=1,\ \ H_{n,0}(\cdot)=0\ \text{ for n} > 0,\ \ H_{n,p}(\cdot)=0\ \text{ for } n<p,\ \text{and} \prod_{i=1}^{q}(\cdot) = \begin{cases} 1 & q=0, p>1 \\ 0 & q=0, p\le 1 \end{cases} \tag{11.1}$$

The nth-order GFRF for (2.11) can be written as (2.26), i.e.,

$$H_n(j\omega_1,\cdots,j\omega_n)=\frac{1}{L_n\left(j\sum_{i=1}^{n}\omega_i\right)}\sum_{q=0}^{n}\sum_{p=0}^{n-q}\sum_{k_1,k_{p+q}=0}^{K}c_{p,q}(k_1,\cdots,k_{p+q}) \times\left(\prod_{i=1}^{q}(j\omega_{n-q+i})^{k_{p+i}}\right)H_{n-q,p}(j\omega_1,\cdots,j\omega_{n-q}) \tag{11.2}$$

The parametric characteristic of the *n*th-order GFRF can be simply computed as (See Corollary 5.1 for details)

$$CE(H_n(j\omega_1,\cdots,j\omega_n))=C_{0,n}\oplus\left(\mathop{\oplus}_{q=1}^{n-1}\mathop{\oplus}_{p=1}^{n-q}C_{p,q}\otimes CE(H_{n-q-p+1}(\cdot))\right) \oplus\left(\mathop{\oplus}_{p=2}^{n}C_{p,0}\otimes CE(H_{n-p+1}(\cdot))\right) \tag{11.3}$$

Moreover, $CE(H_n(j\omega_1,\cdots,j\omega_n))$ can also be determined by following the results in Proposition 5.1, which allows the direct determination of the parameter characteristic vector of the *n*th-order GFRF without recursive computations and provides a sufficient and necessary condition for which nonlinear parameters and how these parameters are included in $CE(H_n(j\omega_1,\cdots,j\omega_n))$.

Based on the parametric characteristic analysis in Chaps. 4–5, the *n*th-order GFRF can be expressed as

$$H_n(j\omega_1,\cdots,j\omega_n)=CE(H_n(j\omega_1,\cdots,j\omega_n))\cdot f_n(j\omega_1,\cdots,j\omega_n) \tag{11.4}$$

where $f_n(j\omega_1,\cdots,j\omega_n)$ is a complex valued function vector with an appropriate dimension, which is referred to as the correlative function of the parametric characteristic $CE(H_n(j\omega_1,\cdots,j\omega_n))$ in this study.

Equation (11.4) provides an explicit expression for the analytical relationship between the GFRFs and system time-domain model parameters. Based on these results, system nonlinear characteristics can be studied in the frequency domain from novel perspectives including frequency characteristics of system output frequency response, nonlinear effect from specific nonlinear parameters, and parametric sensitivity analysis etc as demonstrated in the previous chapters. In this chapter, an algorithm is provided to explicitly determine the correlative function $f_n(j\omega_1,\cdots,j\omega_n)$ in (11.4) directly in terms of the first order GFRF $H_1(j\omega_1)$ based on the parametric characteristic vector $CE(H_n(j\omega_1,\cdots,j\omega_n))$. To achieve this objective, a mapping from $CE(H_n(j\omega_1,\cdots,j\omega_n))$ to $H_n(j\omega_1,\cdots,j\omega_n)$ is established such that the nth-order GFRF can directly be written into the parametric characteristic function (11.4) in an analytical form by using this mapping function, and some new properties of the GFRFs are developed. These results allow higher order GFRFs and consequently the OFRF to be analytically expressed as a functional of the system linear FRF (i.e., the first order GFRF), and thus provide a novel qualitative and quantitative approach to understanding of nonlinear dynamics in the frequency domain (see more discussions in Chap. 12).

11.3 Mapping from the Parametric Characteristic to the *n*th-Order GFRF

The parametric characteristic vector $CE(H_n(j\omega_1,\cdots,j\omega_n))$ of the nth-order GFRF can be recursively determined by (11.3), which has elements of the form $C_{p,q}\otimes C_{p_1,q_1}\otimes C_{p_2,q_2}\otimes\cdots\otimes C_{p_k,q_k}$ $(n-2\geq k\geq 0)$, and each element of which has a corresponding complex valued correlative function in vector $f_n(j\omega_1,\cdots,j\omega_n)$. For example, $c_{0,n}(k_1,\cdots,k_n)$ corresponds to the complex valued function $(j\omega_1)^{k_1}\cdots(j\omega_n)^{k_n}$ in the nth-order GFRF.

From Proposition 5.1, an element in $CE(H_n(j\omega_1,\cdots,j\omega_n))$ is either a single parameter coming from pure input nonlinearity such as $c_{0n}(.)$, or a nonlinear parameter monomial function of the form $C_{p,q}\otimes C_{p_1,q_1}\otimes C_{p_2,q_2}\otimes\cdots\otimes C_{p_k,q_k}$ satisfying (5.15), and the first parameter of $C_{p,q}\otimes C_{p_1,q_1}\otimes C_{p_2,q_2}\otimes\cdots\otimes C_{p_k,q_k}$ must come from pure output nonlinearity or input–output cross nonlinearity, *i.e.*, $c_{pq}(.)$ with $p\geq 1$ and $p+q>1$. For this reason, the following definition is given.

Definition 11.1 A parameter monomial of the form $C_{p,q}\otimes C_{p_1,q_1}\otimes C_{p_2,q_2}\otimes\cdots\otimes C_{p_k,q_k}$ with $k\geq 0$ and $p+q>1$ is said to be effective or an effective combination of the involved nonlinear parameters for $CE(H_n(j\omega_1,\cdots,j\omega_n))$ if $p+q=n(>1)$ for $k=0$, or (5.15) is satisfied for $k>0$. □

From Definition 11.1, it is obvious that all the monomials in $CE(H_n(j\omega_1,\cdots,j\omega_n))$ are effective combinations. The following lemma shows further that what an effective monomial should be in certain order GFRF and how it is generated in this order GFRF.

Lemma 11.1 For a monomial $c_{p_0,q_0}(\cdot)c_{p_1,q_1}(\cdot)\cdots c_{p_k,q_k}(\cdot)$ with k>0, the following statements hold:

(1) it is effective for the Zth-order GFRF if and only if there is at least one parameter $c_{pi,qi}(.)$ with $p_i>0$, where $Z=\sum_{i=0}^{k}(p_i+q_i)-k$.

(2) if there are l different parameters with $p_i>0$, then there are l different cases in which this monomial is produced by the recursive computation of the Zth-order GFRF.

Proof (1) This is directly from Definition 11.1. Z can be computed according to Proposition 5.1, *i.e.*, $p_0+q_0+\sum_{i=1}^{k}(p_i+q_i)=Z+k$, which yields $Z=\sum_{i=0}^{k}(p_i+q_i)-k$. (2) From the second and third terms in the recursive algorithm of (2.19) or (5.1), *i.e.*,

$$\sum_{q=1}^{n-1}\sum_{p=1}^{n-q}\sum_{k_1,k_{p+q}=0}^{K}c_{p,q}(k_1,\cdots,k_{p+q})\left(\prod_{i=1}^{q}(j\omega_{n-q+i})^{k_{p+i}}\right)H_{n-q,p}(j\omega_1,\cdots,j\omega_{n-q})$$
$$+\sum_{p=2}^{n}\sum_{k_1,k_p=0}^{K}c_{p,0}(k_1,\cdots,k_p)H_{n,p}(j\omega_1,\cdots,j\omega_n) \tag{11.5}$$

it can be seen that all the nonlinear parameters with $p>0$ and $p+q\leq n$ are involved in the nth-order GFRF, and each of these parameters must correspond to the initial parameter in an effective monomial of $CE(H_n(j\omega_1,\cdots,j\omega_n))$. Hence, if there are l different parameters with $p_i>0$ in the monomial $c_{p_0,q_0}(\cdot)c_{p_1,q_1}(\cdot)\cdots c_{p_k,q_k}(\cdot)$, then there will be l different cases in which this monomial is produced in the Zth order GFRF. This completes the proof. □

Definition 11.2 A (p,q)-partition of $H_n(j\omega_1,\cdots,j\omega_n)$ is a combination $H_{r_1}(w_{r_1})H_{r_2}(w_{r_2})\cdots H_{r_p}(w_{r_p})$ satisfying $\sum_{i=1}^{p}r_i=n-q$, where $1\leq r_i\leq n-p-q+1$, and w_{r_i} is a set consisting of r_i different frequency variables such that $\bigcup_{i=1}^{p}w_{r_i}=\{\omega_1,\omega_2,\cdots,\omega_n\}$ and $w_{r_i}\cap w_{r_j}=\phi$ for $i\neq j$. □

For example, $H_1(\omega_1)H_1(\omega_2)H_3(\omega_3\cdots\omega_5)$ and $H_1(\omega_1)H_2(\omega_2,\omega_3)H_2(\omega_4,\omega_5)$ are two (3,0)-partitions of $H_5(j\omega_1,\cdots,j\omega_5)$.

Definition 11.3 A p-partition of an effective monomial $c_{p_1,q_1}(\cdot)\cdots c_{p_k,q_k}(\cdot)$ is a combination $s_{x_1}s_{x_2}\cdots s_{x_p}$, where s_{x_i} is a monomial of x_i parameters in $\{c_{p_1,q_1}(\cdot),\cdots,c_{p_k,q_k}(\cdot)\}$, $0\leq x_i\leq k$, $s_0=1$, and each non-unitary s_{x_i} is an effective monomial satisfying $\sum_{i=1}^{p}x_i=k$ and $s_{x_1}s_{x_2}\cdots s_{x_p}=c_{p_1,q_1}(\cdot)\cdots c_{p_k,q_k}(\cdot)$. □

The sub-monomial s_{x_i} in a p-partition of an effective monomial $c_{p_1,q_1}(\cdot)\cdots c_{p_k,q_k}(\cdot)$ is denoted by $s_{x_i}(c_{p_1,q_1}(\cdot)\cdots c_{p_k,q_k}(\cdot))$. Suppose that a p-partition for 1 is still 1, i.e., $\underbrace{1\cdot 1\cdots 1}_{p}=1$. Obviously $c_{p_1,q_1}(\cdot)\cdots c_{p_k,q_k}(\cdot)=s_{x_1}s_{x_2}\cdots s_{x_p}(c_{p_1,q_1}(\cdot)\cdots c_{p_k,q_k}(\cdot))=s_k(c_{p_1,q_1}(\cdot)\cdots c_{p_k,q_k}(\cdot))$. For example, $s_1(c_{1,1}(\cdot))s_2(c_{2,1}(\cdot)c_{3,0}(\cdot))$ and $s_2(c_{1,1}(\cdot)c_{2,1}(\cdot))s_1(c_{3,0}(\cdot))$ are two 2-partitions of $c_{1,1}(\cdot)c_{2,1}(\cdot)c_{3,0}(\cdot)$. Moreover, note that when s_0 appear in a p-partition of a monomial, it means that there is a $H_1(.)$ which appears in the corresponding (p,q)-partition for $H_n(.)$.

For an effective monomial $c_{p,q}(\cdot)c_{p_1,q_1}(\cdot)\cdots c_{p_k,q_k}(\cdot)$ in $CE(H_n(j\omega_1,\cdots,j\omega_n))$, without speciality, suppose the first parameter $c_{p,q}(\cdot)$ is directly generated in the recursive computation of $H_n(j\omega_1,\cdots,j\omega_n)$, then the other parameters must be generated from the lower order GFRFs that are involved in the recursive computation of $H_n(j\omega_1,\cdots,j\omega_n)$. From (2.19)–(2.23) or (5.1)–(5.5) it can be seen that each parameter in a monomial corresponds to a certain order GFRF from which it is generated. The following lemma shows how a monomial is generated in $H_n(j\omega_1,\cdots,j\omega_n)$ by using the new concepts defined above. This provides an important insight into the mapping from a monomial to its correlative function.

Lemma 11.2 If a monomial $c_{p,q}(\cdot)c_{p_1,q_1}(\cdot)\cdots c_{p_k,q_k}(\cdot)$ is effective, and $c_{p,q}(\cdot)$ is the initial parameter directly generated in the xth-order GFRF and $p>0$, then

(1) $c_{p_1,q_1}(\cdot)\cdots c_{p_k,q_k}(\cdot)$ comes from (p,q)-partitions of the xth-order GFRF, where $x=p+q+\sum_{i=1}^{k}(p_i+q_i)-k$;

(2) if additionally s_0 is supposed to be generated from $H_1(.)$, then each p-partition of $c_{p_1,q_1}(\cdot)\cdots c_{p_k,q_k}(\cdot)$ corresponds to a (p,q)-partition of the xth-order GFRF, and each (p,q)-partition of the xth-order GFRF produces at least one p-partition for $c_{p_1,q_1}(\cdot)\cdots c_{p_k,q_k}(\cdot)$;

(3) the correlative function of $c_{p_1,q_1}(\cdot)\cdots c_{p_k,q_k}(\cdot)$ is the summation of the correlative functions from all the (p,q)-partitions of the xth-order GFRF which produces $c_{p_1,q_1}(\cdot)\cdots c_{p_k,q_k}(\cdot)$, and therefore is the summation of the correlative functions corresponding to all the p-partition of $c_{p_1,q_1}(\cdot)\cdots c_{p_k,q_k}(\cdot)$.

Proof See Sect. 11.6 for the proof. □

Remark 11.1 From Lemma 11.2, it can be seen that all the (p,q)-partitions of the xth-order GFRF which produce $c_{p_1,q_1}(\cdot)\cdots c_{p_k,q_k}(\cdot)$ are all the (p,q)-partitions corresponding to all the p-partitions for $c_{p_1,q_1}(\cdot)\cdots c_{p_k,q_k}(\cdot)$. Therefore, to obtain all the (p,q)-partitions of interest, all the p-partitions for $c_{p_1,q_1}(\cdot)\cdots c_{p_k,q_k}(\cdot)$ is needed to be determined. □

Based on the results above, in order to determine the mapping between a parameter monomial $c_{p,q}(\cdot)c_{p_1,q_1}(\cdot)\cdots c_{p_k,q_k}(\cdot)$ and its correlative function in $f_n(j\omega_1,\cdots,j\omega_n)$, the following operator is defined.

Definition 11.4 Let $S_C(n)$ be a set composed of all the elements in $CE(H_n(j\omega_1,\cdots,j\omega_n))$, and let $S_f(n)$ be a set of the complex-valued functions of the frequency variables $j\omega_1,\cdots,j\omega_n$. Then define a mapping

$$\varphi_n : S_C(n) \to S_f(n) \tag{11.6a}$$

such that in $\omega_1,\cdots,\omega_n$

$$H_n^{sym}(j\omega_1,\cdots,j\omega_n) = \frac{1}{n!} \sum_{\substack{\text{all the permutations} \\ \text{of } \{1,2,\ldots,n\}}} CE(H_n(j\omega_1,\cdots,j\omega_n))\cdot\varphi_n(CE(H_n(j\omega_1,\cdots,j\omega_n))) \tag{11.6b}$$

□

That is, by using the mapping function above, an asymmetric GFRF can be obtained as

$$H_n(j\omega_1,\cdots,j\omega_n) = CE(H_n(j\omega_1,\cdots,j\omega_n)) \cdot \varphi_n(CE(H_n(j\omega_1,\cdots,j\omega_n)))$$

The existence of this mapping function is obvious. For example, $\varphi_n(c_{0,n}(k_1,\cdots,k_n)) = (j\omega_1)^{k_1}\cdots(j\omega_n)^{k_n}$. The task is to determine the complex valued correlative function $\varphi_n\left(c_{p,q}(\cdot)c_{p_1,q_1}(\cdot)\cdots c_{p_k,q_k}(\cdot)\right)$ for any nonlinear parameter monomial $c_{p,q}(\cdot)c_{p_1,q_1}(\cdot)\cdots c_{p_k,q_k}(\cdot)$ $(0 \le k \le n-2)$ in $CE(H_n(j\omega_1,\cdots,j\omega_n))$.

Based on Lemma 11.1–11.2, the following result can be obtained.

Proposition 11.1 For an effective nonlinear parameter monomial $c_{p_0,q_0}(\cdot)c_{p_1,q_1}(\cdot)\cdots c_{p_k,q_k}(\cdot)$, let $\bar{s} = c_{p_0,q_0}(\cdot)c_{p_1,q_1}(\cdot)\cdots c_{p_k,q_k}(\cdot)$, $n(s_x(\bar{s})) = \sum_{i=1}^{x}(p_i + q_i) - x + 1$, where x is the number of the parameters in s_x, $\sum_{i=1}^{x}(p_i + q_i)$ is the summation of the subscripts of all the parameters in s_x, $\sum_{i=1}^{x}(.) = 0$ if $x < 1$ and $n(1) = 1$. Then for $0 \le k \le n(\bar{s}) - 2$

$$\begin{aligned}
&\phi_{n(\bar{s})}\left(c_{p_0,q_0}(\cdot)c_{p_1,q_1}(\cdot)\cdots c_{p_k,q_k}(\cdot);\omega_{l(1)}\cdots\omega_{l(n(\bar{s}))}\right)\\
&= \sum_{\substack{\text{all the 2-partitions}\\ \text{for } \bar{s} \text{ satisfying}\\ s_1(\bar{s})=c_{p,q}(\cdot) \text{ and } p>0}} \Bigg\{ f_1\left(c_{p,q}(\cdot),n(\bar{s});\omega_{l(1)}\cdots\omega_{l(n(\bar{s}))}\right)\cdot \sum_{\substack{\text{all the p-partitions}\\ \text{for } \bar{s}/c_{pq}(\cdot)}}\\
&\sum_{\substack{\text{all the different}\\ \text{permutations}\\ \text{of } \{s_{x_1},\cdots,s_{x_p}\}}} \Bigg[f_{2a}\left(s_{\bar{x}_1}\cdots s_{\bar{x}_p}\left(\bar{s}/c_{p,q}(\cdot)\right);\omega_{l(1)}\cdots\omega_{l(n(\bar{s})-q)}\right)\\
&\cdot\prod_{i=1}^{p}\phi_{n\left(s_{\bar{x}_i}\left(\bar{s}/c_{p,q}(\cdot)\right)\right)}\left(s_{\bar{x}_i}\left(\bar{s}/c_{p,q}(\cdot)\right);\omega_{l(\bar{X}(i)+1)}\cdots\omega_{l\left(\bar{X}(i)+n\left(s_{\bar{x}_i}\left(\bar{s}/c_{pq}(\cdot)\right)\right)\right)}\right)\Bigg]\Bigg\}
\end{aligned} \tag{11.7a}$$

or simplified as

$$\begin{aligned}
&\varphi_{n(\bar{s})}\left(c_{p_0,q_0}(\cdot)c_{p_1,q_1}(\cdot)\cdots c_{p_k,q_k}(\cdot);\omega_{l(1)}\cdots\omega_{l(n(\bar{s}))}\right)\\
&= \sum_{\substack{\text{all the 2-partitions}\\ \text{for } \bar{s} \text{ satisfying}\\ s_1(\bar{s})=c_{p,q}(\cdot) \text{ and } p>0}} \Bigg\{ f_1\left(c_{p,q}(\cdot),n(\bar{s});\omega_{l(1)}\cdots\omega_{l(n(\bar{s}))}\right)\\
&\cdot \sum_{\substack{\text{all the p-partitions}\\ \text{for } \bar{s}/c_{p,q}(\cdot)}} \Bigg[f_{2b}\left(s_{x_1}\cdots s_{x_p}\left(\bar{s}/c_{p,q}(\cdot)\right);\omega_{l(1)}\cdots\omega_{l(n(\bar{s})-q)}\right)\\
&\cdot\prod_{i=1}^{p}\varphi_{n\left(s_{x_i}\left(\bar{s}/c_{p,q}(\cdot)\right)\right)}\left(s_{x_i}\left(\bar{s}/c_{p,q}(\cdot)\right);\omega_{l(X(i)+1)}\cdots\omega_{l\left(X(i)+n\left(s_{x_i}\left(\bar{s}/c_{p,q}(\cdot)\right)\right)\right)}\right)\Bigg]\Bigg\}
\end{aligned} \tag{11.7b}$$

the terminating condition is $k=0$ and $\varphi_1(1;\omega_i)=H_1(j\omega_i)$, where,

$$\bar{X}(i)=\sum_{j=1}^{i-1} n\left(s_{\bar{x}_j}\left(\bar{s}/c_{pq}(\cdot)\right)\right) \text{ or } X(i)=\sum_{j=1}^{i-1} n\left(s_{x_j}\left(\bar{s}/c_{pq}(\cdot)\right)\right) \tag{11.8a}$$

$$f_1\left(c_{p,q}(\cdot),n(\bar{s});\omega_{l(1)}\cdots\omega_{l(n(\bar{s}))}\right)=\left(\prod_{i=1}^{q}\left(j\omega_{l(n(\bar{s})-q+i)}\right)^{k_{p+i}}\middle/ L_{n(\bar{s})}\left(j\sum_{i=1}^{n(\bar{s})}\omega_{l(i)}\right)\right. \tag{11.8b}$$

$$f_{2a}\left(s_{\overline{x}_1}\cdots s_{\overline{x}_p}\left(\overline{s}/c_{p,q}(\cdot)\right);\omega_{l(1)}\cdots\omega_{l(n(\overline{s})-q)}\right)$$
$$=\prod_{i=1}^{p}\left(j\omega_{l(\overline{X}(i)+1)}+\cdots+j\omega_{l(\overline{X}(i)+n(s_{\overline{x}_i}(\overline{s}/c_{pq}(\cdot))))}\right)^{k_i} \tag{11.8c}$$

$$f_{2b}\left(s_{x_1}\cdots s_{x_p}\left(\overline{s}/c_{p,q}(\cdot)\right);\omega_{l(1)}\cdots\omega_{l(n(\overline{s})-q)}\right)$$
$$=\frac{n_x^*}{n_k^*}\sum_{\substack{\text{all the different}\\ \text{permutations}\\ \text{of }\{k_1,\cdots,k_p\}}}\prod_{i=1}^{p}\left(j\omega_{l(X(i)+1)}+\cdots+j\omega_{l(X(i)+n(s_{x_i}(\overline{s}/c_{pq}(\cdot))))}\right)^{k_i} \tag{11.8d}$$

Moreover, $\{s_{\overline{x}_1},\cdots s_{\overline{x}_p}\}$ is a permutation of $\{s_{x_1},\cdots s_{x_p}\}$, $\omega_{l(1)}\cdots\omega_{l(n(\overline{s}))}$ represents the frequency variables involved in the corresponding functions, $li)$ for $i=1\ldots n(\overline{s})$ is a positive integer representing the index of the frequency variables, $n_k^*=\frac{p!}{n_1!n_2!\cdots n_c!}$, $n_1+\ldots+n_c=p$, c is the number of distinct differentials k_i appearing in the combination, n_i is the number of repetitions of the ith distinct differential k_i, and a similar definition holds for n_x^*. □

Proof See Sect. 11.6 for the proof. □

Remark 11.2 Equations (11.7a,b) are recursive. The terminating condition is $k=0$, which is also included in (11.7a,b). For k=0, it can be derived from (11.7b) that

$$\phi_{n(\overline{s})}\left(c_{p,q}(\cdot);\omega_{l(1)}\cdots\omega_{l(n(\overline{s}))}\right)=\phi_{p+q}\left(c_{p,q}(\cdot);\omega_{l(1)}\cdots\omega_{l(p+q)}\right)$$
$$=f_1\left(c_{p,q}(\cdot),p+q;\omega_{l(1)}\cdots\omega_{l(p+q)}\right)$$
$$\cdot\sum_{\substack{\text{all the }p-\text{partitions}\\ \text{for }1}} f_{2b}\left(s_{x_1}\cdots s_{x_p}(1);\omega_{l(1)}\cdots\omega_{l(p+q-q)}\right)\prod_{i=1}^{p}\phi_{n(s_{x_i}(1))}\left(s_{x_i}(1);\omega_{l(X(i)+1)}\cdots\omega_{l(X(i)+n(s_{x_i}(1)))}\right)$$
$$=f_1\left(c_{p,q}(\cdot),p+q;\omega_{l(1)}\cdots\omega_{l(p+q)}\right)\cdot f_{2b}\left(\underbrace{11\cdots1}_{p};\omega_{l(1)}\cdots\omega_{l(p)}\right)\cdot\prod_{i=1}^{p}\phi_1(1;\omega_i)$$
$$=\frac{1}{L_{p+q}\left(j\sum_{i=1}^{p+q}\omega_{l(i)}\right)}\left(\prod_{i=1}^{q}(j\omega_{l(p+i)})^{k_{p+i}}\cdot\prod_{i=1}^{p}(j\omega_{l(i)})^{k_i}\cdot\prod_{i=1}^{p}H_1(j\omega_{l(i)})\right) \tag{11.9}$$

Note that in this case, p+q=$n(\overline{s})$ from (5.15), and $\overline{s}=c_{p,q}(\cdot)$ corresponding to a specific recursive terminal. Hence, (11.9) can be written as

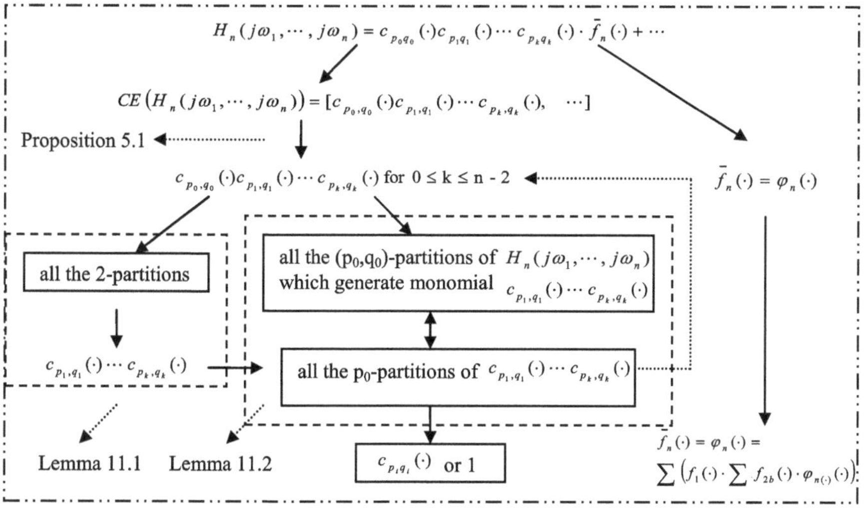

Fig. 11.1 An illustration of the relationships in Proposition 11.1 (Jing et al. 2008e)

$$\varphi_{n(\bar{s})}\left(c_{p,q}(\cdot);\omega_{l(1)}\cdots\omega_{l(n(\bar{s}))}\right) = \frac{1}{L_{n(\bar{s})}\left(j\sum_{i=1}^{n(\bar{s})}\omega_{l(i)}\right)}\left(\prod_{i=1}^{q}\left(j\omega_{l(p+i)}\right)^{k_{p+i}}\right.$$
$$\cdot\prod_{i=1}^{p}\left(j\omega_{l(i)}\right)^{k_i}\cdot\prod_{i=1}^{p}H_1\left(j\omega_{l(i)}\right) \qquad (11.10)$$

In order to verify this result, let $n = n(\bar{s}) = \mathrm{p}+\mathrm{q}$, it can be obtained from (11.2) that for a parameter $c_{p,q}(\cdot)$, its correlative function is

$$\frac{1}{L_{n(\bar{s})}\left(j\sum_{i=1}^{n(\bar{s})}\omega_i\right)}\prod_{i=1}^{q}\left(j\omega_{p+i}\right)^{k_{p+i}}\Big)H_{p,p}\left(j\omega_1,\cdots,j\omega_p\right)$$

From (5.5), $H_{p,p}\left(j\omega_1,\cdots,j\omega_p\right) = \prod_{i=1}^{p}\left(j\omega_i\right)^{k_i}\cdot\prod_{i=1}^{p}H_1\left(j\omega_i\right)$. This is consistent with (11.10). To further understand the results in Proposition 11.1, the following figure can be referred, which demonstrates the recursive process in the new mapping function and the structure of the theoretical results above (See Fig. 11.1). □

To further demonstrate the results, the following example is given.

Example 11.1 Consider the fourth-order GFRF. The parametric characteristic of the fourth-order GFRF can be obtained from Proposition 5.1 that

$$
\begin{aligned}
CE(H_4(j\omega_1,\cdots,j\omega_4)) = {} & \mathrm{C}_{0,4} \oplus \mathrm{C}_{1,3} \oplus \mathrm{C}_{3,1} \oplus \mathrm{C}_{2,2} \oplus \mathrm{C}_{4,0} \oplus \mathrm{C}_{1,1} \otimes \mathrm{C}_{0,3} \oplus \mathrm{C}_{1,1} \\
& \otimes \mathrm{C}_{1,2} \oplus \mathrm{C}_{1,1} \otimes \mathrm{C}_{2,1} \oplus \mathrm{C}_{1,1} \otimes \mathrm{C}_{3,0} \oplus \mathrm{C}_{1,2} \otimes \mathrm{C}_{0,2} \oplus \mathrm{C}_{1,2} \\
& \otimes \mathrm{C}_{2,0} \oplus \mathrm{C}_{2,0} \otimes \mathrm{C}_{0,3} \oplus \mathrm{C}_{2,0} \otimes \mathrm{C}_{2,1} \oplus \mathrm{C}_{2,0} \otimes \mathrm{C}_{3,0} \oplus \mathrm{C}_{2,1} \\
& \otimes \mathrm{C}_{0,2} \oplus \mathrm{C}_{3,0} \otimes \mathrm{C}_{0,2} \oplus \mathrm{C}_{1,1} \otimes \mathrm{C}_{0,2}{}^2 \oplus \mathrm{C}_{1,1}{}^2 \otimes \mathrm{C}_{0,2} \oplus \mathrm{C}_{1,1} \\
& \otimes \mathrm{C}_{0,2} \otimes \mathrm{C}_{2,0} \oplus \mathrm{C}_{1,1}{}^3 \oplus \mathrm{C}_{1,1}{}^2 \otimes \mathrm{C}_{2,0} \oplus \mathrm{C}_{1,1} \otimes \mathrm{C}_{2,0}{}^2 \oplus \mathrm{C}_{2,0} \\
& \otimes \mathrm{C}_{0,2}{}^2 \oplus \mathrm{C}_{2,0}{}^2 \otimes \mathrm{C}_{0,2} \oplus \mathrm{C}_{2,0}{}^3
\end{aligned}
$$

By using Proposition 11.1, the correlative function of each term in $CE(H_4(j\omega_1,\cdots, j\omega_4))$ can all be obtained. For example, for the term $c_{1,1}(.)c_{0,2}(.)c_{2,0}(.)$, it can be derived that

$$
\begin{aligned}
& \varphi_{n(\bar{s})}\left(c_{1,1}(\cdot)c_{0,2}(\cdot)c_{2,0}(\cdot);\omega_{l(1)}\cdots\omega_{l(n(\bar{s}))}\right) = \varphi_4(c_{1,1}(\cdot)c_{0,2}(\cdot)c_{2,0}(\cdot);\omega_1\cdots\omega_4) \\
& = f_1(c_{1,1}(\cdot),4;\omega_1\cdots\omega_4) \\
& \cdot\left[f_{2b}(s_2(c_{0,2}(\cdot)c_{2,0}(\cdot));\omega_1\cdots\omega_3)\cdot\varphi_{n(s_2(c_{0,2}(\cdot)c_{2,0}(\cdot)))}\left(s_2(c_{0,2}(\cdot)c_{2,0}(\cdot));\omega_{X(1)+1}\cdots\omega_{X(1)+n(s_2(c_{0,2}(\cdot)c_{2,0}(\cdot)))}\right)\right] \\
& + f_1(c_{2,0}(\cdot),4;\omega_1\cdots\omega_4) \\
& \quad \cdot\left[f_{2b}(s_0s_2(c_{1,1}(\cdot)c_{0,2}(\cdot));\omega_1\cdots\omega_4)\cdot\varphi_{n(s_0(c_{1,1}(\cdot)c_{0,2}(\cdot)))}\left(s_0(c_{1,1}(\cdot)c_{0,2}(\cdot));\omega_{X(1)+1}\cdots\omega_{X(1)+n(s_0(c_{1,1}(\cdot)c_{0,2}(\cdot)))}\right)\right. \\
& \qquad \cdot\varphi_{n(s_2(c_{1,1}(\cdot)c_{0,2}(\cdot)))}\left(s_2(c_{1,1}(\cdot)c_{0,2}(\cdot));\omega_{X(2)+1}\cdots\omega_{X(2)+n(s_2(c_{1,1}(\cdot)c_{0,2}(\cdot)))}\right) \\
& \quad + f_{2b}(s_1s_1(c_{1,1}(\cdot)c_{0,2}(\cdot));\omega_1\cdots\omega_4)\cdot\varphi_{n(s_1(c_{1,1}(\cdot)))}\left(s_1(c_{1,1}(\cdot));\omega_{X(1)+1}\cdots\omega_{X(1)+n(s_1(c_{1,1}(\cdot)))}\right) \\
& \qquad \left.\cdot\varphi_{n(s_1(c_{0,2}(\cdot)))}\left(s_1(c_{0,2}(\cdot));\omega_{X(2)+1}\cdots\omega_{X(2)+n(s_1(c_{0,2}(\cdot)))}\right)\right] \\
& = f_1(c_{1,1}(\cdot),4;\omega_1\cdots\omega_4) \\
& \cdot\left[f_{2b}(c_{0,2}(\cdot)c_{2,0}(\cdot);\omega_1\cdots\omega_3)\cdot\varphi_{n(c_{0,2}(\cdot)c_{2,0}(\cdot))}\left(c_{0,2}(\cdot)c_{2,0}(\cdot);\omega_{0+1}\cdots\omega_{0+n(c_{0,2}(\cdot)c_{2,0}(\cdot))}\right)\right] \\
& + f_1(c_{2,0}(\cdot),4;\omega_1\cdots\omega_4) \\
& \cdot\left[f_{2b}(s_0s_2(c_{1,1}(\cdot)c_{0,2}(\cdot));\omega_1\cdots\omega_4)\cdot\varphi_{n(1)}(1;\omega_1\cdots\omega_{n(1)})\varphi_{n(c_{1,1}(\cdot)c_{0,2}(\cdot))}\left(c_{1,1}(\cdot)c_{0,2}(\cdot);\omega_{n(1)+1}\cdots\omega_{n(1)+n(c_{1,1}(\cdot)c_{0,2}(\cdot))}\right)\right. \\
& + f_{2b}(s_1s_1(c_{1,1}(\cdot)c_{0,2}(\cdot));\omega_1\cdots\omega_4)\cdot\varphi_2\left(c_{1,1}(\cdot);\omega_{X(1)+1}\cdots\omega_{X(1)+n(s_1(c_{1,1}(\cdot)))}\right) \\
& \qquad \left.\cdot\varphi_2\left(c_{0,2}(\cdot);\omega_{X(2)+1}\cdots\omega_{X(2)+n(s_1(c_{0,2}(\cdot)))}\right)\right] \\
& = f_1(c_{1,1}(\cdot),4;\omega_1\cdots\omega_4)\cdot[f_{2b}(c_{0,2}(\cdot)c_{2,0}(\cdot);\omega_1\cdots\omega_3)\cdot\phi_3(c_{0,2}(\cdot)c_{2,0}(\cdot);\omega_1\cdots\omega_3)] \\
& + f_1(c_{2,0}(\cdot),4;\omega_1\cdots\omega_4)\cdot[f_{2b}(s_0s_2(c_{1,1}(\cdot)c_{0,2}(\cdot));\omega_1\cdots\omega_4)\cdot\phi_1(1;\omega_1)\phi_3(c_{1,1}(\cdot)c_{0,2}(\cdot);\omega_2\cdots\omega_4) \\
& \qquad\qquad + f_{2b}(s_1s_1(c_{1,1}(\cdot)c_{0,2}(\cdot));\omega_1\cdots\omega_4)\cdot\phi_2(c_{1,1}(\cdot);\omega_1,\omega_2)\cdot\phi_2(c_{0,2}(\cdot);\omega_3,\omega_4)]
\end{aligned}
\tag{11.11}
$$

To proceed with the recursive computation, it can be derived that

$$
f_1(c_{1,1}(\cdot),4;\omega_1\cdots\omega_4) = \prod_{i=1}^{1}(j\omega_{3+i})^{k_{1+i}} \Big/ L_4\left(j\sum_{i=1}^{4}\omega_i\right) = (j\omega_4)^{k_2} \Big/ L_4\left(j\sum_{i=1}^{4}\omega_i\right) \tag{11.12a}
$$

$$
f_1(c_{2,0}(\cdot),4;\omega_1\cdots\omega_4) = 1 \Big/ L_4\left(j\sum_{i=1}^{4}\omega_i\right) \tag{11.12b}
$$

$$
f_{2b}(s_{x_1}(c_{2,0}(\cdot)c_{0,2}(\cdot));\omega_1\cdots\omega_3) = (j\omega_1 + \cdots + j\omega_3)^{k_1} \tag{11.12c}
$$

$$f_{2b}(s_0s_2(c_{1,1}(\cdot)c_{0,2}(\cdot));\omega_1\cdots\omega_4)=\sum_{\substack{\text{all the different}\\ \text{permutations}\\ \text{of } \{k_1,\cdots,k_p\}}}\prod_{i=1}^{2}\left(j\omega_{X(i)+1}+\cdots+j\omega_{X(i)+n\left(s_{x_i}\left(\bar{s}/c_{pq}(\cdot)\right)\right)}\right)^{k_i}$$

$$=(j\omega_1)^{k_1}(j\omega_2+\cdots+j\omega_4)^{k_2}+(j\omega_2+\cdots+j\omega_4)^{k_1}(j\omega_1)^{k_2} \tag{11.12d}$$

$$\begin{aligned}
&\phi_3(c_{0,2}(\cdot)c_{2,0}(\cdot);\omega_1\cdots\omega_3)\\
&=f_1(c_{2,0}(\cdot),3;\omega_1\cdots\omega_3)\cdot f_{2b}(s_{x_1}s_{x_2}(c_{0,2}(\cdot));\omega_1\cdots\omega_3)\\
&\qquad\prod_{i=1}^{2}\phi_{n\left(s_{x_i}\left(\bar{s}/c_{pq}(\cdot)\right)\right)}\left(s_{x_i}(c_{0,2}(\cdot));\omega_{X(i)+1}\cdots\omega_{X(i)+n\left(s_{x_i}\left(c_{0,2}(\cdot)\right)\right)}\right)\\
&=f_1(c_{2,0}(\cdot),3;\omega_1\cdots\omega_3)\cdot f_{2b}(s_{x_1}s_{x_2}(c_{0,2}(\cdot));\omega_1\cdots\omega_3)\phi_1(1;\omega_1)\phi_2(c_{0,2}(\cdot);\omega_2,\omega_3)\\
&=\frac{1}{L_3\left(j\sum_{i=1}^{3}\omega_i\right)}\cdot\left((j\omega_1)^{k_1}(j\omega_2+j\omega_3)^{k_2}+(j\omega_3+j\omega_2)^{k_1}(j\omega_1)^{k_2}\right)\\
&\qquad\cdot H_1(j\omega_1)\cdot\frac{1}{L_2(j\omega_2+j\omega_3)}(j\omega_2)^{k_1}(j\omega_3)^{k_2}
\end{aligned} \tag{11.12e}$$

$$\begin{aligned}
&\varphi_3(c_{1,1}(\cdot)c_{0,2}(\cdot);\omega_2\cdots\omega_4)\\
&=f_1(c_{1,1}(\cdot),3;\omega_2\cdots\omega_4)\cdot f_{2b}(s_{x_1}(c_{0,2}(\cdot));\omega_2,\omega_3)\cdot\varphi_{n\left(s_{x_1}\left(c_{0,2}(\cdot)\right)\right)}(s_{x_1}(c_{0,2}(\cdot));\omega_2,\omega_3)\\
&=f_1(c_{1,1}(\cdot),3;\omega_2\cdots\omega_4)\cdot f_{2b}(c_{0,2}(\cdot);\omega_2,\omega_3)\cdot\varphi_2(c_{0,2}(\cdot);\omega_2,\omega_3)\\
&=\frac{(j\omega_4)^{k_2}}{L_3(j\omega_2+\cdots+j\omega_4)}\cdot(j\omega_2+j\omega_3)^{k_1}\cdot\frac{1}{L_2(j\omega_2+j\omega_3)}(j\omega_2)^{k_1}(j\omega_3)^{k_2}
\end{aligned} \tag{11.12f}$$

Using (11.12a–f) in (11.11) yields

$$\begin{aligned}
&\varphi_4(c_{1,1}(\cdot)c_{0,2}(\cdot)c_{2,0}(\cdot);\omega_1\cdots\omega_4)\\
&=f_1(c_{1,1}(\cdot),4;\omega_1\cdots\omega_4)\cdot[f_{2b}(c_{0,2}(\cdot)c_{2,0}(\cdot);\omega_1\cdots\omega_3)\cdot\varphi_3(c_{0,2}(\cdot)c_{2,0}(\cdot);\omega_1\cdots\omega_3)]\\
&+f_1(c_{2,0}(\cdot),4;\omega_1\cdots\omega_4)\cdot[f_{2b}(s_0s_2(c_{1,1}(\cdot)c_{0,2}(\cdot));\omega_1\cdots\omega_4)\cdot\varphi_1(1;\omega_1)\varphi_3(c_{1,1}(\cdot)c_{0,2}(\cdot);\omega_2\cdots\omega_4)\\
&\qquad+f_{2b}(s_1s_1(c_{1,1}(\cdot)c_{0,2}(\cdot));\omega_1\cdots\omega_4)\cdot\varphi_2(c_{1,1}(\cdot);\omega_1,\omega_2)\cdot\varphi_2(c_{0,2}(\cdot);\omega_3,\omega_4)]\\
&=\frac{(j\omega_4)^{k_2}(j\omega_1+\cdots+j\omega_3)^{k_1}\left((j\omega_1)^{k_1}(j\omega_2+j\omega_3)^{k_2}+(j\omega_3+j\omega_2)^{k_1}(j\omega_1)^{k_2}\right)(j\omega_2)^{k_1}(j\omega_3)^{k_2}}{L_4(j\omega_1+\cdots+j\omega_4)L_3(j\omega_1+j\omega_2+j\omega_3)L_2(j\omega_2+j\omega_3)}\cdot H_1(j\omega_1)\\
&+\frac{\left((j\omega_1)^{k_1}(j\omega_2+\cdots+j\omega_4)^{k_2}+(j\omega_2+\cdots+j\omega_4)^{k_1}(j\omega_1)^{k_2}\right)(j\omega_4)^{k_2}(j\omega_2+j\omega_3)^{k_1}(j\omega_2)^{k_1}(j\omega_3)^{k_2}}{L_4(j\omega_1+\cdots+j\omega_4)L_3(j\omega_2+\cdots+j\omega_4)L_2(j\omega_2+j\omega_3)}H_1(j\omega_1)\\
&+\frac{\left((j\omega_1+j\omega_2)^{k_1}(j\omega_3+j\omega_4)^{k_2}+(j\omega_3+j\omega_4)^{k_1}(j\omega_1+j\omega_2)^{k_2}\right)(j\omega_4)^{k_2}(j\omega_3)^{k_1}(j\omega_1)^{k_1}(j\omega_2)^{k_2}}{L_4(j\omega_1+\cdots+j\omega_4)L_2(j\omega_3+j\omega_4)L_2(j\omega_2+j\omega_1)}H_1(j\omega_1)
\end{aligned} \tag{11.13}$$

Therefore, the correlative function of the parameter monomial $c_{1,1}(\cdot)c_{0,2}(\cdot)c_{2,0}(\cdot)$ is obtained. It can be verified that the same result can be obtained by using the recursive algorithm in (11.2), (5.2)–(5.3), (11.1). For the sake of brevity, this is omitted. By following the same method, the whole correlative function vector $\varphi_4(CE(H_4(j\omega_1,\cdots,j\omega_4)))$ can be determined. Thus the fourth-order GFRF $H_4(j\omega_1,\cdots,j\omega_4)$ can directly be written into a parametric characteristic form which can provide a straightforward and meaningful insight into the relationship between $H_4(j\omega_1,\cdots,j\omega_4)$ and nonlinear parameters, and also between $H_4(j\omega_1,\cdots,j\omega_4)$ and $H_1(j\omega_1)$. □

Remark 11.3 From Example 11.1, it can be seen that Proposition 11.1 provides an effective method to determine the correlative function for an effective monomial $c_{p_0,q_0}(\cdot)c_{p_1,q_1}(\cdot)\cdots c_{p_k,q_k}(\cdot)$, and the computation process should be able to be carried out automatically without manual intervention. Therefore, Proposition 11.1 provides a simplified method to determine the nth-order GFRF directly into a more meaningful form as (11.4) which can demonstrate the parametric characteristic clearly and describe the nth-order GFRF in terms of the first order GFRF $H_1(j\omega)$ and nonlinear parameters. This reveals a more straightforward insight into the relationships between $H_n(j\omega_1,\cdots,j\omega_n)$ and nonlinear parameters, and between $H_n(j\omega_1,\cdots,j\omega_n)$ and $H_1(j\omega)$. Note that the high order GFRFs can represent system nonlinear frequency response characteristics (Billings and Peyton-Jones1990; Yue et al. 2005) and $H_1(j\omega)$ represents the linear part of the system model. Hence, the results in Proposition 11.1 not only facilitate the analysis of the connection between system frequency response characteristics and model linear and nonlinear parameters, but also provide a new perspective on the understanding of the GFRFs and on the analysis of nonlinear systems based on the GFRFs. □

11.4 Some New Properties

Based on the mapping function φ_n established in the last section, some new properties of the nth-order GFRF are discussed in this section.

11.4.1 Determination of FRFs Based on Parametric Characteristics

There are several relationships involved in this paper. $H_n(j\omega_1,\cdots,j\omega_n)$ is determined from the NDE model in terms of the model parameters. The CE operator is a mapping from $H_n(j\omega_1,\cdots,j\omega_n)$ to its parametric characteristic, which can also be regarded as a mapping from the nonlinear parameters of the NDE model to the parametric characteristics of $H_n(j\omega_1,\cdots,j\omega_n)$. Thus there is a bijective mapping between $H_n(j\omega_1,\cdots,j\omega_n)$ and the NDE model. The function φ_n can be regarded as an inverse mapping of the CE operator such that the nth-order GFRF can be

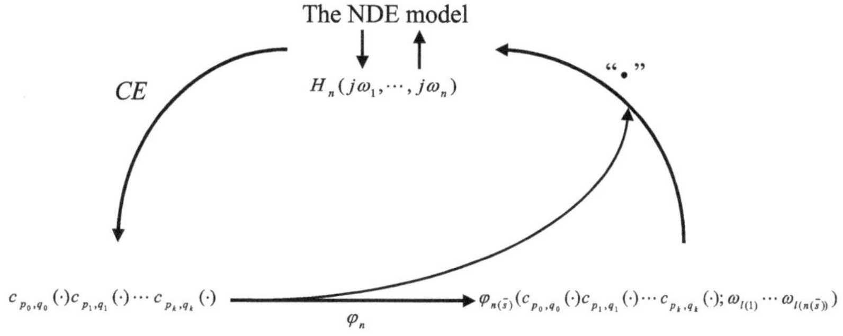

Fig. 11.2 Relationship between φ_n and CE (Jing et al. 2008e)

reconstructed from its parametric characteristic, which can also be regarded as a mapping from the nonlinear parameters of the NDE model to $H_n(j\omega_1,\cdots,j\omega_n)$. This can refer to Fig. 11.2, where "•" represents the point multiplication between the parametric monomial and its correlative function.

It can be seen from Fig. 11.2 that

$$H_n(j\omega_1,\cdots,j\omega_n) = CE(H_n(\cdot)) \cdot \varphi_n(CE(H_n(\cdot))) \tag{11.14}$$

From (11.14), the inverse of the operator CE can simply be written as ($x=CE(H_n(\cdot))$)

$$CE^{-1}(x) = x \cdot \varphi_n(x)$$

which constructs a mapping directly from the parametric characteristic of the nth-order GFRF to the nth-order GFRF itself. Note that $CE(H_n(\cdot))$ includes all the nonlinear parameters of degree from 2 to n of the nonlinear system of interest, and $\varphi_n(CE(H_n(\cdot)))$ is a complex valued function vector including the effect of the complicated nonlinear characteristics and also the effect of the linear part of the nonlinear system. Hence, (11.14) reveals a new perspective on the computation and understanding of the GFRFs as discussed in Sect. 11.3, and also provides a new insight into the frequency domain analysis of nonlinear systems based on the GFRFs.

From the results in Chaps. 5–6, the output spectrum of model (2.11) can now be determined as

$$Y(j\omega) = \sum_{n=1}^{N} CE(H_n(j\omega_1,\cdots,j\omega_n)) \cdot \hat{F}_n(j\omega) \tag{11.15a}$$

when the input is a general input $U(j\omega)$,

$$\hat{F}_n(j\omega) = \frac{1}{\sqrt{n}(2\pi)^{n-1}} \int\limits_{\omega_1+\cdots+\omega_n=\omega} \varphi_n(CE(H_n(j\omega_1,\cdots,j\omega_n))) \cdot \prod_{i=1}^{n} U(j\omega_i)d\sigma_\omega \tag{11.15b}$$

when the input is a multi-tone function $u(t) = \sum_{i=1}^{K} |F_i| \cos(\omega_i t + \angle F_i)$,

$$\hat{F}_n(j\omega) = \frac{1}{2^n} \sum_{\omega_{k_1}+\cdots+\omega_{k_n}=\omega} \varphi_n(CE(H_n(j\omega_{k_1},\cdots,j\omega_{k_n}))) \cdot F(\omega_{k_1})\cdots F(\omega_{k_n}) \tag{11.15c}$$

It is obvious that (11.15a) is an explicit analytical polynomial functions with coefficients in $S_C(1)\cup\cdots\cup S_C(N)$ and the corresponding correlative functions in $S_f(1)\cup\cdots\cup S_f(N)$. This demonstrates a direct analytical relationship between system output spectrum and system time-domain model parameters. The effects on system output spectrum from the linear parameters are included in $S_f(1)\cup\cdots\cup S_f(N)$, and the effects from the nonlinear parameters are included in $S_C(1)\cup\cdots\cup S_C(N)$ and also embodied in $S_f(1)\cup\cdots\cup S_f(N)$. This will facilitate the analysis of output frequency response characteristics of nonlinear systems. For example, for any parameters of model (2.11) of interest, which may represent some specific physical characteristics, the output spectrum can therefore directly be written as a polynomial in terms of these parameters. Then how these parameters affect the system output spectrum needs only to be investigated by studying the frequency characteristics of the new mapping functions involved in the polynomial and simultaneously optimizing the values of these nonlinear parameters. Further study in this topic will be introduced in the next chapter.

11.4.2 Magnitude of the nth-Order GFRF

Based on (11.14), the magnitude of the nth-order GFRF can be expressed in terms of its parametric characteristic.

Corollary 11.1 Let $CE_n=CE(H_n(\cdot))$, $\Theta_n=\varphi_n(CE(H_n(\cdot)))\cdot\varphi_n(CE(H_n(\cdot)))^*$, $\varphi_n=\varphi_n(CE(H_n(\cdot)))$, and $\Lambda_n=CE(H_n(\cdot))^T CE(H_n(\cdot))$, then

$$|H_n(j\omega_1,\cdots,j\omega_n)|^2 = CE_n\Theta_n CE_n^T \tag{11.16a}$$

$$|H_n(j\omega_1,\cdots,j\omega_n)|^2 = \varphi_n^*\Lambda_n\varphi_n \tag{11.16b}$$

Proof It can be derived from (11.14) that

$$
\begin{aligned}
&|H_n(j\omega_1,\cdots,j\omega_n)|^2 = H_n(j\omega_1,\cdots,j\omega_n)\cdot H_n^*(j\omega_1,\cdots,j\omega_n)\\
&= CE(H_n(\cdot))\cdot\varphi_n(CE(H_n(\cdot)))\cdot(CE(H_n(\cdot))\cdot\varphi_n(CE(H_n(\cdot))))^*\\
&= CE(H_n(\cdot))\cdot(\varphi_n(CE(H_n(\cdot)))\cdot\varphi_n(CE(H_n(\cdot)))^*)\cdot CE(H_n(\cdot))^T = CE_n\Theta_n CE_n^T
\end{aligned}
$$

The result in (11.16b) can also be achieved by following the same method. This completes the proof. □

From Corollary 11.1, the square of the magnitude of the *n*th-order GFRF is proportional to a quadratic function of the parametric characteristic and also proportional to a quadratic function of the corresponding correlative function. Corollary 11.1 provides a new property of the *n*th-order GFRF, which reveals the relationship between the magnitude of $H_n(j\omega_1,\cdots,j\omega_n)$ and its nonlinear parametric characteristic, and also the relationship between the magnitude of $H_n(j\omega_1,\cdots,j\omega_n)$ and the correlative functions which involve both the system linear and nonlinear characteristics. Given a requirement on $|\, H_n(j\omega_1,\cdots,j\omega_n) \,|$, the condition on model parameters can be derived by using (11.16a,b). This may provide a new technique for the analysis and design of nonlinear systems based on the *n*th-order GFRF in the frequency domain.

Moreover, it can be seen that the frequency characteristic matrix Θ_n is a Hermitian matrix, whose eigenvalues are the positive real valued functions of the system linear parameters but invariant to the values of the system nonlinear parameters in $CE(H_n(\cdot))$. Thus different nonlinearities may result in different frequency characteristic matrix Θ_n, but the same nonlinearities will have an invariant matrix Θ_n. This property of the nth-order GFRF provides a new insight into the nonlinear effect on the high order GFRFs from different nonlinearities. For this purpose, define a new function

$$
\overline{\lambda}_n(\omega_1,\cdots,\omega_n) = \lambda_{\max}(\Theta_n) \tag{11.17}
$$

which is the maximum eigenvalue of the frequency characteristic matrix Θ_n. As mentioned, the frequency spectrum of this function can act as a novel insight into the nonlinear effect on the GFRFs from different nonlinearities, since this function is only dependent on different nonlinearities but independent of their values. However, the frequency response spectrum of the GFRFs will change greatly with the values of the involved nonlinear parameters, which cannot provide a clear insight into the nonlinear effects between different nonlinearities.

Moreover, the following results can be obtained for the bound evaluation for the nth-order GFRF based on the results above.

Proposition 11.2

$$\sup_{\omega_1,\cdots,\omega_n} |H_n(j\omega_1,\cdots,j\omega_n)| \le \int_{-\infty}^{\infty}\cdots\int_{-\infty}^{\infty} |h_n(\tau_1,\cdots,\tau_n)| d\tau_1\cdots d\tau_n$$
$$\le \sqrt{\sup_{\omega_1,\cdots,\omega_n} (\lambda_{\max}(\Theta_n))} \cdot \|CE_n\| \tag{11.18a}$$

$$\sup_{\omega_1,\cdots,\omega_n} |H_n(j\omega_1,\cdots,j\omega_n)| \le \int_{-\infty}^{\infty}\cdots\int_{-\infty}^{\infty} |h_n(\tau_1,\cdots,\tau_n)| d\tau_1\cdots d\tau_n$$
$$\le \sqrt{\lambda_{\max}(\Lambda_n)} \cdot \sup_{\omega_1,\cdots,\omega_n} (\|\varphi_n\|) \tag{11.18b}$$

Proof See Sect. 11.6 for the proof. □

From (11.18a,b), it can be seen that the magnitude of the nth-order GFRF is proportional to a quadratic function of the parametric characteristics and also proportional to a quadratic function of the corresponding correlative function. These results demonstrate a new property of the nth-order GFRF, which reveals the relationship between the magnitude of $H_n(j\omega_1,\cdots,j\omega_n)$ and its nonlinear parametric characteristics, and also the relationship between the magnitude of $H_n(j\omega_1,\cdots,j\omega_n)$ and the correlative functions which include the linear (the first order GFRF) and nonlinear characteristics. Given a requirement on $|H_n(j\omega_1,\cdots,j\omega_n)|$, the condition on model parameters or the first order GFRF can be derived by using these results. Proposition 11.2 also shows that the absolute integration of the nth-order Volterra kernel function in the time domain is bounded by a quadratic function of the parametric characteristics. This reveals the relationship between the model parameters and the stability of Volterra series. Obviously, these may provide a new insight into the analysis and design of nonlinear systems based on the nth-order GFRF in the frequency domain.

11.4.3 Relationship Between $H_n(j\omega_1,\cdots,j\omega_n)$ *and* $H_1(j\omega_1)$

As illustrated in Example 11.1, $H_n(j\omega_1,\cdots,j\omega_n)$ can directly be determined in terms of the first order GFRF $H_1(j\omega)$ based on the novel mapping function φ_n according to its parametric characteristic. The following results can be concluded.

Corollary 11.2 For an effective parametric monomial $c_{p_0,q_0}(\cdot)c_{p_1,q_1}(\cdot)\cdots c_{p_k,q_k}(\cdot)$, its correlative function is a ρ-degree function of $H_1(j\omega_{l(1)})$ which can be written as a symmetric form

$$\varphi_{n(\bar{s})}\left(c_{p_0,q_0}(\cdot)c_{p_1,q_1}(\cdot)\cdots c_{p_k,q_k}(\cdot);\omega_{l(1)}\cdots\omega_{l(n(\bar{s}))}\right)$$
$$=\frac{(n(\bar{s})-\rho)!\rho!}{n(\bar{s})!}\sum_{\substack{\text{all the combinations of } \rho \text{ integers } \{r_1,r_2,\cdots,r_\rho\} \\ \text{taken from} \{1,2,\cdots,n(\bar{s})\} \text{without repetition} \\ j \text{ is for different combination}}}\mu_j\left(\omega_{l(1)}\cdots\omega_{l(n(\bar{s}))}\right)\prod_{i=1}^{\rho}H_1\left(j\omega_{\bar{l}(i)}\right)$$

where $\rho=n(\bar{s})-\sum_{i=0}^{k}q_i=\sum_{i=0}^{k}p_i-k$, $\bar{l}=\left[r_1,r_2,\cdots,r_\rho\right]$, and $\mu_j\left(\omega_{l(1)}\cdots\omega_{l(n(\bar{s}))}\right)$ can be determined by (11.7a–11.8d). Therefore, the nth-order GFRF can be regarded as an n-degree polynomial function of $H_1(j\omega_{l(1)})$. □

Proof See Sect. 11.6 for the proof. □

Corollary 11.2 demonstrates the relationship between $H_n(j\omega_1,\cdots,j\omega_n)$ and $H_1(j\omega)$, and reveals how the first order GFRF, which represents the linear part of system model, affects the higher order GFRFs, together with the nonlinear dynamics. Note that for any specific parameters of interest, the polynomial structure of the FRFs is explicitly determined in terms of these parameters, thus the property of this polynomial function is greatly dependent on the "coefficients" of these parameter monomials in the polynomial, which correspond to the correlative functions of the parametric characteristics of the polynomial and are determined by the new mapping function. Hence, Corollary 11.2 is important for the qualitative analysis of the connection between $H_n(j\omega_1,\cdots,j\omega_n)$ and $H_1(j\omega)$, and also between nonlinear parameters and high order GFRFs.

Example 11.2 To demonstrate the theoretical results above, consider a simple mechanical system shown in Fig. 11.3.

The output property of the spring satisfies $F=Ky+c_1y^3$, and the damper $F=B\dot{y}+c_2\dot{y}^3$. u(t) is the external input force. The system dynamics can be described by

$$m\ddot{y}=-Ky-B\dot{y}-c_1y^3-c_2\dot{y}^3+u(t) \tag{11.19}$$

which can be written into the form of NDE model (2.11) with $M=3$, $K=2$, $c_{1,0}(2)=m$, $c_{1,0}(1)=B$, $c_{1,0}(0)=K$, $c_{3,0}(000)=c_1$, $c_{3,0}(111)=c_2$, $c_{0,1}(0)=-1$, and all the other parameters are zero.

There are two nonlinear terms $c_{3,0}(000)=c_1$ and $c_{3,0}(111)=c_2$ in model (11.19), which are all pure output nonlinearity and can be written as $C_{3,0}=[c_1,c_2]$. The parametric characteristics of the GFRFs of model (11.19) with respect to nonlinear parameter $C_{3,0}$ can be obtained according to (11.3) or Proposition 5.1 as

$CE(H_{2i+1}(.))=C_{3,0}{}^{i}$ for $i=0,1,2,\ldots,$, otherwise $CE(H_{2i}(.))=0$ for $i=1,2,3,\ldots$

Therefore,

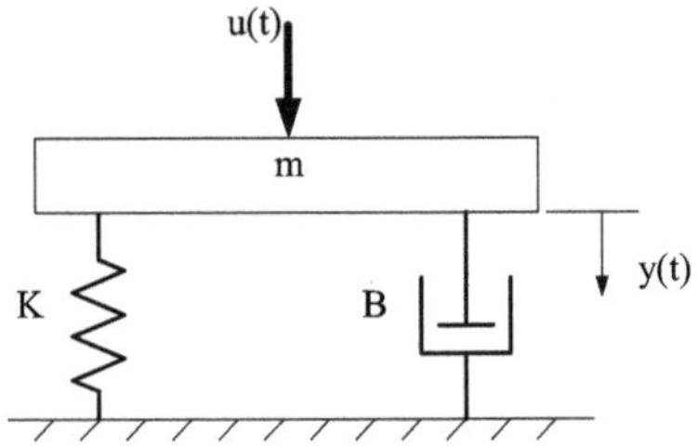

Fig. 11.3 A mechanical system

$$
\begin{aligned}
&CE(H_1(.)) = 1;\\
&CE(H_3(.)) = C_{3,0} = \begin{bmatrix} c_1 & c_2 \end{bmatrix};\\
&CE(H_5(.)) = C_{3,0} \otimes C_{3,0} = \begin{bmatrix} c_1{}^2 & c_1c_2 & c_2{}^2 \end{bmatrix};\\
&CE(H_7(.)) = C_{3,0} \otimes C_{3,0} \otimes C_{3,0} = \begin{bmatrix} c_1{}^3 & c_1{}^2c_2 & c_1c_2{}^2 & c_2{}^3 \end{bmatrix} \ldots\ldots
\end{aligned}
$$

By using (11.7a)–(11.10), it can be obtained that

$$
\begin{aligned}
\varphi_3(c_{3,0}(000);\omega_1,\omega_2,\omega_3) &= \frac{1}{L_3\left(j\sum_{i=1}^{3}\omega_i\right)}\cdot\prod_{i=1}^{3}(j\omega_i)^0\cdot\prod_{i=1}^{3}H_1(j\omega_i)\\
&= \frac{1}{L_3\left(j\sum_{i=1}^{3}\omega_i\right)}\cdot\prod_{i=1}^{3}H_1(j\omega_i)
\end{aligned}
$$

$$
\begin{aligned}
\varphi_3(c_{3,0}(111);\omega_1,\omega_2,\omega_3) &= \frac{1}{L_3\left(j\sum_{i=1}^{3}\omega_i\right)}\cdot\prod_{i=1}^{3}(j\omega_i)\cdot\prod_{i=1}^{3}H_1(j\omega_i)\\
&= \frac{\prod_{i=1}^{3}(j\omega_i)}{L_3\left(j\sum_{i=1}^{3}\omega_i\right)}\cdot\prod_{i=1}^{3}H_1(j\omega_i)
\end{aligned}
$$

$$
\begin{aligned}
&\varphi_5(c_{3,0}(000)c_{3,0}(000);\omega_1,\cdots,\omega_5)\\
&=f_1(c_{3,0}(000),5;\omega_1,\cdots,\omega_5)\cdot\sum_{\substack{\text{all the } 3-\text{partitions}\\ \text{for } c_{3,0}(000)}}\ \sum_{\substack{\text{all the different}\\ \text{permutations of } \{0,0,1\}}}\Big[f_{2a}(s_{\bar{x}_1}\cdots s_{\bar{x}_p}(c_{3,0}(000));\omega_1\cdots\omega_5)\\
&\quad\cdot\prod_{i=1}^{3}\varphi_{n\left(s_{\bar{x}_i}(\bar{s}/c_{pq}(\cdot))\right)}\left(s_{\bar{x}_i}(c_{3,0}(000));\omega_{l(\bar{X}(i)+1)}\cdots\omega_{l\left(\bar{X}(i)+n\left(s_{\bar{x}_i}(\bar{s}/c_{p,q}(\cdot))\right)\right)}\right)\Big]\\
&=f_1(c_{3,0}(000),5;\omega_1,\cdots,\omega_5)\cdot\begin{pmatrix} f_{2a}(s_0s_0s_1(c_{3,0}(000));\omega_1\cdots\omega_5)\varphi_1(1;\omega_1)\varphi_1(1;\omega_2)\varphi_3(c_{3,0}(000);\omega_3\cdots\omega_5)\\ +f_{2a}(s_0s_1s_0(c_{3,0}(000));\omega_1\cdots\omega_5)\varphi_1(1;\omega_1)\varphi_3(c_{3,0}(000);\omega_2\cdots\omega_4)\varphi_1(1;\omega_5)\\ +f_{2a}(s_1s_0s_0(c_{3,0}(000));\omega_1\cdots\omega_5)\varphi_3(c_{3,0}(000);\omega_1\cdots\omega_3)\varphi_1(1;\omega_4)\varphi_1(1;\omega_5)\end{pmatrix}
\end{aligned}
$$

$$=\frac{1}{L_5\left(j\sum_{i=1}^{5}\omega_i\right)}\cdot\left(\begin{array}{l}H_1(\omega_1)H_1(\omega_2)\prod_{i=3}^{5}H_1(j\omega_i)/L_3\left(j\sum_{i=3}^{5}\omega_i\right)\\+H_1(\omega_1)\prod_{i=2}^{4}H_1(j\omega_i)H_1(\omega_5)/L_3\left(j\sum_{i=2}^{4}\omega_i\right)\\+\prod_{i=1}^{3}H_1(j\omega_i)H_1(\omega_4)H_1(\omega_5)/L_3\left(j\sum_{i=1}^{3}\omega_i\right)\end{array}\right)$$

$$=\frac{1}{L_5\left(j\sum_{i=1}^{5}\omega_i\right)}\cdot\left(\frac{1}{L_3\left(j\sum_{i=3}^{5}\omega_i\right)}+\frac{1}{L_3\left(j\sum_{i=2}^{4}\omega_i\right)}+\frac{1}{L_3\left(j\sum_{i=1}^{3}\omega_i\right)}\right)\cdot\prod_{i=1}^{5}H_1(j\omega_i)$$

$$\varphi_5(c_{3,0}(111)c_{3,0}(111);\omega_1,\cdots,\omega_5)$$
$$=f_1(c_{3,0}(111),5;\omega_1,\cdots,\omega_5)\cdot\sum_{\substack{\text{all the }3-\text{partitions}\\ \text{for } c_{3,0}(111)}}\ \sum_{\substack{\text{all the different}\\ \text{permutations of }\{0,0,1\}}}\left[f_{2a}\left(s_{\overline{x}_1}\cdots s_{\overline{x}_p}(c_{3,0}(111));\omega_1\cdots\omega_5\right)\right.$$
$$\left.\cdot\prod_{i=1}^{3}\varphi_{n\left(s_{\overline{x}_i}\left(\overline{s}/c_{p,q}(\cdot)\right)\right)}\left(s_{\overline{x}_i}(c_{3,0}(111));\omega_{l\left(\overline{X}(i)+1\right)}\cdots\omega_{l\left(\overline{X}(i)+n\left(s_{\overline{x}_i}\left(\overline{s}/c_{p,q}(\cdot)\right)\right)\right)}\right)\right]$$

$$=f_1(c_{3,0}(111),5;\omega_1,\cdots,\omega_5)\cdot\left(\begin{array}{l}f_{2a}(s_0s_0s_1(c_{3,0}(111));\omega_1\cdots\omega_5)\varphi_1(1;\omega_1)\varphi_1(1;\omega_2)\varphi_3(c_{3,0}(111);\omega_3\cdots\omega_5)\\+f_{2a}(s_0s_1s_0(c_{3,0}(111));\omega_1\cdots\omega_5)\varphi_1(1;\omega_1)\varphi_3(c_{3,0}(111);\omega_2\cdots\omega_4)\varphi_1(1;\omega_5)\\+f_{2a}(s_1s_0s_0(c_{30}(111));\omega_1\cdots\omega_5)\varphi_3(c_{30}(111);\omega_1\cdots\omega_3)\varphi_1(1;\omega_4)\varphi_1(1;\omega_5)\end{array}\right)$$

$$=\frac{1}{L_5\left(j\sum_{i=1}^{5}\omega_i\right)}\cdot\left(\frac{\left(j\sum_{i=3}^{5}\omega_i\right)\prod_{i=1}^{5}(j\omega_i)}{L_3\left(j\sum_{i=3}^{5}\omega_i\right)}+\frac{\left(j\sum_{i=2}^{4}\omega_i\right)\prod_{i=1}^{5}(j\omega_i)}{L_3\left(j\sum_{i=2}^{4}\omega_i\right)}+\frac{\left(j\sum_{i=1}^{3}\omega_i\right)\prod_{i=1}^{5}(j\omega_i)}{L_3\left(j\sum_{i=1}^{3}\omega_i\right)}\right)$$
$$\cdot\prod_{i=1}^{5}H_1(j\omega_i)$$

$$\varphi_5(c_{3,0}(000)c_{3,0}(111);\omega_1,\cdots,\omega_5)$$
$$=f_1(c_{3,0}(000),5;\omega_1,\cdots,\omega_5)$$
$$\cdot\sum_{\substack{\text{all the }3-\text{partitions}\\ \text{for } c_{3,0}(111)}}\ \sum_{\substack{\text{all the different}\\ \text{permutations of }\{0,0,1\}}}\left[f_{2a}\left(s_{\overline{x}_1}\cdots s_{\overline{x}_p}(c_{3,0}(111));\omega_1\cdots\omega_5\right)\right.$$
$$\left.\cdot\prod_{i=1}^{3}\varphi_{n\left(s_{\overline{x}_i}\left(\overline{s}/c_{p,q}(\cdot)\right)\right)}\left(s_{\overline{x}_i}(c_{3,0}(111));\omega_{l\left(\overline{X}(i)+1\right)}\cdots\omega_{l\left(\overline{X}(i)+n\left(s_{\overline{x}_i}\left(\overline{s}/c_{p,q}(\cdot)\right)\right)\right)}\right)\right]$$
$$+f_1(c_{3,0}(111),5;\omega_1,\cdots,\omega_5)$$

$$\cdot \sum_{\substack{\text{all the 3-partitions}\\ \text{for } c_{3,0}(000)}} \sum_{\substack{\text{all the different}\\ \text{permutations of } \{0,0,1\}}} \Big[f_{2a}(s_{\bar{x}_1}\cdots s_{\bar{x}_p}(c_{3,0}(000));\omega_1\cdots\omega_5)$$

$$\cdot \prod_{i=1}^{3} \varphi_{n(s_{\bar{x}_i}(\bar{s}/c_{p,q}(\cdot)))}\left(s_{\bar{x}_i}(c_{3,0}(000));\omega_{l(\overline{X}(i)+1)}\cdots\omega_{l(\overline{X}(i)+n(s_{\bar{x}_i}(\bar{s}/c_{p,q}(\cdot))))}\right)\Big]$$

$$= f_1(c_{3,0}(000),5;\omega_1,\cdots,\omega_5)$$

$$\cdot\begin{pmatrix} f_{2a}(s_0s_0s_1(c_{3,0}(111));\omega_1\cdots\omega_5)\varphi_1(1;\omega_1)\varphi_1(1;\omega_2)\varphi_3(c_{3,0}(111);\omega_3\cdots\omega_5) \\ +f_{2a}(s_0s_1s_0(c_{3,0}(111));\omega_1\cdots\omega_5)\varphi_1(1;\omega_1)\varphi_3(c_{3,0}(111);\omega_2\cdots\omega_4)\varphi_1(1;\omega_5) \\ +f_{2a}(s_1s_0s_0(c_{3,0}(111));\omega_1\cdots\omega_5)\varphi_3(c_{3,0}(111);\omega_1\cdots\omega_3)\varphi_1(1;\omega_4)\varphi_1(1;\omega_5) \end{pmatrix}$$

$$+ f_1(c_{3,0}(111),5;\omega_1,\cdots,\omega_5)$$

$$\cdot\begin{pmatrix} f_{2a}(s_0s_0s_1(c_{3,0}(000));\omega_1\cdots\omega_5)\varphi_1(1;\omega_1)\varphi_1(1;\omega_2)\varphi_3(c_{3,0}(000);\omega_3\cdots\omega_5) \\ +f_{2a}(s_0s_1s_0(c_{3,0}(000));\omega_1\cdots\omega_5)\varphi_1(1;\omega_1)\varphi_3(c_{3,0}(000);\omega_2\cdots\omega_4)\varphi_1(1;\omega_5) \\ +f_{2a}(s_1s_0s_0(c_{3,0}(000));\omega_1\cdots\omega_5)\varphi_3(c_{3,0}(000);\omega_1\cdots\omega_3)\varphi_1(1;\omega_4)\varphi_1(1;\omega_5) \end{pmatrix}$$

$$= \frac{1}{L_5\left(j\sum_{i=1}^{5}\omega_i\right)}$$

$$\cdot\left(\frac{1+\left(j\sum_{i=3}^{5}\omega_i\right)\prod_{i=1}^{5}(j\omega_i)}{L_3\left(j\sum_{i=3}^{5}\omega_i\right)} + \frac{1+\left(j\sum_{i=2}^{4}\omega_i\right)\prod_{i=1}^{5}(j\omega_i)}{L_3\left(j\sum_{i=2}^{4}\omega_i\right)} + \frac{1+\left(j\sum_{i=1}^{3}\omega_i\right)\prod_{i=1}^{5}(j\omega_i)}{L_3\left(j\sum_{i=1}^{3}\omega_i\right)} \right)$$

$$\cdot \prod_{i=1}^{5} H_1(j\omega_i)$$

Hence, it can be obtained that

$$\varphi_3(CE(H_3(\cdot))) = \frac{1}{L_3\left(j\sum_{i=1}^{3}\omega_i\right)} \cdot \left[\prod_{i=1}^{3}(j\omega_i)\right]^1 \cdot \prod_{i=1}^{3} H_1(j\omega_i)$$

$$\varphi_5(CE(H_5(\omega_1,\cdots,\omega_5))) = \frac{1}{L_5\left(j\sum_{i=1}^{5}\omega_i\right)}$$

$$\cdot\begin{bmatrix} \dfrac{1}{L_3\left(j\sum_{i=3}^{5}\omega_i\right)}+\dfrac{1}{L_3\left(j\sum_{i=2}^{4}\omega_i\right)}+\dfrac{1}{L_3\left(j\sum_{i=1}^{3}\omega_i\right)} \\ \dfrac{1+\left(j\sum_{i=3}^{5}\omega_i\right)\prod_{i=1}^{5}(j\omega_i)}{L_3\left(j\sum_{i=3}^{5}\omega_i\right)}+\dfrac{1+\left(j\sum_{i=2}^{4}\omega_i\right)\prod_{i=1}^{5}(j\omega_i)}{L_3\left(j\sum_{i=2}^{4}\omega_i\right)}+\dfrac{1+\left(j\sum_{i=1}^{3}\omega_i\right)\prod_{i=1}^{5}(j\omega_i)}{L_3\left(j\sum_{i=1}^{3}\omega_i\right)} \\ \dfrac{\left(j\sum_{i=3}^{5}\omega_i\right)\prod_{i=1}^{5}(j\omega_i)}{L_3\left(j\sum_{i=3}^{5}\omega_i\right)}+\dfrac{\left(j\sum_{i=2}^{4}\omega_i\right)\prod_{i=1}^{5}(j\omega_i)}{L_3\left(j\sum_{i=2}^{4}\omega_i\right)}+\dfrac{\left(j\sum_{i=1}^{3}\omega_i\right)\prod_{i=1}^{5}(j\omega_i)}{L_3\left(j\sum_{i=1}^{3}\omega_i\right)} \end{bmatrix}$$

$$\cdot\prod_{i=1}^{5}H_1(j\omega_i)$$

By using (11.14), the GFRFs for $n=3$ and 5 of system (11.19) can be obtained. Proceeding with the computation process above, any higher order GFRFs of system (11.19) can be derived and written in a much more meaningful form. It can be seen that, the correlative function of a monomial in the parametric characteristic of the nth-order GFRF is an n-degree polynomial of the first order GFRF as stated in Corollary 11.2, and so the nth-order GFRF is. Based on (11.14), the first order parametric sensitivity of the GFRFs with respect to any nonlinear parameter can be studied as

$$\frac{\partial H_n(j\omega_1,\cdots,j\omega_n)}{\partial c}=\frac{\partial CE(H_n(\cdot))}{\partial c}\cdot\varphi_n(CE(H_n(\cdot)))$$

For example,

$$\begin{aligned}\frac{\partial H_3(j\omega_1,\cdots,j\omega_3)}{\partial c_1}&=\frac{\partial CE(H_3(\cdot))}{\partial c_1}\cdot\varphi_3(CE(H_3(\cdot)))=[1,0]\cdot\varphi_3(CE(H_3(\cdot)))\\&=\prod_{i=1}^{3}H_1(j\omega_i)/L_3\left(j\sum_{i=1}^{3}\omega_i\right)\end{aligned}$$

Similarly,

$$\frac{\partial H_5(j\omega_1,\cdots,j\omega_5)}{\partial c_1}=\frac{\partial CE(H_5(\cdot))}{\partial c_1}\cdot\varphi_5(CE(H_5(\cdot)))=[2c_1,c_2,0]\cdot\varphi_5(CE(H_5(\cdot))).$$

Similar results can also be obtained for parameter c_2. It can be seen that the sensitivity of the third order GFRF with respect to the nonlinear spring c_1 and nonlinear damping c_2 is constant which is dependent on linear parameters, but the sensitivity of the higher order GFRFs will be a function of these nonlinearities and the linear parameters. Note that for a Volterra system, the system output is usually dominated by its several low order GFRFs (Boyd and Chua 1985). Hence, in order to make the system less sensitive to these nonlinearities, the linear parameters should properly be designed.

Moreover, the magnitude of $H_n(j\omega_1,\cdots,j\omega_n)$ can also be evaluated readily according to Corollary 11.1. For example, for $n=3$

$$\begin{aligned}|H_3(j\omega_1,\cdots,j\omega_3)|^2 &= CE_3\Theta_3 CE_3^T\\ &= \left|\frac{\prod_{i=1}^{3} H_1(j\omega_i)}{L_3\left(j\sum_{i=1}^{3}\omega_i\right)}\right|^2 \cdot \begin{bmatrix} c_1\\ c_2\end{bmatrix}^T \begin{bmatrix} 1 & \prod_{i=1}^{3}(j\omega_i)\\ -\prod_{i=1}^{3}(j\omega_i) & \prod_{i=1}^{3}(\omega_i^2)\end{bmatrix}\begin{bmatrix} c_1\\ c_2\end{bmatrix}\end{aligned}$$

As mentioned above, instead of studying the Bode diagram of $H_3(j\omega_1,\cdots,j\omega_3)$, the frequency response spectrum of the maximum eigenvalue of the third order frequency characteristic matrix defined in Corollary 11.1 can be investigated. See Fig. 11.4. Different values of the linear parameters will result in a different view. An increase of the linear damping enables the magnitude to increase for higher $\omega_1+\omega_2+\omega_3$ along the line $\omega_1+\omega_3=0$. Note that the system output spectrum (11.15a–c) involves the computation of the GFRFs along a super-plane$\omega_1+\cdots+\omega_n=\omega$. The frequency response spectra of the maximum eigenvalue on the plane $\omega_1+\cdots+\omega_3=\omega$ with different output frequency ω are given in Fig. 11.5. The peak and valley in the figures can represent special properties of the system. Understanding of these diagrams can follow the method in Yue et al. (2005), and further results are under study.

The system output spectrum can also be studied. For example, suppose the system is subject to a harmonic input $u(t)=F_d\sin(\omega_0 t)$ $(F_d>0)$, then the magnitude of the third order output spectrum can be evaluated as (Jing et al. 2007a)

$$\begin{aligned}|Y_3(j\omega)| &\le \frac{1}{2^3}\sum_{\omega_{k_1}+\cdots+\omega_{k_3}=\omega} |H_3(j\omega_{k_1},\cdots,j\omega_{k_3})||F(\omega_{k_1})\cdots F(\omega_{k_3})|\\ &\le \frac{F_d^3}{2^3}\sum_{\omega_{k_1}+\cdots+\omega_{k_3}=\omega} |H_3(j\omega_{k_1},\cdots,j\omega_{k_3})|\end{aligned}$$

From corollary 11.1, $|H_3(j\omega_1,\cdots,j\omega_3)| \le \sqrt{\lambda_3(j\omega_1,\cdots,j\omega_n)}\left\|CE_3^T\right\|$. Therefore,

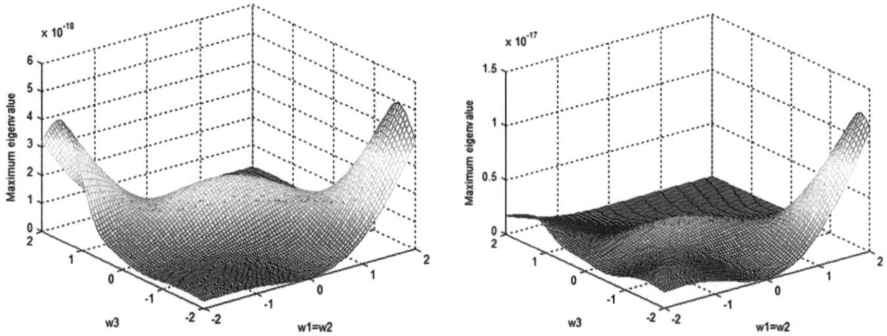

Fig. 11.4 Frequency response spectrum of the maximum eigenvalue when m=24, B=2.96 (*left*) or 29.6 (*right*), K=160 (Jing et al. 2008e)

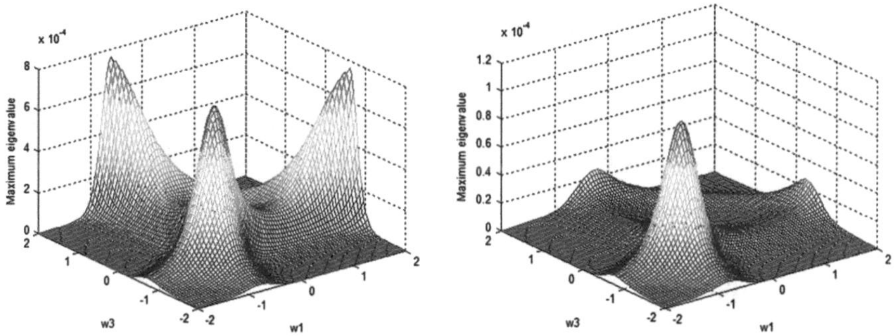

Fig. 11.5 Frequency response spectrum of the maximum eigenvalue when m=2.4, B=2.96, K= 1.6 and $\omega_1+\omega_2+\omega_3=0.8$ (*left*) or 1.5 (*right*) (Jing et al. 2008e)

$$
\begin{aligned}
|Y_3(j\omega)| &\leq \frac{F_d^3}{2^3} \sum_{\omega_{k_1}+\cdots+\omega_{k_3}=\omega} \sqrt{\lambda_3(j\omega_1,\cdots,j\omega_n)}\left\|CE_3^T\right\| \\
&= \frac{F_d^3}{2^3}\sqrt{c_1^2+c_2^2} \sum_{\omega_{k_1}+\cdots+\omega_{k_3}=\omega} \sqrt{\lambda_3(j\omega_1,\cdots,j\omega_n)}
\end{aligned}
$$

For $\omega=0.8$ and m=2.4, B=29.6, K=1.6, it can be obtained that $\sqrt{\lambda_3(j\omega_1,\cdots,j\omega_n)}$ ≤ 0.006055896. Hence, in this case

$$
|Y_3(j\omega)| \leq 0.00227096 F_d^3 \sqrt{c_1^2+c_2^2}
$$

Obviously, given a requirement on the bound of $|Y_3(j\omega)|$, the design restriction on the nonlinear parameters c_1 and c_2 can be further derived. □

11.5 Conclusions

A mapping function from the parametric characteristics to the GFRFs is established. The *n*th-order GFRF can directly be written into a more straightforward and meaningful form in terms of the first order GFRF and model parameters based on the parametric characteristic, which explicitly unveils the linear and nonlinear factors included in the GFRFs and can be regarded as an n-degree polynomial function of the first order GFRF. These results demonstrate some new properties of the GFRFs, which can reveal clearly the relationship between the nth-order GFRF and its parametric characteristic, and also the relationship between the higher order GFRF and the first order GFRF. These provide a novel and useful insight into the frequency domain analysis and design of nonlinear systems based on the GFRFs. Note that the results of this study are established for nonlinear systems described by the NDE model, similar results can be extended to discrete time nonlinear systems described by NARX model (Jing and Lang 2009a). As shown, the results provide a useful tool for investigation of nonlinear dynamics in the frequency domain and thus a useful insight into frequency domain analysis and design of nonlinear systems based on the GFRFs. With the mapping function established, nonlinear influence on system output spectrum can be studied, which is discussed in the next chapter.

11.6 Proofs

A. Proof of Lemma 11.2

(1) From Proposition 5.1, it can be computed that $c_{pq}(\cdot)c_{p_1q_1}(\cdot)\cdots c_{p_kq_k}(\cdot)$ comes from the *x*th-order GFRF, where $x=p+q+\sum_{i=1}^{k}(p_i+q_i)-k$. It is obvious that $c_{p_1q_1}(\cdot)\cdots c_{p_kq_k}(\cdot)$ comes from the correlative function of the parameter $c_{pq}(\cdot)$ in (5.1) or (11.2) for the *x*th-order GFRF, *i.e.*, $\left(\prod_{i=1}^{q}(j\omega_{x-q+i})^{k_{p+i}}\right)H_{x-q,p}(j\omega_1,\cdots,j\omega_{x-q})$, that is, it comes from $H_{x-q,p}(j\omega_1,\cdots,j\omega_{n-q})$. From (5.5), it follows that

$$H_{x-q,p}(j\omega_1,\cdots,j\omega_{x-q})=\sum_{\substack{r_1\cdots r_p=1\\ \sum r_i=x-q}}^{x-p-q+1}\prod_{i=1}^{p}H_{r_i}\left(j\omega_{rX+1},\cdots,j\omega_{rX+r_i}\right)\left(j\omega_{rX+1}+\cdots+j\omega_{rX+r_i}\right)^{k_i} \quad \text{(B1)}$$

Obviously, $\prod_{i=1}^{p}H_{r_i}\left(j\omega_{rX+1},\cdots,j\omega_{rX+r_i}\right)$ is a (p,q)-partition for the *x*th-order GFRF

(2) Supposing that s_0 comes from $H_1(.)$, each monomial s_{x_i} in a p-partition for $c_{p_1q_1}(\cdot)\cdots c_{p_kq_k}(\cdot)$ comes from the $\left(\sum_{j=1}^{x_i}(p_j+q_j)-x_i+1\right)$th-order GFRF if $x_i>0$, therefore, each p-partition for $c_{p_1q_1}(\cdot)\cdots c_{p_kq_k}(\cdot)$ corresponds to a combination of $H_{r_1}(w_{r_1})H_{r_2}(w_{r_2})\cdots H_{r_p}(w_{r_p})$ which must included in (B1) since (B1) includes all the possible (*p*,*q*)-partitions, where $r_i=\sum_{j=1}^{x_i}(p_j+q_j)-x_i+1$. That is, each p-partition for $c_{p_1q_1}(\cdot)\cdots c_{p_kq_k}(\cdot)$ corresponds to a (*p*,*q*)-partition for the *x*th-order GFRF. On the other hand, each (*p*,*q*)-partition in (B1) which produces $c_{p_1q_1}(\cdot)\cdots c_{p_kq_k}(\cdot)$ must correspond to at least one p-partition for $c_{p_1q_1}(\cdot)\cdots c_{p_kq_k}(\cdot)$.

(3) Equation (B1) includes all the (*p*,*q*)-partitions for the *x*th-order GFRF which produce $c_{p_1q_1}(\cdot)\cdots c_{p_kq_k}(\cdot)$, thus the correlative function of $c_{p_1q_1}(\cdot)\cdots c_{p_kq_k}(\cdot)$ are the summation of all the correlative functions of each (*p*,*q*)-partition. Note that each (*p*,*q*)-partition may produce more than one p-partition for $c_{p_1q_1}(\cdot)\cdots c_{p_kq_k}(\cdot)$. This implies there are more than one cases in the same (*p*, *q*)-partition to produce $c_{p_1q_1}(\cdot)\cdots c_{p_kq_k}(\cdot)$. Therefore, the correlative function of $c_{p_1q_1}(\cdot)\cdots c_{p_kq_k}(\cdot)$ should be the summation of the correlative functions corresponding to all the cases where $c_{p_1q_1}(\cdot)\cdots c_{p_kq_k}(\cdot)$ are produced.

This completes the proof. □

B. Proof of Proposition 11.1

Considering the recursive equation (11.2), the recursive structure in (11.7a) is directly followed from Lemma 11.1 (2) and Lemma 11.2 (3). That is, the correlative function of $c_{p_1q_1}(\cdot)\cdots c_{p_kq_k}(\cdot)$ are the summation of the correlative functions with respect to all the cases by which this monomial is produced in the same $n(\bar{s})$th-order GFRF, in each case it should include all the correlative functions corresponding to all the p-partition for $c_{p_1q_1}(\cdot)\cdots c_{p_kq_k}(\cdot)$, and for each p-partition of $c_{p_1q_1}(\cdot)\cdots c_{p_kq_k}(\cdot)$, the correlative function should include all the permutations of $x_1x_2...x_p$, since the correlative function $f_{2a}(s_{\bar{x}_1}\cdots s_{\bar{x}_p}(\bar{s}/c_{pq}(\cdot));\omega_{l(1)}\cdots\omega_{l(n(\bar{s})-q)})$ is different with each different permutation which can be seen from (5.5). $f_1(c_{p,q}(\cdot),n(\bar{s});\omega_{l(1)}\cdots\omega_{l(n(\bar{s}))})$ is a part of the correlative function for $c_{p,q}(k_1,\cdots,k_{p+q})$ except for $H_{n(\bar{s})-q,p}(j\omega_1,\cdots,j\omega_{n(\bar{s})-q})$, which directly follows from (11.2). $f_{2a}(s_{\bar{x}_1}\cdots s_{\bar{x}_p}(\bar{s}/c_{pq}(\cdot));\omega_{l(1)}\cdots\omega_{l(n(\bar{s})-q)})$ is a part of the correlative function with respect to a permutation of a p-partition $s_{x_1}\cdots s_{x_p}(\bar{s}/c_{pq}(\cdot))$ of the monomial $\bar{s}/c_{pq}(\cdot)$ which corresponds to a (*p*,*q*)-partition for the $n(\bar{s})$th-order GFRF, and it is followed from (5.5). Equation (11.7b) has a similar structure with (11.7a), and is an optimised one which simplifies the computation of (11.7a) for the reason that $\prod_{i=1}^{p}\varphi_{n(s_{x_i}(\bar{s}/c_{pq}(\cdot)))}\left(s_{x_i}(\bar{s}/c_{pq}(\cdot));\omega_{l(X(i)+1)}\cdots\omega_{l(X(i)+n(s_{x_i}(\bar{s}/c_{pq}(\cdot))))}\right)$ is identical to each other under each permutation of a p-partition for the monomial $\bar{s}/c_{pq}(\cdot)$, and therefore the contribution from each permutation is included in $f_{2b}\left(s_{x_1}\cdots s_{x_p}\left(\overline{s}/c_{pq}(\cdot)\right);\omega_{l(1)}\cdots\omega_{l(n(\overline{s})-q)}\right)$ which can be obtained from (5.5) and is also given in Peyton Jones (2007). This completes the proof. □

C. Proof of Proposition 11.2

From (2.2), it can be obtained that

$$|H_n(j\omega_1,\cdots,j\omega_n)| \le \int_{-\infty}^{\infty}\cdots\int_{-\infty}^{\infty}|h_n(\tau_1,\cdots,\tau_n)\exp(-j(\omega_1\tau_1+\cdots+\omega_n\tau_n))|d\tau_1\cdots d\tau_n$$

which further gives

$$\sup_{\omega_1,\cdots,\omega_n}|H_n(j\omega_1,\cdots,j\omega_n)| \le \int_{-\infty}^{\infty}\cdots\int_{-\infty}^{\infty}|h_n(\tau_1,\cdots,\tau_n)|d\tau_1\cdots d\tau_n$$

Suppose at point $(\omega_1^*,\cdots,\omega_n^*)$, it holds that

$$\begin{aligned}\sup_{\omega_1,\cdots,\omega_n}|H_n(j\omega_1,\cdots,j\omega_n)| &= \left|H_n\left(j\omega_1^*,\cdots,j\omega_n^*\right)\right|\\ &= \int_{-\infty}^{\infty}\cdots\int_{-\infty}^{\infty}|h_n(\tau_1,\cdots,\tau_n)|d\tau_1\cdots d\tau_n\end{aligned}$$

From (11.16a), it can be obtained that

$$|H_n(j\omega_1,\cdots,j\omega_n)|^2 \le \lambda_{\max}(\Theta_n)\cdot CE_nCE_n^T$$

Thus it holds that

$$\left|H_n\left(j\omega_1^*,\cdots,j\omega_n^*\right)\right|^2 \le \lambda_{\max}\left(\Theta_n\left(\omega_1^*,\cdots,\omega_n^*\right)\right)\cdot CE_nCE_n^T$$

Hence, $\int_{-\infty}^{\infty}\cdots\int_{-\infty}^{\infty}|h_n(\tau_1,\cdots,\tau_n)|d\tau_1\cdots d\tau_n \le \sqrt{\sup_{\omega_1,\cdots,\omega_n}(\lambda_{\max}(\Theta_n))}\cdot\|CE_n\|$.
Following a similar process, (11.18b) can be obtained. This completes the proof. □

D. Proof of Corollary 11.2

From (11.10), for a parameter corresponding to a pure input nonlinear term $c_{0,q}(.)$, it can be derived that

$$\varphi_{n(\overline{s})}\left(c_{0q}(\cdot);\omega_{l(1)}\cdots\omega_{l(n(\overline{s}))}\right) = \frac{1}{L_{n(\overline{s})}\left(j\sum_{i=1}^{n(\overline{s})}\omega_{l(i)}\right)}\left(\prod_{i=1}^{q}\left(j\omega_{l(i)}\right)^{k_i}\right.$$

There is no $H_1(j\omega_{l(1)})$ appearing in the correlative function. That is, the degree of $H_1(j\omega_{l(1)})$ in the correlative function of this kind of nonlinear parameters is zero. For a parameter corresponding to a pure output nonlinear term $c_{p,0}(.)$, it can be derived that $\varphi_{n(\bar{s})}\left(c_{p0}(\cdot);\omega_{l(1)}\cdots\omega_{l(n(\bar{s}))}\right)=\varphi_{n(\bar{s})}\left(c_{n(\bar{s})0}\right.$ $\left.(\cdot);\omega_{l(1)}\cdots\omega_{l(n(\bar{s}))}\right)=\frac{1}{L_{n(\bar{s})}\left(j\sum_{i=1}^{n(\bar{s})}\omega_{l(i)}\right)}\prod_{i=1}^{n(\bar{s})}\left(j\omega_{l(i)}\right)^{k_i}\cdot\prod_{i=1}^{n(\bar{s})}H_1\left(j\omega_{l(i)}\right).$

The degree of $H_1(j\omega_{l(1)})$ in the correlative function of this kind of nonlinear parameters is $n(\bar{s})$. For a parameter corresponding to a pure input–output nonlinear term $c_{p,q}(.)$, it can be seen from (11.10) that the degree of $H_1(j\omega_{l(1)})$ in the correlative function of this kind of nonlinear parameters is $n(\bar{s})-q$. Hence, after recursive computation, for a monomial $c_{p_0q_0}(\cdot)c_{p_1q_1}(\cdot)\cdots c_{p_kq_k}(\cdot)$, the degree of $H_1(j\omega_{l(1)})$ in the correlative function is $n(\bar{s})-\sum_{i=0}^{k}q_i=\sum_{i=0}^{k}(p_i+q_i)-k-\sum_{i=0}^{k}q_i=\sum_{i=0}^{k}p_i-k$. It is also noted that the largest order is $n(\bar{s})$ when all $q_i=0$ corresponding to the parametric monomial whose parameters are all from pure output nonlinearity, and the smallest order is zero when $n(\bar{s})=\sum_{i=0}^{k}q_i$ corresponding to the parametric monomial whose parameters are all from pure input nonlinearity. Therefore, $H_n(j\omega_1,\cdots,j\omega_n)$ can be regarded as an n-degree polynomial function of $H_1(j\omega_{l(1)})$. This completes the proof. □

Chapter 12
The Alternating Series Approach to Nonlinear Influence in the Frequency Domain

12.1 Introduction

It is known that the transfer function of a linear system provides a coordinate-free and equivalent description for system dynamics, which greatly facilitates the analysis and design of system output response. Although the frequency-domain theory for linear systems is well established in the literature, the corresponding methods for linear systems cannot directly be used for frequency domain analysis of nonlinear systems. Nonlinear systems usually have very complicated output frequency characteristics such as harmonics and inter-modulation. Investigation of these nonlinear phenomena in the frequency domain is far from full development.

In this study, understanding of nonlinear effect in the frequency domain is investigated from a novel viewpoint for the Volterra class of nonlinear systems. The system output spectrum is shown to be an alternating series with respect to some model parameters that define system nonlinearities. The output spectrum can therefore be suppressed by exploiting the alternating properties to design corresponding parameters. The concept of alternating series provides a novel insight into the nonlinear influence on system output response in the frequency domain. The sufficient (and necessary) conditions in which the output spectrum can be transformed into an alternating series are studied. To illustrate the new results, several examples are given, which investigated a single degree of freedom (SDOF) mass-spring-damper system with a cubic nonlinear damper. All these results demonstrate a novel insight into the analysis and design of nonlinearities in the frequency domain.

The content of this chapter is organised as follows. Section 12.2 provides a simple explanation for the background of this study. The novel nonlinear characteristic and its influence are discussed in Sect. 12.3. Section 12.4 gives a sufficient and necessary condition under which system output spectrum can be transformed into an alternating series. A conclusion is given in Sect. 12.5.

X. Jing, Z. Lang, *Frequency Domain Analysis and Design of Nonlinear Systems based on Volterra Series Expansion*, Understanding Complex Systems,
DOI 10.1007/978-3-319-12391-2_12

12.2 An Outline of Frequency Response Functions of Nonlinear Systems

For convenience, an outline is given in this section for some results discussed in the previous chapters relating to frequency response functions that form the basis of this study. As mentioned, a wide class of nonlinear systems can be approximated by the Volterra series up to a maximum order N around the zero equilibrium (Boyd and Chua 1985) described by (2.1a,b). In this Chapter, consider nonlinear systems described by the NDE model (2.11). The computation of the nth-order generalized frequency response function (GFRF) for the NDE model (2.11) can be conducted by following (2.19–2.26). The output spectrum of model (2.11) can be evaluated by (3.1), i.e.,

$$Y(j\omega) = \sum_{n=1}^{N} \frac{1}{\sqrt{n}(2\pi)^{n-1}} \int_{\omega_1+\cdots+\omega_n=\omega} H_n(j\omega_1,\cdots,j\omega_n)\prod_{i=1}^{n} U(j\omega_i)d\sigma_\omega \tag{12.1}$$

where,

$$H_n(j\omega_1,\cdots,j\omega_n) = \int_{-\infty}^{\infty}\cdots\int_{-\infty}^{\infty} h_n(\tau_1,\cdots,\tau_n)\exp(-j(\omega_1\tau_1+\cdots+\omega_n\tau_n))d\tau_1\cdots d\tau_n \tag{12.2}$$

is known as the nth-order GFRF defined in George (1959). When the system input is a multi-tone function described by (3.2), the system output frequency response can be described as:

$$Y(j\omega) = \sum_{n=1}^{N} \frac{1}{2^n} \sum_{\omega_{k_1}+\cdots+\omega_{k_n}=\omega} H_n(j\omega_{k_1},\cdots,j\omega_{k_n})F(\omega_{k_1})\cdots F(\omega_{k_n}) \tag{12.3}$$

where $F(\omega_{k_i})$ can be explicitly written as

$$F(\omega_{k_i}) = \left|F_{|k_i|}\right| e^{j\angle F_{|k_i|}\cdot \mathrm{sig}(k_i)} \text{ for } k_i \in \{\pm 1,\cdots,\pm \overline{K}\} \tag{12.4}$$

where $\mathrm{sgn}(a) = \begin{cases} 1 & a \geq 0 \\ -1 & a < 0 \end{cases}$, and $\omega_{k_i} \in \{\pm\omega_1,\cdots,\pm\omega_{\overline{K}}\}$.

In order to reveal the relationship between the system frequency response functions and model parameters, the parametric characteristics of the GFRFs and output spectrum are studied in Chaps. 5 and 6. The results show that the nth-order GFRF can be expressed as a more straightforward polynomial function of the system nonlinear parameters, i.e.,

$$H_n(j\omega_1,\cdots,j\omega_n) = CE(H_n(j\omega_1,\cdots,j\omega_n)) \cdot f_n(j\omega_1,\cdots,j\omega_n) \tag{12.5}$$

where, $CE(H_n(j\omega_1,\cdots,j\omega_n))$ is referred to as the parametric characteristic of the nth-order GFRF $H_n(j\omega_1,\cdots,j\omega_n)$, which can recursively be determined by (5.17) or (5.18), e.g.,

$$CE(H_n(j\omega_1,\cdots,j\omega_n)) = C_{0,n} \oplus \left(\mathop{\oplus}_{q=1}^{n-1} \mathop{\oplus}_{p=1}^{n-q} C_{p,q} \otimes CE\left(H_{n-q-p+1}(\cdot)\right)\right)$$
$$\oplus \left(\mathop{\oplus}_{p=2}^{n} C_{p,0} \otimes CE\left(H_{n-p+1}(\cdot)\right)\right)$$

Note that CE is a new operator with two operations "$\otimes$" and "$\oplus$" defined in Chap. 4, and $C_{p,q}$ is a vector consisting of all the $(p+q)$th degree nonlinear parameters, i.e.,

$$C_{p,q} = [c_{p,q}(0,\cdots,0), c_{p,q}(0,\cdots,1),\cdots, c_{p,q}(\underbrace{K,\cdots,K}_{p+q=m})]$$

In (12.5), $f_n(j\omega_1,\cdots,j\omega_n)$ is a complex valued vector with the same dimension as $CE(H_n(j\omega_1,\cdots,j\omega_n))$. In Chap. 11, a mapping $\varphi_n(CE(H_n(\cdot)); \omega_1,\cdots,\omega_n)$ from the parametric characteristic $CE(H_n(j\omega_1,\cdots,j\omega_n))$ to its corresponding correlative function $f_n(j\omega_1,\cdots,j\omega_n)$ is established as

$$\begin{aligned}
&\phi_{n(\bar{s})}\left(c_{p_0q_0}(\cdot)c_{p_1q_1}(\cdot)\cdots c_{p_kq_k}(\cdot);\omega_{l(1)}\cdots\omega_{l(n(\bar{s}))}\right)\\
&= \sum_{\substack{\text{all the 2-partitions}\\ \text{for } \bar{s} \text{ satisfying}\\ s_1(\bar{s})=c_{pq}(\cdot) \text{ and } p>0}} \left\{ f_1\left(c_{p,q}(\cdot), n(\bar{s}); \omega_{l(1)}\cdots\omega_{l(n(\bar{s}))}\right)\right.\\
&\cdot \sum_{\substack{\text{all the p-partitions}\\ \text{for } \bar{s}/c_{pq}(\cdot)}} \sum_{\substack{\text{all the different}\\ \text{permutations}\\ \text{of } \{s_{x_1},\cdots,s_{x_p}\}}} \left[f_{2a}\left(s_{\bar{x}_1}\cdots s_{\bar{x}_p}\left(\bar{s}/c_{pq}(\cdot)\right);\omega_{l(1)}\cdots\omega_{l(n(\bar{s})-q)}\right)\right.\\
&\left.\left.\cdot \prod_{i=1}^{p} \phi_{n\left(s_{\bar{x}_i}\left(\bar{s}/c_{pq}(\cdot)\right)\right)}\left(s_{\bar{x}_i}\left(\bar{s}/c_{pq}(\cdot)\right);\omega_{l(\bar{X}(i)+1)}\cdots\omega_{l\left(\bar{X}(i)+n\left(s_{\bar{x}_i}\left(\bar{s}/c_{pq}(\cdot)\right)\right)\right)}\right)\right]\right\}
\end{aligned} \tag{12.6a}$$

where the terminating condition is $k=0$ and $\varphi_1(1;\omega_i)=H_1(j\omega_i)$ (which is the first order GFRF, i.e., transfer function when all nonlinear parameters are zero), $\left\{s_{\bar{x}_1},\cdots s_{\bar{x}_p}\right\}$ is a permutation of $\left\{s_{x_1},\cdots s_{x_p}\right\}$, $\omega_{l(1)}\cdots\omega_{l(n(\bar{s}))}$ represents the frequency variables involved in the corresponding functions, $li)$ for $i=1\ldots n(\bar{s})$ is a positive integer representing the index of the frequency variables, $\bar{s}=c_{p_0q_0}(\cdot)c_{p_1q_1}(\cdot)\cdots c_{p_kq_k}(\cdot)$, $n(s_x(\bar{s}))=\sum_{i=1}^{x}(p_i+q_i)-x+1$, x is the number of the parameters in s_x, $\sum_{i=1}^{x}(p_i+q_i)$ is the summation of the subscripts of all the parameters in s_x. Moreover,

$$\overline{X}(i)=\sum_{j=1}^{i-1}n\left(s_{\overline{x}_j}\left(\overline{s}/c_{pq}(\cdot)\right)\right),$$

$$L_n(j\omega_1+\cdots+j\omega_n)=-\sum_{r_1=0}^{K}c_{1,0}(r_1)(j\omega_1+\cdots+j\omega_n)^{r_1},$$

$$f_1\left(c_{p,q}(\cdot),n(\overline{s});\omega_{l(1)}\cdots\omega_{l(n(\overline{s}))}\right)=\left(\prod_{i=1}^{q}\left(j\omega_{l(n(\overline{s})-q+i)}\right)^{r_{p+i}}\Big/L_{n(\overline{s})}\left(j\sum_{i=1}^{n(\overline{s})}\omega_{l(i)}\right)\right)$$

(12.6b)

$$f_{2a}\left(s_{\overline{x}_1}\cdots s_{\overline{x}_p}\left(\overline{s}/c_{pq}(\cdot)\right);\omega_{l(1)}\cdots\omega_{l(n(\overline{s})-q)}\right)$$
$$=\prod_{i=1}^{p}\left(j\omega_{l(\overline{X}(i)+1)}+\cdots+j\omega_{l\left(\overline{X}(i)+n\left(s_{\overline{x}_i}\left(\overline{s}/c_{pq}(\cdot)\right)\right)\right)}\right)^{r_i} \tag{12.6c}$$

The mapping function $\varphi_n(CE(H_n(\cdot));\omega_1,\cdots,\omega_n)$ enables the complex valued function $f_n(j\omega_1,\cdots,j\omega_n)$ to be analytically and directly determined in terms of the first order GFRF and model nonlinear parameters. Therefore, the nth-order GFRF can directly be written into a more straightforward and meaningful polynomial function in terms of the first order GFRF and model parameters by using the mapping function $\varphi_n(CE(H_n(\cdot));\omega_1,\cdots,\omega_n)$ as

$$H_n(j\omega_1,\cdots,j\omega_n)=CE(H_n(j\omega_1,\cdots,j\omega_n))\cdot\varphi_n(CE(H_n(\cdot));\omega_1,\cdots,\omega_n) \tag{12.7}$$

Note that although the recursive expression of $\varphi_n(CE(H_n(\cdot));\omega_1,\cdots,\omega_n)$ seems complicated, both $CE(H_n(j\omega_1,\cdots,j\omega_n))$ and $\varphi_n(CE(H_n(\cdot));\omega_1,\cdots,\omega_n)$ can be computed through the symbolic manipulation using some available computer software such as Matlab. Therefore, given any nonlinear model as (2.11), a clear polynomial expression as (12.7) can be obtained readily.

Using (12.7), (12.1) can be written as

$$Y(j\omega)=\sum_{n=1}^{N}CE(H_n(j\omega_1,\cdots,j\omega_n))\cdot\overline{F}_n(j\omega) \tag{12.8a}$$

where $\overline{F}_n(j\omega)=\frac{1}{\sqrt{n}(2\pi)^{n-1}}\int_{\omega_1+\cdots+\omega_n=\omega}\varphi_n(CE(H_n(\cdot);\omega_1,\cdots,\omega_n)\cdot\prod_{i=1}^{n}U(j\omega_i)d\sigma_\omega.$

Similarly, (12.3) can be written as

$$Y(j\omega) = \sum_{n=1}^{N} CE(H_n(j\omega_{k_1}, \cdots, j\omega_{k_n})) \cdot \widetilde{F}_n(\omega) \tag{12.8b}$$

where $\widetilde{F}_n(j\omega) = \frac{1}{2^n} \sum\limits_{\omega_{k_1}+\cdots+\omega_{k_n}=\omega} \varphi_n(CE(H_n(\cdot)); \omega_{k_1}, \cdots, \omega_{k_n}) \cdot F(\omega_{k_1}) \cdots F(\omega_{k_n})$.

Note that (12.8a) or (12.8b) is also a polynomial function of model parameters whose structure is determined by the parametric characteristics $CE(H_n(j\omega_1, \cdots, j\omega_n))$ and truncated at the largest order N. The significance of the expressions in (12.7, 12.8a,b) is that, the explicit relationship between any model parameters and the frequency response functions can be demonstrated and thus it is convenient to be used for system analysis and design. For example, if one wants to know how a parameter c is related to the GFRFs and output spectrum, one can directly find the polynomial expansions of the GFRFs and output spectrum in terms of the parameter c using the method above. Usually, in this polynomial expansion, the first several orders take a dominant part in the overall effect of parameter c. Then for an analysis and design purpose, one needs only to study the first several coefficients of the polynomial which are determined by $\varphi_n(CE(H_n(\cdot)); \omega_1, \cdots, \omega_n)$.

Example 12.1 Consider a simple example to demonstrate the results above. Suppose all the other nonlinear parameters in (2.11) are zero except $c_{1,1}(1,1)$, $c_{0,2}(1,1)$, $c_{2,0}(1,1)$. For convenience, $c_{1,1}(1,1)$ is written as $c_{1,1}$ and so on. Consider the parametric characteristic of $H_3(.)$, which can be derived from (5.8),

$$\begin{aligned} &CE(H_3(j\omega_1, \cdots, j\omega_3)) \\ &= C_{0,3} \oplus C_{1,1} \otimes C_{0,2} \oplus C_{1,1}^2 \oplus C_{1,1} \otimes C_{2,0} \oplus C_{2,1} \oplus C_{1,2} \oplus C_{2,0} \otimes C_{0,2} \oplus C_{2,0}^2 \oplus C_{3,0} \\ &= C_{1,1} \otimes C_{0,2} \oplus C_{1,1}^2 \oplus C_{1,1} \otimes C_{2,0} \oplus C_{2,0} \otimes C_{0,2} \oplus C_{2,0}^2 \end{aligned}$$

Note that $C_{1,1}=c_{1,1}$, $C_{0,2}=c_{0,2}$, $C_{2,0}=c_{2,0}$. Thus,

$$CE(H_3(j\omega_1, \cdots, j\omega_3)) = \left[c_{1,1}c_{0,2}, c_{1,1}^2, c_{1,1}c_{2,0}, c_{2,0}c_{0,2}, c_{2,0}c_{1,1}, c_{2,0}^2\right]$$

Using (12.6a–c), the correlative functions of each term in $CE(H_3(j\omega_1, \cdots, j\omega_3))$ can all be obtained. For example, for the term $c_{1,1}c_{0,2}$, it can be derived directly from (12.6a–c) that

$$\begin{aligned} &\varphi_{n(\overline{s})}\left(c_{1,1}(\cdot)c_{0,2}(\cdot); \omega_{l(1)} \cdots \omega_{l(n(\overline{s}))}\right) = \varphi_3(c_{1,1}(\cdot)c_{0,2}(\cdot); \omega_1 \cdots \omega_3) \\ &= f_1(c_{1,1}(\cdot), 3; \omega_1 \cdots \omega_3) \cdot f_{2a}(s_1(c_{1,1}(\cdot)c_{0,2}(\cdot)/c_{1,1}(\cdot)); \omega_1, \omega_2) \cdot \varphi_2(s_1(c_{0,2}(\cdot)); \omega_1, \omega_2) \\ &= f_1(c_{1,1}(\cdot), 3; \omega_1 \cdots \omega_3) \cdot f_{2a}(c_{0,2}(\cdot); \omega_1, \omega_2) \cdot \varphi_2(c_{0,2}(\cdot); \omega_1, \omega_2) \\ &= \frac{j\omega_3}{L_3(j\omega_1 + \cdots + j\omega_3)} \cdot (j\omega_1 + j\omega_2) \cdot \frac{j\omega_1 j\omega_2}{L_2(j\omega_1 + j\omega_2)} = \frac{j\omega_1 j\omega_2 j\omega_3 (j\omega_1 + j\omega_2)}{L_3(j\omega_1 + \cdots + j\omega_3) L_2(j\omega_1 + j\omega_2)} \end{aligned}$$

Proceed with the process above, the whole correlative function of $CE(H_3(j\omega_1, \cdots, j\omega_3))$ can be obtained, and then (12.7, 12.8a,b) can be determined. The process above demonstrates a new way to analytically compute the high order GFRFs, and the final results can directly be written into a polynomial form as (12.7, 12.8a,b), for example

$$\begin{aligned} H_3(j\omega_1,\cdots,j\omega_3) &= \left[c_{1,1}c_{0,2}, c_{1,1}^2, c_{1,1}c_{2,0}, c_{2,0}c_{0,2}, c_{2,0}c_{1,1}, c_{2,0}^2\right] \\ &\quad \cdot \varphi_3(CE(H_3(j\omega_1,\cdots,j\omega_3)); \omega_1,\cdots,\omega_3) \\ &= c_{1,1}c_{0,2}\cdot\varphi_3(c_{1,1}c_{0,2};\omega_1,\cdots,\omega_3) + c_{1,1}^2\cdot\varphi_3\left(c_{1,1}^2;\omega_1,\cdots,\omega_3\right) \\ &\quad + \ldots + c_{2,0}^2\cdot\varphi_3\left(c_{2,0}^2;\omega_1,\cdots,\omega_3\right) \end{aligned}$$

Compared with the recursive computation of the GFRFs in Appendix E, the expression above demonstrates the polynomial relationship between model parameters and the GFRFs in a more straightforward manner.

As discussed in Chap. 11, it can be seen from (12.7, 12.8a,b) and Example 12.1 that the mapping function $\varphi_n(CE(H_n(\cdot)); \omega_1,\cdots,\omega_n)$ can facilitate the frequency domain analysis of nonlinear systems such that the relationship between the frequency response functions and model parameters, and the relationship between the frequency response functions and $H_1(j\omega_{l(1)})$ can be demonstrated explicitly, and some new properties of the GFRFs and output spectrum can be revealed. As revealed in those previous studies, the output spectrum of a nonlinear system can be expanded as a power series with respect to a specific model parameter (e.g., c) of interest by using (12.8a,b) for $N\rightarrow\infty$. The nonlinear effect on system output spectrum incurred by this model parameter c, which may represent the physical characteristic of a structural unit in the system, can then be analysed and designed by studying the resulting power series in the frequency domain. Note that the fundamental properties of this power series (e.g. convergence) are to a large extent dominated by the properties of its coefficients, which are explicitly determined by the mapping function $\varphi_n(CE(H_n(\cdot)); \omega_1,\cdots,\omega_n)$. Thus studying the properties of this power series is now equivalent to studying the properties of the mapping function $\varphi_n(CE(H_n(\cdot)); \omega_1,\cdots,\omega_n)$. Therefore, the mapping function $\varphi_n(CE(H_n(\cdot)); \omega_1,\cdots,\omega_n)$ introduced above provides an important and significant technique for this frequency domain analysis to study the nonlinear influence on system output spectrum.

In this Chapter, a novel property of the nonlinear influence on system output spectrum is revealed by exploring the new mapping function $\varphi_n(CE(H_n(\cdot)); \omega_1,\cdots,\omega_n)$ and frequency response functions defined in (12.7, 12.8a,b). It will be shown that the nonlinear terms in a system can drive the system output spectrum to be an alternating series under certain conditions when the system is subjected to a sinusoidal input, and the system output spectrum will be shown to have some interesting properties when it can be expanded into an alternating series with respect to a specific model parameter of interest. This provides a novel insight into the nonlinear effect on the system output spectrum incurred by corresponding nonlinear terms in a nonlinear system.

It should be noted that the alternating series is an important concept adopted in this study, which might not be something surprising in practice. It can be seen that a stable root of a linear system is an alternating series in Taylor series expansion (for example the Taylor series expansion of $\frac{1}{s+2}$). Therefore, the alternating series might be a natural characteristic related to system dynamics and a potentially promising way to understand nonlinear behaviours in the frequency domain. However, all these are yet to be developed and this is the first study in this direction.

12.3 Alternating Phenomenon in the Output Spectrum and Its Influence

The alternating phenomena and its influence are firstly discussed in this section to point out the significance of this novel property, and then the conditions under which system output spectrum can be expressed into an alternating series are studied in the following section.

For any specific nonlinear parameter c in model (2.11), the output spectrum (12.8a,b) can be expanded with respect to this parameter into a power series as

$$Y(j\omega) = F_0(j\omega) + cF_1(j\omega) + c^2F_2(j\omega) + \cdots + c^{\rho}F_{\rho}(j\omega) + \cdots \tag{12.9}$$

Note that when c represents a pure input nonlinearity, (12.9) may be a finite series; in other cases, it is definitely an infinite series, and if only the first ρ terms in the series (12.9) are considered, there is a truncation error denoted by $o(\rho)$. $F_i(j\omega)$ for i=0,1,2,... can be obtained from $\overline{F}_i(j\omega)$ or $\widetilde{F}_i(j\omega)$ in (12.8a,b) by using the mapping $\varphi_n(CE(H_n(\cdot)); \omega_1, \cdots, \omega_n)$. Clearly, $F_i(j\omega)$ dominate the property of this power series. Thus the property of this power series can be revealed by studying the property of $\varphi_n(CE(H_n(\cdot)); \omega_1, \cdots, \omega_n)$. This will be discussed in detail in the next section. In this section, the alternating phenomenon of this power series and its influence are discussed.

For any $\upsilon \in \mathbb{C}$, define an operator as

$$\mathrm{sgn}_c(\upsilon) = [\,\mathrm{sgn}_r(\mathrm{Re}(\upsilon)) \quad \mathrm{sgn}_r(\mathrm{Im}(\upsilon))\,] \tag{12.10}$$

where $\mathrm{sgn}_r(x) = \begin{cases} +1 & x > 0 \\ 0 & x = 0 \\ -1 & x < 0 \end{cases}$ for $x \in \mathbb{R}$.

Definition 12.1 (Alternating Series) Consider a power series of form (12.9) with c>0. If $\mathrm{sgn}_c(F_i(j\omega)) = -\,\mathrm{sgn}_c(F_{i+1}(j\omega))$ for i=0,1,2,3,..., then the series is an alternating series.

The series (12.9) can be written into two series as

$$\begin{aligned} Y(j\omega) &= \mathrm{Re}(Y(j\omega)) + j(\mathrm{Im}(Y(j\omega))) \\ &= \mathrm{Re}(F_0(j\omega)) + c\mathrm{Re}(F_1(j\omega)) + c^2\mathrm{Re}(F_2(j\omega)) + \cdots + c^\rho \mathrm{Re}\left(F_\rho(j\omega)\right) + \cdots \\ &\quad + j\left(\mathrm{Im}(F_0(j\omega)) + c\mathrm{Im}(F_1(j\omega)) + c^2\mathrm{Im}(F_2(j\omega)) + \cdots + c^\rho \mathrm{Im}\left(F_\rho(j\omega)\right) + \cdots\right) \end{aligned} \tag{12.11}$$

From Definition 12.1, if $Y(j\omega)$ is an alternating series, then $\mathrm{Re}(Y(j\omega))$ and $\mathrm{Im}(Y(j\omega))$ are both alternating. When (12.9) is an alternating series, there are some interesting properties summarized in Proposition 12.1. Denote

$$Y(j\omega)_{1\to\rho} = F_0(j\omega) + cF_1(j\omega) + c^2F_2(j\omega) + \cdots + c^\rho F_\rho(j\omega) \tag{12.12}$$

Proposition 12.1 Suppose (12.9) is an alternating series for $c>0$, then:

(1) if there exist $T>0$ and $R>0$ such that for $i>T$

$$\min\left\{-\frac{\mathrm{Re}(F_i(j\omega))}{\mathrm{Re}(F_{i+1}(j\omega))}, -\frac{\mathrm{Im}(F_i(j\omega))}{\mathrm{Im}(F_{i+1}(j\omega))}\right\} > R$$

then (12.9) has a radius of convergence R, the truncation error for a finite order $\rho > T$ is $|o(\rho)| \le c^{\rho+1}|F_{\rho+1}(j\omega)|$, and for all $n \ge 0$,

$$|Y(j\omega)| \in \Pi_n = \left[\left|Y(j\omega)_{1\to T+2n+1}\right|, \left|Y(j\omega)_{1\to T+2n}\right|\right] \text{ and } \Pi_{n+1} \subset \Pi_n;$$

(2) $|Y(j\omega)|^2 = Y(j\omega)Y(-j\omega)$ is also an alternating series with respect to parameter c; Furthermore, $|Y(j\omega)|^2 = Y(j\omega)Y(-j\omega)$ is alternating only if $\mathrm{Re}(Y(j\omega))$ is alternating;

(3) there exists a $\bar{c} > 0$ such that $\frac{\partial|Y(j\omega)|}{\partial c} < 0$ for $0 < c < \bar{c}$.

Proof

(1) $Y(j\omega)$ is convergent if and only if $\mathrm{Re}(Y(j\omega))$ and $\mathrm{Im}(Y(j\omega))$ are both convergent. Since $Y(j\omega)$ is an alternating series, $\mathrm{Re}(Y(j\omega))$ and $\mathrm{Im}(Y(j\omega))$ are both alternating from Definition 12.1. Then according to Bromwich (1991), $\mathrm{Re}(Y(j\omega))$ is convergent if $|\mathrm{Re}(c^iF_i(j\omega))| > |\mathrm{Re}(c^{i+1}F_{i+1}(j\omega))|$ and $\lim_{i\to\infty}|\mathrm{Re}(c^iF_i(j\omega))| = 0$. Therefore, if there exists $T>0$ such that $|\mathrm{Re}(c^iF_i(j\omega))| > |\mathrm{Re}(c^{i+1}F_{i+1}(j\omega))|$ for i>T and $\lim_{i\to\infty}|\mathrm{Re}(c^iF_i(j\omega))| = 0$, the alternating series $\mathrm{Re}(Y(j\omega))$ is also convergent. Now since there exist $T>0$ and $R>0$ such that $-\frac{\mathrm{Re}(F_i(j\omega))}{\mathrm{Re}(F_{i+1}(j\omega))} > R$ for $i>T$ and note $c<R$, it can be obtained that for $i>T$

$$-\frac{\mathrm{Re}(c^{i+1}F_{i+1}(j\omega))}{\mathrm{Re}(c^iF_i(j\omega))} = -\frac{\mathrm{Re}(cF_{i+1}(j\omega))}{\mathrm{Re}(F_i(j\omega))} = \left|\frac{\mathrm{Re}(cF_{i+1}(j\omega))}{\mathrm{Re}(F_i(j\omega))}\right| < \frac{c}{R} < 1$$

i.e., $|\mathrm{Re}(c^iF_i(j\omega))| > |\mathrm{Re}(c^{i+1}F_{i+1}(j\omega))|$ for $i>T$ and c<R. Moreover, it can also be obtained that for n>0

$$|\mathrm{Re}(F_{T+n}(j\omega))| < \frac{1}{R^n}|\mathrm{Re}(F_T(j\omega))|$$

It further yields that

$$\left|\mathrm{Re}\left(c^{T+n}F_{T+n}(j\omega)\right)\right| < \left(\frac{c}{R}\right)^n c^T|\mathrm{Re}(F_T(j\omega))|$$

That is, $\lim_{n\to\infty}\left|\mathrm{Re}(c^{T+n}F_{T+n}(j\omega))\right| = 0$. Therefore, $\mathrm{Re}(Y(j\omega))$ is convergent. Similarly, it can be proved that $\mathrm{Im}(Y(j\omega))$ is convergent. This proves that $Y(j\omega)$ is convergent. The truncation errors for the real convergent alternating series $\mathrm{Re}(Y(j\omega))$ and $\mathrm{Im}(Y(j\omega))$ are

$$|o_R(\rho)| \le c^{\rho+1}\left|\mathrm{Re}\left(F_{\rho+1}(j\omega)\right)\right| \text{ and } |o_I(\rho)| \le c^{\rho+1}\left|\mathrm{Im}\left(F_{\rho+1}(j\omega)\right)\right|$$

Therefore, the truncation error for the series $Y(j\omega)$ is

$$|o(\rho)| = \sqrt{o_R(\rho)^2 + o_I(\rho)^2} \le c^{\rho+1}\left|F_{\rho+1}(j\omega)\right|$$

It can be shown that for $\mathrm{Re}(Y(j\omega))$ and $\mathrm{Im}(Y(j\omega))$, for $n\ge0$

$$\begin{aligned}\left|\mathrm{Re}\left(Y(j\omega)_{1\to T+1}\right)\right| &< \cdots < \left|\mathrm{Re}\left(Y(j\omega)_{1\to T+2n+1}\right)\right| < |\mathrm{Re}(Y(j\omega))| \\ &< \left|\mathrm{Re}\left(Y(j\omega)_{1\to T+2n}\right)\right| < \cdots < \left|\mathrm{Re}\left(Y(j\omega)_{1\to T}\right)\right| \\ \left|\mathrm{Im}\left(Y(j\omega)_{1\to T+1}\right)\right| &< \cdots < \left|\mathrm{Im}\left(Y(j\omega)_{1\to T+2n+1}\right)\right| < |\mathrm{Im}(Y(j\omega))| \\ &< \left|\mathrm{Im}\left(Y(j\omega)_{1\to T+2n}\right)\right| < \cdots < \left|\mathrm{Im}\left(Y(j\omega)_{1\to T}\right)\right|\end{aligned}$$

Therefore, $|Y(j\omega)_{1\to T+1}| < \cdots < |Y(j\omega)_{1\to T+2n+1}| < |Y(j\omega)| < |Y(j\omega)_{1\to T+2n}| < \cdots < |Y(j\omega)_{1\to T}|$.

(2) $|Y(j\omega)|^2 = Y(j\omega)Y(-j\omega)$

$$\begin{aligned}&= \left(F_0(j\omega) + cF_1(j\omega) + c^2F_2(j\omega) + \cdots\right)\left(F_0(-j\omega) + cF_1(-j\omega), +c^2F_2(-j\omega) + \cdots\right) \\ &= \sum_{n=0,1,2,\ldots} c^n\sum_{i=0}^{n}F_i(j\omega)F_{n-i}(-j\omega)\end{aligned}$$

It can be verified that the (2k)th terms in the series are positive and the (2k+1) th terms are negative. Moreover, it needs only the real parts of the terms in $Y(j\omega)$ to be alternating for $|Y(j\omega)|^2 = Y(j\omega)Y(-j\omega)$ to be alternating.

(3) $$\frac{\partial|Y(j\omega)|}{\partial c}=\frac{1}{2|Y(j\omega)|}\frac{\partial|Y(j\omega)|^2}{\partial c}$$
$$=\frac{1}{2|Y(j\omega)|}\left\{\mathrm{Re}(F_0(j\omega)F_1(-j\omega))+c\sum_{n=1,2,\ldots}nc^{n-1}\sum_{i=0}^{n}F_i(j\omega)F_{n-i}(-j\omega)\right\}$$

Since $\mathrm{Re}(F_0(j\omega)F_1(-j\omega))<0$, there must exist $\bar{c}>0$ such that $\frac{\partial|Y(j\omega)|}{\partial c}<0$ for $0<c<\bar{c}$. This completes the proof. □

Proposition 12.1 shows that if the system output spectrum can be expressed as an alternating series with respect to a specific parameter c, it is always easier to find a c such that the output spectrum is convergent and its magnitude can always be suppressed by a properly designed c. Moreover, it is also shown that the low limit of the magnitude of the output spectrum that can be reached is larger than $|Y(j\omega)_{1\to T+2}|$ and the truncation error can also be easily evaluated, if the output spectrum can be expressed into an alternating series.

An example is given to illustrate these results.

Example 12.2 Consider a SDOF spring-damping system with a cubic nonlinear damping which can be described by the following differential equation,

$$m\ddot{y}=-k_0y-B\dot{y}-c\dot{y}^3+u(t) \tag{12.13}$$

Note that k_0 represents the spring characteristic, B the damping characteristic and c is the cubic nonlinear damping characteristic. This system is a simple case of NDE model (2.11) and can be written into the form of NDE model with $M=3$, $K=2$, $c_{10}(2)=m$, $c_{10}(1)=B$, $c_{10}(0)=k_0$, $c_{30}(111)=c$, $c_{01}(0)=-1$ and all the other parameters are zero.

Note that there is only one output nonlinear term in this case, the nth-order GFRF for system (12.13) can be derived according to the algorithm in (2.19–2.26), which can recursively be given as

$$H_n(j\omega_1,\cdots,j\omega_n)=\frac{c_{3,0}(1,1,1)H_{n,3}(j\omega_1,\cdots,j\omega_n)}{L_n(j\omega_1+\cdots+j\omega_n)}$$

$$H_{n,3}(\cdot)=\sum_{i=1}^{n-2}H_i(j\omega_1,\cdots,j\omega_i)H_{n-i,2}(j\omega_{i+1},\cdots,j\omega_n)(j\omega_1+\cdots+j\omega_i)$$

$$H_{n,1}(j\omega_1,\cdots,j\omega_n)=H_n(j\omega_1,\cdots,j\omega_n)(j\omega_1+\cdots+j\omega_n)$$

Proceeding with the recursive computation above, it can be seen that $H_n(j\omega_1,\cdots,j\omega_n)$ is a polynomial of $c_{30}(111)$, and substituting these equations above into (12.8a, b) gives another polynomial for the output spectrum. By using the relationship (12.7) and the mapping function $\varphi_n(CE(H_n(\cdot));\omega_1,\cdots,\omega_n)$, these results can be obtained directly as follows.

For simplicity, let $u(t) = F_d \sin(\Omega t)$ $(F_d > 0)$. Then $F(\omega_{k_l}) = -jk_l F_d$, for $k_l = \pm 1$, $\omega_{k_l} = k_l \Omega$, and $l = 1, \cdots, n$ in (12.8b). By using (5.15) or Property 5.3, it can be obtained that for n=0,1,2,3,...

$$CE\left(H_{2n+1}\left(j\omega_{k_1}, \cdots, j\omega_{k_{2n+1}}\right)\right) = \left(c_{3,0}(1,1,1)\right)^n \text{ and } CE\left(H_{2n}\left(j\omega_{k_1}, \cdots, j\omega_{k_{2n}}\right)\right) = 0 \tag{12.14}$$

Therefore, for n=0,1,2,3,...

$$H_{2n+1}\left(j\omega_{k_1}, \cdots, j\omega_{k_{2n+1}}\right) = c^n \cdot \varphi_{2n+1}\left(CE(H_{2n+1}(\cdot)); \omega_{k_1}, \cdots, \omega_{k_{2n+1}}\right) \text{ and}$$
$$H_{2n}\left(j\omega_{k_1}, \cdots, j\omega_{k_{2n}}\right) = 0 \tag{12.15}$$

Then the output spectrum at frequency Ω can be computed as

$$Y(j\Omega) = \sum_{n=0}^{\lfloor N-1/2 \rfloor} c^n \cdot \widetilde{F}_{2n+1}(\Omega) \tag{12.16}$$

where $\widetilde{F}_{2n+1}(j\Omega)$ can be computed as

$$\begin{aligned}
\widetilde{F}_{2n+1}(j\Omega) &= \frac{1}{2^{2n+1}} \sum_{\omega_{k_1}+\cdots+\omega_{k_{2n+1}}=\Omega} \varphi_{2n+1}\left(CE(H_{2n+1}(\cdot)); \omega_{k_1}, \cdots, \omega_{k_{2n+1}}\right) \\
&\quad \cdot (-jF_d)^{2n+1} \cdot k_1 k_2 \cdots k_{2n+1} \\
&= \frac{1}{2^{2n+1}} \sum_{\omega_{k_1}+\cdots+\omega_{k_{2n+1}}=\Omega} \varphi_{2n+1}\left(CE(H_{2n+1}(\cdot)); \omega_{k_1}, \cdots, \omega_{k_{2n+1}}\right) \\
&\quad \cdot (-1)^{n+1} j (F_d)^{2n+1} \cdot (-1)^n \\
&= -j\left(\frac{F_d}{2}\right)^{2n+1} \sum_{\omega_{k_1}+\cdots+\omega_{k_{2n+1}}=\Omega} \varphi_{2n+1}\left(CE(H_{2n+1}(\cdot)); \omega_{k_1}, \cdots, \omega_{k_{2n+1}}\right)
\end{aligned} \tag{12.17}$$

and $\varphi_{2n+1}\left(CE(H_{2n+1}(\cdot)); \omega_{k_1}, \cdots, \omega_{k_{2n+1}}\right) = \varphi_{2n+1}\left(c_{30}(1,1,1)^n; \omega_{k_1}, \cdots, \omega_{k_{2n+1}}\right)$ can be obtained according to (12.6a–c). For example,

$$\varphi_3(c_{30}(111);\omega_{k_1},\omega_{k_2},\omega_{k_3}) = \frac{1}{L_3\left(j\sum_{i=1}^{3}\omega_{k_i}\right)}\cdot\prod_{i=1}^{3}(j\omega_{k_i})\cdot\prod_{i=1}^{3}H_1(j\omega_{k_i})$$

$$= \frac{\prod_{i=1}^{3}(j\omega_{k_i})}{L_3\left(j\sum_{i=1}^{3}\omega_{k_i}\right)}\cdot\prod_{i=1}^{3}H_1(j\omega_{k_i})$$

$$\begin{aligned}
&\phi_5\left(c_{3,0}(111)c_{3,0}(111);\omega_{k_1},\cdots,\omega_{k_5}\right)\\
&=f_1\left(c_{3,0}(111),5;\omega_{k_1},\cdots,\omega_{k_5}\right)\\
&\cdot\sum_{\substack{\text{all the }3-\text{partitions}\\ \text{for } c_{3,0}(111)}}\ \sum_{\substack{\text{all the different}\\ \text{permutations of }\{0,0,1\}}}\left[f_{2a}\left(s_{\overline{x}_1}\cdots s_{\overline{x}_p}\left(c_{3,0}(111)\right);\omega_{k_1},\cdots,\omega_{k_5}\right)\right.\\
&\left.\cdot\prod_{i=1}^{3}\phi_{n\left(s_{\overline{x}_i}(\overline{s}/c_{p,q}(\cdot))\right)}\left(s_{\overline{x}_i}\left(c_{3,0}(111)\right);\omega_{l(\overline{X}(i)+1)}\cdots\omega_{l\left(\overline{X}(i)+n\left(s_{\overline{x}_i}(\overline{s}/c_{p,q}(\cdot))\right)\right)}\right)\right]\\
&=f_1\left(c_{3,0}(111),5;\omega_{k_1},\cdots,\omega_{k_5}\right)\\
&\cdot\begin{pmatrix}
f_{2a}\left(s_0s_0s_1\left(c_{3,0}(111)\right);\omega_{k_1},\cdots,\omega_{k_5}\right)\phi_1\left(1;\omega_{k_1}\right)\phi_1\left(1;\omega_{k_2}\right)\phi_3\left(c_{3,0}(111);\omega_{k_3}\cdots\omega_{k_5}\right)\\
+f_{2a}\left(s_0s_1s_0\left(c_{3,0}(111)\right);\omega_{k_1},\cdots,\omega_{k_5}\right)\phi_1\left(1;\omega_{k_1}\right)\phi_3\left(c_{3,0}(111);\omega_{k_2}\cdots\omega_{k_4}\right)\phi_1\left(1;\omega_{k_5}\right)\\
+f_{2a}\left(s_1s_0s_0(c_{30}(111));\omega_{k_1},\cdots,\omega_{k_5}\right)\phi_3\left(c_{30}(111);\omega_{k_1}\cdots\omega_{k_3}\right)\phi_1\left(1;\omega_{k_4}\right)\phi_1\left(1;\omega_{k_5}\right)
\end{pmatrix}\\
&=\frac{1}{L_5\left(j\sum_{i=1}^{5}\omega_{k_i}\right)}\\
&\cdot\left(\frac{\left(j\sum_{i=3}^{5}\omega_{k_i}\right)\prod_{i=1}^{5}(j\omega_{k_i})}{L_3\left(j\sum_{i=3}^{5}\omega_{k_i}\right)}+\frac{\left(j\sum_{i=2}^{4}\omega_{k_i}\right)\prod_{i=1}^{5}(j\omega_{k_i})}{L_3\left(j\sum_{i=2}^{4}\omega_{k_i}\right)}+\frac{\left(j\sum_{i=1}^{3}\omega_{k_i}\right)\prod_{i=1}^{5}(j\omega_{k_i})}{L_3\left(j\sum_{i=1}^{3}\omega_{k_i}\right)}\right)\\
&\cdot\prod_{i=1}^{5}H_1(j\omega_{k_i})
\end{aligned}$$

where $\omega_{k_i}\in\{\Omega,-\Omega\}$, and so on. Substituting these results into (12.16), the output spectrum is clearly a power series with respect to the parameter c. When there are more nonlinear terms, it is obvious that the computation process above can directly result in a straightforward multivariate power series with respect to these nonlinear parameters. To check the alternating phenomenon of the output spectrum, consider the following values for each linear parameter: m=240, k_0=16,000, B=296, F_d=100, and $\Omega = 8.165$. Then it is obtained that

$$\begin{aligned} Y(j\Omega) &= \widetilde{F}_1(\Omega) + c\widetilde{F}_3(\Omega) + c^2\widetilde{F}_5(\Omega) + \cdots \\ &= -j\left(\frac{F_d}{2}\right)H_1(j\Omega) + 3\left(\tfrac{F_d}{2}\right)^3 \frac{\Omega^3|H_1(j\Omega)|^2 H_1(j\Omega)}{L_1(j\Omega)} \\ &\quad + 3\left(\tfrac{F_d}{2}\right)^5 \frac{\Omega^5|H_1(j\Omega)|^4 H_1(j\Omega)}{L_1(j\Omega)}\left(\frac{j6\Omega}{L_1(j\Omega)} + \frac{j3\Omega}{L_1(j3\Omega)} + \frac{-j3\Omega}{L_1(-j\Omega)}\right) + \cdots \\ &= (-0.02068817126756 + 0.000001147041 16\mathrm{i}) \\ &\quad + (5.982851578532449\text{e-}006 - 12.634300276113922\text{e-}010i)\mathrm{c} \\ &\quad + (-5.192417616715994\text{e-}009 + 3.323565122085705\text{e-}011\mathrm{i})\mathrm{c}^2 + \ldots \end{aligned} \tag{12.18a}$$

The series is alternating. In order to check the series further, computation of $\varphi_{2n+1}\left(c_{3,0}(1,1,1)^n; \omega_{k_1}, \cdots, \omega_{k_{2n+1}}\right)$ can be carried out for higher orders. It can also be verified that the magnitude square of the output spectrum (12.18a) is still an alternating series, *i.e.*,

$$\begin{aligned} |Y(j\Omega)|^2 &= (4.280004317115985\text{e-}004) - (2.475485177721052\text{e-}007)\mathrm{c} \\ &\quad + (2.506378395908398\text{e-}010)\mathrm{c}^2 - \ldots \end{aligned} \tag{12.18b}$$

As pointed in Proposition 12.1, it is easy to find a c such that (12.8a,b) are convergent and their limits are decreased. From (12.18b) and according to Proposition 12.1, it can be computed that $0.01671739 < |Y(j\Omega)| < 0.0192276 < 0.0206882$ for c=600. This can be verified by Fig. 12.1. Figure 12.1 is a result from simulation tests, and shows that the magnitude of the output spectrum decreases when c increases. This property is of great significance in practical engineering systems for output suppression through structural characteristic design or feedback control.

12.4 Alternating Conditions

In this section, the conditions under which the output spectrum described by (12.9) can be expressed as an alternating series with respect to a specific nonlinear parameter are studied. Suppose the system is subjected to a harmonic input $u(t) = F_d \sin(\Omega t)$ $(F_d > 0)$ (The results for this case can be extended to the general input) and only the output nonlinearities (*i.e.*, $c_{p,0}(.)$ with $p \geq 2$) are considered. For convenience, assume that there is only one nonlinear parameter $c_{p,0}(.)$ in model (2.11) and all the other nonlinear parameters are zero.

Under the assumptions above, it can be obtained from the parametric characteristic analysis in Chaps. 5 and 6 as demonstrated in Example 12.2 and (12.8b) that

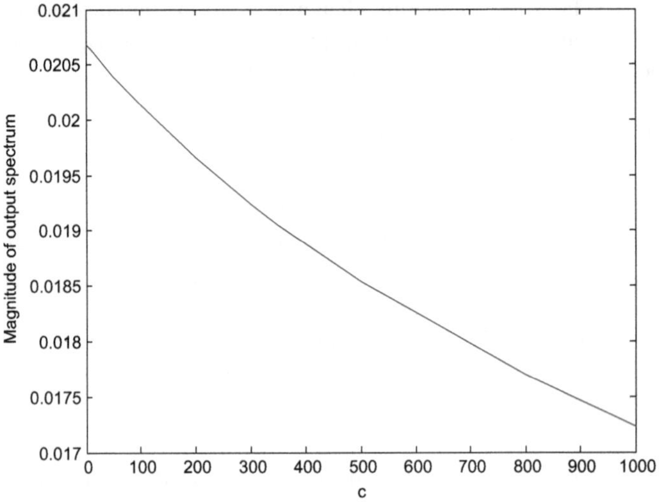

Fig. 12.1 Magnitude of output spectrum (Jing et al. 2011)

$$
\begin{aligned}
Y(j\Omega) &= Y_1(j\Omega) + Y_p(j\Omega) + \cdots + Y_{(p-1)n+1}(j\Omega) + \cdots \\
&= \widetilde{F}_1(\Omega) + c_{p,0}(\cdot)\widetilde{F}_p(\Omega) + \cdots + c_{p,0}(\cdot)^n \widetilde{F}_{(p-1)n+1}(\Omega) + \cdots \\
&= \widetilde{F}_1(\Omega) + c_{p,0}(\cdot)\widetilde{F}_p(\Omega) + \cdots + c_{p,0}(\cdot)^n \widetilde{F}_{(p-1)n+1}(\Omega) + \cdots
\end{aligned}
\tag{12.19a}
$$

where $\omega_{k_i} \in \{\pm\Omega\}, \widetilde{F}_{(p-1)n+1}(j\Omega)$ can be computed from (12.8b), and n is a positive integer. Noting that $F(\omega_{k_l}) = -jk_l F_d,\ k_l = \pm 1,\ \omega_{k_l} = k_l\Omega$, and $l = 1, \cdots, n$ in (12.8b),

$$
\begin{aligned}
\widetilde{F}_{(p-1)n+1}(j\Omega) = &\frac{1}{2^{(p-1)n+1}} \sum_{\omega_{k_1}+\cdots+\omega_{k_{(p-1)n+1}}=\Omega} \varphi_{(p-1)n+1}\left(c_{p,0}(\cdot)^n; \omega_{k_1}, \cdots, \omega_{k_{(p-1)n+1}}\right) \\
&\cdot (-jF_d)^{(p-1)n+1} \cdot k_1 k_2 \cdots k_{(p-1)n+1}
\end{aligned}
\tag{12.19b}
$$

If p is an odd integer, then (p−1)n+1 is also an odd integer. Thus there should be (p−1)n/2 frequency variables being $-\Omega$ and (p−1)n/2+1 frequency variables being Ω such that $\omega_{k_1} + \cdots + \omega_{k_{(p-1)n+1}} = \Omega$. In this case,

$$
\begin{aligned}
(-jF_d)^{(p-1)n+1} \cdot k_1 k_2 \cdots k_{(p-1)n+1} &= (-1) \cdot j \cdot \left(j^2\right)^{(p-1)n/2} \cdot (F_d)^{(p-1)n+1} \\
&\quad \cdot (-1)^{(p-1)n/2} \\
&= -j(F_d)^{(p-1)n+1}
\end{aligned}
$$

If p is an even integer, then (p−1)n+1 is an odd integer for n=2k (k=1,2,3,. . .) and an even integer for n=2k−1 (k=1,2,3,. . .). When n is an odd integer, $\omega_{k_1} + \cdots + \omega_{k_{(p-1)n+1}} \neq \Omega$ for $\omega_{k_l} \in \{\pm\Omega\}$. This gives that $\widetilde{F}_{(p-1)n+1}(j\Omega) = 0$. When n is an even integer, (p−1)n+1 is an odd integer. In this case, it is similar to that p is an odd integer. Therefore, for n>0

$$\widetilde{F}_{(p-1)n+1}(j\Omega) = \begin{cases} -j\left(\frac{F_d}{2}\right)^{(p-1)n+1} \sum\limits_{\omega_{k_1}+\cdots+\omega_{k_{(p-1)n+1}}=\Omega} \varphi_{(p-1)n+1}\left(c_{p,0}(\cdot)^n; \omega_{k_1}, \cdots, \omega_{k_{(p-1)n+1}}\right) & \text{if p is odd or n is even} \\ 0 & \text{else} \end{cases} \tag{12.19c}$$

From (12.19a–c) it is obvious that the property of the new mapping $\varphi_{(p-1)n+1}\left(c_{p,0}(\cdot)^n; \omega_{k_1}, \cdots, \omega_{k_{(p-1)n+1}}\right)$ plays a key role in the series. To develop the alternating conditions for series (12.19a), the following results can be obtained.

Lemma 12.1 That $\varphi_{(p-1)n+1}\left(c_{p,0}(\cdot)^n; \omega_{k_1}, \cdots, \omega_{k_{(p-1)n+1}}\right)$ is symmetric or asymmetric has no influence on $\widetilde{F}_{(p-1)n+1}(j\Omega)$.

This lemma is obvious since $\sum\limits_{\omega_{k_1}+\cdots+\omega_{k_{(p-1)n+1}}=\Omega} (\cdot)$ includes all the possible permutations of $\left(\omega_{k_1}, \cdots, \omega_{k_{2n+1}}\right)$. Although there are many choices to obtain the asymmetric $\varphi_{(p-1)n+1}\left(c_{p,0}(\cdot)^n; \omega_{k_1}, \cdots, \omega_{k_{(p-1)n+1}}\right)$ which may be different at different permutation $\left(\omega_{k_1}, \cdots, \omega_{k_{(p-1)n+1}}\right)$, they have no different effect on the analysis of $\widetilde{F}_{(p-1)n+1}(j\Omega)$.

Lemma 12.2 Consider parameter $c_{p,q}(r_1,r_2,\ldots,r_{p+q})$.

(a1) If $p \geq 2$ and q=0, then

$$\phi_{n(\bar{s})}\left(c_{p,0}(\cdot)^{n};\omega_{l(1)}\cdots\omega_{l(n(\bar{s}))}\right)=\phi_{(p-1)n+1}\left(c_{p,0}(\cdot)^{n};\omega_{l(1)}\cdots\omega_{l((p-1)n+1)}\right)$$

$$=\frac{(-1)^{n-1}\prod_{i=1}^{(p-1)n+1}H_1\left(j\omega_{l(i)}\right)}{L_{(p-1)n+1}\left(j\omega_{l(1)}+\cdots+j\omega_{l((p-1)n+1)}\right)}\cdot\sum_{\substack{\text{all the different combinations}\\ \text{of }\{\bar{x}_1,\bar{x}_2,\ldots,\bar{x}_p\}\text{ satisfying}\\ \bar{x}_1+\cdots+\bar{x}_p=n-1,\ 0\le\bar{x}_i\le n-1}}\left[\prod_{i=1}^{p}\phi'_{(p-1)\bar{x}_i+1}\left(c_{p,0}(\cdot)^{\bar{x}_i};\omega_{l(\bar{X}(i)+1)}\cdots\omega_{l(\bar{X}(i)+(p-1)\bar{x}_i+1)}\right)\right.$$

$$\left.\cdot\frac{n_x{}^*\left(\bar{x}_1,\cdots,\bar{x}_p\right)}{n_r{}^*\left(r_1,\cdots,r_p\right)}\cdot\sum_{\substack{\text{all the different}\\ \text{permutations of}\\ \{k_1,\cdots,k_p\}}}\prod_{i=1}^{p}\left(j\omega_{l(\bar{X}(i)+1)}+\cdots+j\omega_{l(\bar{X}(i)+(p-1)\bar{x}_i+1)}\right)^{r_i}\right]$$

where,

$$\phi'_{(p-1)n+1}\left(c_{p,0}(\cdot)^{n};\omega_{l(1)}\cdots\omega_{l((p-1)n+1)}\right)$$

$$=\frac{-1}{L_{(p-1)n+1}\left(j\omega_{l(1)}+\cdots+j\omega_{l((p-1)n+1)}\right)}\cdot\sum_{\substack{\text{all the different combinations}\\ \text{of }\{\bar{x}_1,\bar{x}_2,\ldots,\bar{x}_p\}\text{ satisfying}\\ \bar{x}_1+\cdots+\bar{x}_p=n-1,\ 0\le\bar{x}_i\le n-1}}\left[\prod_{i=1}^{p}\phi'_{(p-1)\bar{x}_i+1}\left(c_{p,0}(\cdot)^{\bar{x}_i};\omega_{l(\bar{X}(i)+1)}\cdots\omega_{l(\bar{X}(i)+(p-1)\bar{x}_i+1)}\right)\right.$$

$$\left.\cdot\frac{n_x{}^*\left(\bar{x}_1,\cdots,\bar{x}_p\right)}{n_r{}^*\left(r_1,\cdots,r_p\right)}\cdot\sum_{\substack{\text{all the different}\\ \text{permutations of}\\ \{r_1,\cdots,r_p\}}}\prod_{i=1}^{p}\left(j\omega_{l(\bar{X}(i)+1)}+\cdots+j\omega_{l(\bar{X}(i)+(p-1)\bar{x}_i+1)}\right)^{r_i}\right]$$

the termination is $\varphi'_1(1;\omega_i)=1$. $n_r^*\left(r_1,\cdots,r_p\right)=\frac{p!}{n_1!n_2!\cdots n_e!}$, $n_1+\ldots+n_e=p$, e is the number of distinct differentials r_i appearing in the combination, n_i is the number of repetitions of r_i, and a similar definition holds for $n_x^*\left(\bar{x}_1,\cdots,\bar{x}_p\right)$.

(a2) If $p \geq 2$, q=0 and $r_1=r_2=\ldots=r_p=r$, then

$$\varphi_{(p-1)n+1}\left(c_{p,0}(\cdot)^n;\omega_{l(1)}\cdots\omega_{l((p-1)n+1)}\right)$$

$$=\frac{(-1)^{n-1}\prod_{i=1}^{(p-1)n+1}\left[\left(j\omega_{l(i)}\right)^r H_1\left(j\omega_{l(i)}\right)\right]}{L_{(p-1)n+1}\left(j\omega_{l(1)}+\cdots+j\omega_{l((p-1)n+1)}\right)}$$

$$\cdot \sum_{\substack{\text{all the different combinations}\\ \text{of } \{\overline{x}_1,\overline{x}_2,\ldots,\overline{x}_p\} \text{ satisfying}\\ \overline{x}_1+\cdots+\overline{x}_p=n-1,\ 0\leq \overline{x}_i\leq n-1}} n_x^*\left(\overline{x}_1,\cdots,\overline{x}_p\right)$$

$$\cdot\prod_{i=1}^{p}\varphi''_{(p-1)\overline{x}_i+1}\left(c_{p,0}(\cdot)^{\overline{x}_i};\omega_{l(\overline{X}(i)+1)}\cdots\omega_{l(\overline{X}(i)+(p-1)\overline{x}_i+1)}\right)$$

where,

if $\overline{x}_i=0,\ \varphi''_{(p-1)\overline{x}_i+1}\left(c_{p,0}(\cdot)^{\overline{x}_i};\omega_{l(\overline{X}(i)+1)}\cdots\omega_{l(\overline{X}(i)+(p-1)\overline{x}_i+1)}\right)=1,$

otherwise,

$$\varphi''_{(p-1)\overline{x}_i+1}\left(c_{p,0}(\cdot)^{\overline{x}_i};\omega_{l(\overline{X}(i)+1)}\cdots\omega_{l(\overline{X}(i)+(p-1)\overline{x}_i+1)}\right)$$

$$=\frac{\left(j\omega_{l(\overline{X}(i)+1)}+\cdots+j\omega_{l(\overline{X}(i)+(p-1)\overline{x}_i+1)}\right)^r}{-L_{(p-1)\overline{x}_i+1}\left(j\omega_{l(\overline{X}(i)+1)}+\cdots+j\omega_{l(\overline{X}(i)+(p-1)\overline{x}_i+1)}\right)}$$

$$\cdot \sum_{\substack{\text{all the different combinations}\\ \text{of } \{x_1,x_2,\ldots,x_p\} \text{ satisfying}\\ x_1+\cdots+x_p=\overline{x}_i-1,\ 0\leq x_j\leq \overline{x}_i-1}} n_x^*\left(x_1,\cdots,x_p\right)$$

$$\cdot\prod_{j=1}^{p}\varphi''_{(p-1)x_j+1}\left(c_{p,0}(\cdot)^{x_j};\omega_{l(\overline{X}'(j)+1)}\cdots\omega_{l(\overline{X}'(j)+(p-1)x_j+1)}\right)$$

The recursive terminal of $\varphi''_{(p-1)\overline{x}_i+1}\left(c_{p,0}(\cdot)^{\overline{x}_i};\omega_{l(\overline{X}(i)+1)}\cdots\omega_{l(\overline{X}(i)+(p-1)\overline{x}_i+1)}\right)$ is $\overline{x}_i=1$.

Proof

$$\varphi_{n(\bar{s})}\left(c_{p,0}(\cdot)^{n};\omega_{l(1)}\cdots\omega_{l(n(\bar{s}))}\right)=\varphi_{(p-1)n+1}\left(c_{p,0}(\cdot)c_{p,0}(\cdot)\cdots c_{p,0}(\cdot);\omega_{l(1)}\cdots\omega_{(l(p-1)n+1)}\right)$$

$$=\sum_{\substack{\text{all the 2-partitions}\\ \text{for } \bar{s} \text{ satisfying}\\ s_1(\bar{s})=c_{p,0}(\cdot)}}\left\{f_1\left(c_{p,0}(\cdot),(p-1)n+1;\omega_{l(1)}\cdots\omega_{l((p-1)n+1)}\right)\right.$$

$$\cdot\sum_{\substack{\text{all the p-partitions}\\ \text{for } \bar{s}/c_{3,0}(\cdot)}}\;\sum_{\substack{\text{all the different}\\ \text{permutations}\\ \text{of } \{s_{X_1},\cdots,s_{X_p}\}}}\left[f_{2a}\left(s_{\bar{X}_1}\cdots s_{\bar{X}_p}\left(c_{p0}(\cdot)^{n-1}\right);\omega_{l(1)}\cdots\omega_{l(n(\bar{s}))}\right)\right.$$

$$\left.\left.\cdot\prod_{i=1}^{p}\varphi_{n\left(s_{\bar{X}_i}\left(c_{p,0}(\cdot)^{n-1}\right)\right)}\left(s_{\bar{X}_i}\left(c_{p,0}(\cdot)^{n-1}\right);\omega_{l(\bar{X}(i)+1)}\cdots\omega_{l\left(\bar{X}(i)+n\left(s_{\bar{X}_i}\left(c_{p,0}(\cdot)^{n-1}\right)\right)\right)}\right)\right]\right\}$$

$$=\frac{1}{L_{(p-1)n+1}\left(j\omega_{l(1)}+\cdots+j\omega_{l((p-1)n+1)}\right)}$$

$$\cdot\sum_{\substack{\text{all the p-partitions}\\ \text{for } \bar{s}/c_{p,0}(\cdot)}}\;\sum_{\substack{\text{all the different}\\ \text{permutations}\\ \text{of } \{s_{X_1},\cdots,s_{X_p}\}}}\left[\prod_{i=1}^{p}\left(j\omega_{l(\bar{X}(i)+1)}+\cdots+j\omega_{l\left(\bar{X}(i)+n\left(s_{\bar{X}_i}\left(c_{p,0}(\cdot)^{n-1}\right)\right)\right)}\right)^{r_i}\right.$$

$$\left.\cdot\prod_{i=1}^{p}\varphi_{n\left(s_{\bar{X}_i}\left(c_{p,0}(\cdot)^{n-1}\right)\right)}\left(s_{\bar{X}_i}\left(c_{p,0}(\cdot)^{n-1}\right);\omega_{l(\bar{X}(i)+1)}\cdots\omega_{l\left(\bar{X}(i)+n\left(s_{\bar{X}_i}\left(c_{p,0}(\cdot)^{n-1}\right)\right)\right)}\right)\right]$$

$$=\frac{1}{L_{(p-1)n+1}\left(j\omega_{l(1)}+\cdots+j\omega_{l((p-1)n+1)}\right)}$$

$$\cdot\sum_{\substack{\text{all the different combinations}\\ \text{of } \{\bar{x}_1,\bar{x}_2,\ldots,\bar{x}_p\} \text{ satisfying}\\ \bar{x}_1+\cdots+\bar{x}_p=n-1,\ 0\le\bar{x}_i\le n-1}}\;\sum_{\substack{\text{all the different}\\ \text{permutations of}\\ \text{each combination}}}\left[\prod_{i=1}^{p}\left(j\omega_{l(\bar{X}(i)+1)}+\cdots+j\omega_{l(\bar{X}(i)+(p-1)\bar{x}_i+1)}\right)^{r_i}\right.$$

$$\left.\cdot\prod_{i=1}^{p}\varphi_{(p-1)\bar{x}_i+1}\left(c_{p,0}(\cdot)^{\bar{x}_i};\omega_{l(\bar{X}(i)+1)}\cdots\omega_{l(\bar{X}(i)+(p-1)\bar{x}_i+1)}\right)\right]$$

Note that different permutations in each combination have no difference to $\prod_{i=1}^{p}\varphi_{(p-1)\bar{x}_i+1}\left(c_{p,0}(\cdot)^{\bar{x}_i};\omega_{l(\bar{X}(i)+1)}\cdots\omega_{l(\bar{X}(i)+(p-1)\bar{x}_i+1)}\right)$, thus $\varphi_{(p-1)n+1}(c_{p,0}(\cdot)^{n};\omega_1\cdots\omega_{(p-1)n+1})$ can be written as

$$\varphi_{(p-1)n+1}\left(c_{p,0}(\cdot)^n;\omega_1\cdots\omega_{(p-1)n+1}\right)$$

$$=\frac{1}{L_{(p-1)n+1}\left(j\omega_{l(1)}+\cdots+j\omega_{l((p-1)n+1)}\right)}$$

$$\cdot\sum_{\substack{\text{all the different combinations}\\ \text{of }\{\overline{x}_1,\overline{x}_2,\ldots,\overline{x}_p\}\text{ satisfying}\\ \overline{x}_1+\cdots+\overline{x}_3=n-1,\ 0\le\overline{x}_i\le n-1}}\prod_{i=1}^{p}\varphi_{(p-1)\overline{x}_i+1}\left(c_{p,0}(\cdot)^{\overline{x}_i};\omega_{l(\overline{X}(i)+1)}\cdots\omega_{l(\overline{X}(i)+(p-1)\overline{x}_i+1)}\right)$$

$$\cdot\sum_{\substack{\text{all the different}\\ \text{permutations of}\\ \text{each combination}}}\prod_{i=1}^{p}\left(j\omega_{l(\overline{X}(i)+1)}+\cdots+j\omega_{l(\overline{X}(i)+(p-1)\overline{x}_i+1)}\right)^{r_i}$$

$$=\frac{1}{L_{(p-1)n+1}\left(j\omega_{l(1)}+\cdots+j\omega_{l((p-1)n+1)}\right)}$$

$$\cdot\sum_{\substack{\text{all the different combinations}\\ \text{of }\{\overline{x}_1,\overline{x}_2,\ldots,\overline{x}_p\}\text{ satisfying}\\ \overline{x}_1+\cdots+\overline{x}_p=n-1,\ 0\le\overline{x}_i\le n-1}}\prod_{i=1}^{p}\varphi_{(p-1)\overline{x}_i+1}\left(c_{p,0}(\cdot)^{\overline{x}_i};\omega_{l(\overline{X}(i)+1)}\cdots\omega_{l(\overline{X}(i)+(p-1)\overline{x}_i+1)}\right)$$

$$\cdot\frac{n_x{}^*\left(\overline{x}_1,\cdots,\overline{x}_p\right)}{n_r{}^*\left(r_1,\cdots,r_p\right)}\cdot\sum_{\substack{\text{all the different}\\ \text{permutations of}\\ \{r_1,\cdots,r_p\}}}\prod_{i=1}^{p}\left(j\omega_{l(\overline{X}(i)+1)}+\cdots+j\omega_{l(\overline{X}(i)+(p-1)\overline{x}_i+1)}\right)^{r_i}$$

$n_x^*\left(\overline{x}_1,\cdots,\overline{x}_p\right)$ and $n_r^*(r_1,\cdots,r_p)$ are the numbers of the corresponding combinations involved, which can be obtained from the combination theory and can also be referred to Peyton Jones (2007). From an inspection of the recursive relationship in the equation above, it can be seen that there are $(p-1)n+1$ $H_1(j\omega_i)$ with different frequency variable at the end of the recursive relationship. Thus they can be brought out as a common factor. This gives

$$\varphi_{(p-1)n+1}\left(c_{p,0}(\cdot)^n;\omega_{l(1)}\cdots\omega_{l((p-1)n+1)}\right)$$

$$=(-1)^n\prod_{i=1}^{(p-1)n+1}H_1\left(j\omega_{l(i)}\right)\cdot\varphi'_{(p-1)n+1}\left(c_{p,0}(\cdot)^n;\omega_{l(1)}\cdots\omega_{l((p-1)n+1)}\right)\quad(12.20a)$$

where,

$$\varphi'_{(p-1)n+1}\left(c_{p,0}(\cdot)^n;\omega_{l(1)}\cdots\omega_{l((p-1)n+1)}\right)=\frac{-1}{L_{(p-1)n+1}\left(j\omega_{l(1)}+\cdots+j\omega_{l((p-1)n+1)}\right)}$$

$$\cdot\sum_{\substack{\text{all the different combinations}\\ \text{of }\{\bar{x}_1,\bar{x}_2,\ldots,\bar{x}_p\}\text{ satisfying}\\ \bar{x}_1+\cdots+\bar{x}_p=n-1,\ 0\le\bar{x}_i\le n-1}}\prod_{i=1}^{p}\varphi'_{(p-1)\bar{x}_i+1}\left(c_{p,0}(\cdot)^{\bar{x}_i};\omega_{l(\bar{X}(i)+1)}\cdots\omega_{l(\bar{X}(i)+(p-1)\bar{x}_i+1)}\right)$$

$$\cdot\frac{n_x{}^*\left(\bar{x}_1,\cdots,\bar{x}_p\right)}{n_r{}^*\left(r_1,\cdots,r_p\right)}$$

$$\cdot\sum_{\substack{\text{all the different}\\ \text{permutations of}\\ \{r_1,\cdots,r_p\}}}\prod_{i=1}^{p}\left(j\omega_{l(\bar{X}(i)+1)}+\cdots+j\omega_{l(\bar{X}(i)+(p-1)\bar{x}_i+1)}\right)^{r_i}$$

(12.20b)

the termination is $\varphi'_1(1;\omega_i)=1$. Note that when $\bar{x}_i=0$, there is a term $\left(j\omega_{l(\bar{X}(i)+1)}\right)^{r_i}$ appearing from $\frac{n_x{}^*(\bar{x}_1,\cdots,\bar{x}_p)}{n_r{}^*(r_1,\cdots,r_p)}\cdot\sum_{\substack{\text{all the different}\\ \text{permutations of}\\ \{r_1,\cdots,r_p\}}}\prod_{i=1}^{p}\left(j\omega_{l(\bar{X}(i)+1)}+\cdots+j\omega_{l(\bar{X}(i)+(p-1)\bar{x}_i+1)}\right)^{r_i}$.

It can be verified that in each recursion of $\varphi'_{(p-1)n+1}(c_{p,0}(\cdot)^n;\omega_{l(1)}\cdots\omega_{l((p-1)n+1)})$, there may be some frequency variables appearing individually in the form of $\left(j\omega_{l(\bar{X}(i)+1)}\right)^{r_i}$, and these variables will not appear individually in the same form in the subsequent recursion. At the end of the recursion, all the frequency variables should have appeared in this form. Thus these terms can also be brought out as common factors if $r_1=r_2=\ldots=r_p$. In the case of $r_1=r_2=\ldots=r_p=r$,

$$\frac{n_x{}^*\left(\bar{x}_1,\cdots,\bar{x}_p\right)}{n_r{}^*\left(r_1,\cdots,r_p\right)}\cdot\sum_{\substack{\text{all the different}\\ \text{permutations of}\\ \{r_1,\cdots,r_p\}}}\prod_{i=1}^{p}\left(j\omega_{l(\bar{X}(i)+1)}+\cdots+j\omega_{l(\bar{X}(i)+(p-1)\bar{x}_i+1)}\right)^{r_i}$$

$$=n_x{}^*\left(\bar{x}_1,\cdots,\bar{x}_p\right)\cdot\prod_{i=1}^{p}\left(j\omega_{l(\bar{X}(i)+1)}+\cdots+j\omega_{l(\bar{X}(i)+(p-1)\bar{x}_i+1)}\right)^{r_i}$$

Therefore (12.20a,b) can be written, if $r_1=r_2=\ldots=r_p$, as

$$\varphi_{(p-1)n+1}\left(c_{p,0}(\cdot)^n;\omega_{l(1)}\cdots\omega_{l((p-1)n+1)}\right)$$
$$=(-1)^n\prod_{i=1}^{(p-1)n+1}\left[\left(j\omega_{l(i)}\right)^r H_1\left(j\omega_{l(i)}\right)\right]\cdot\varphi'_{(p-1)n+1}\left(c_{p,0}(\cdot)^n;\omega_{l(1)}\cdots\omega_{l((p-1)n+1)}\right)\tag{12.21a}$$

$$\phi'_{(p-1)n+1}\left(c_{p,0}(\cdot)^n;\omega_{l(1)}\cdots\omega_{l((p-1)n+1)}\right)$$
$$=\frac{-1}{L_{(p-1)n+1}\left(j\omega_{l(1)}+\cdots+j\omega_{l((p-1)n+1)}\right)}$$
$$\cdot\sum_{\substack{\text{all the different combinations}\\ \text{of }\{\bar{x}_1,\bar{x}_2,\ldots,\bar{x}_p\}\text{ satisfying}\\ \bar{x}_1+\cdots+\bar{x}_p=n-1,\ 0\le\bar{x}_i\le n-1}}\prod_{i=1}^{p}\phi'_{(p-1)\bar{x}_i+1}\left(c_{p,0}(\cdot)^{\bar{x}_i};\omega_{l(\bar{X}(i)+1)}\cdots\omega_{l(\bar{X}(i)+(p-1)\bar{x}_i+1)}\right)$$
$$\cdot n_x{}^*\left(\bar{x}_1,\cdots,\bar{x}_p\right)\cdot\prod_{i=1}^{p}\left(j\omega_{l(\bar{X}(i)+1)}+\cdots+j\omega_{l(\bar{X}(i)+(p-1)\bar{x}_i+1)}\right)^{r_i(1-\delta(\bar{x}_i))}\tag{12.21b}$$

(12.21b) can be further written as

$$\phi'_{(p-1)n+1}\left(c_{p,0}(\cdot)^n;\omega_{l(1)}\cdots\omega_{l((p-1)n+1)}\right)$$
$$=\frac{-1}{L_{(p-1)n+1}\left(j\omega_{l(1)}+\cdots+j\omega_{l((p-1)n+1)}\right)}$$
$$\cdot\sum_{\substack{\text{all the different combinations}\\ \text{of }\{\bar{x}_1,\bar{x}_2,\ldots,\bar{x}_p\}\text{ satisfying}\\ \bar{x}_1+\cdots+\bar{x}_p=n-1,\ 0\le\bar{x}_i\le n-1}} n_x{}^*\left(\bar{x}_1,\cdots,\bar{x}_p\right)$$
$$\cdot\prod_{i=1}^{p}\phi''_{(p-1)\bar{x}_i+1}\left(c_{p,0}(\cdot)^{\bar{x}_i};\omega_{l(\bar{X}(i)+1)}\cdots\omega_{l(\bar{X}(i)+(p-1)\bar{x}_i+1)}\right)\tag{12.22a}$$

where, if $\bar{x}_i=0$,

$$\varphi''_{(p-1)\bar{x}_i+1}\left(c_{p,0}(\cdot)^{\bar{x}_i};\omega_{l(\bar{X}(i)+1)}\cdots\omega_{l(\bar{X}(i)+(p-1)\bar{x}_i+1)}\right)=1,$$

otherwise,

$$\varphi''_{(p-1)\bar{x}_i+1}\left(c_{p,0}(\cdot)^{\bar{x}_i};\omega_{l(\overline{X}(i)+1)}\cdots\omega_{l(\overline{X}(i)+(p-1)\bar{x}_i+1)}\right)$$
$$=\left(j\omega_{l(\overline{X}(i)+1)}+\cdots+j\omega_{l(\overline{X}(i)+(p-1)x_i+1)}\right)^r$$

$$\varphi'_{(p-1)\bar{x}_i+1}\left(c_{p,0}(\cdot)^{\bar{x}_i};\omega_{l(\overline{X}(i)+1)}\cdots\omega_{l(\overline{X}(i)+(p-1)\bar{x}_i+1)}\right)$$
$$=\frac{\left(j\omega_{l(\overline{X}(i)+1)}+\cdots+j\omega_{l(\overline{X}(i)+(p-1)\bar{x}_i+1)}\right)^r}{-L_{(p-1)\bar{x}_i+1}\left(j\omega_{l(\overline{X}(i)+1)}+\cdots+j\omega_{l(\overline{X}(i)+(p-1)\bar{x}_i+1)}\right)}$$
$$\cdot\sum_{\substack{\text{all the different combinations}\\ \text{of } \{\mathrm{x}_1,\mathrm{x}_2,\ldots,\mathrm{x}_\mathrm{p}\} \text{ satisfying}\\ \mathrm{x}_1+\cdots+\mathrm{x}_\mathrm{p}=\bar{x}_i-1,\ 0\le \mathrm{x}_\mathrm{i}\le \bar{x}_i-1}} n_x{}^*(\mathrm{x}_1,\cdots,\mathrm{x}_\mathrm{p})$$
$$\cdot\prod_{i=1}^{p}\left(j\omega_{l(\overline{X}'(i)+1)}+\cdots+j\omega_{l(\overline{X}'(i)+(p-1)x_i+1)}\right)^{r_i(1-\delta(x_i))}$$

$$\varphi'_{(p-1)x_i+1}\left(c_{p,0}(\cdot)^{x_i};\omega_{l(\overline{X}'(i)+1)}\cdots\omega_{l(\overline{X}'(i)+(p-1)x_i+1)}\right)$$
$$=\frac{\left(j\omega_{l(\overline{X}(i)+1)}+\cdots+j\omega_{l(\overline{X}(i)+(p-1)\bar{x}_i+1)}\right)^r}{-L_{(p-1)\bar{x}_i+1}\left(j\omega_{l(\overline{X}(i)+1)}+\cdots+j\omega_{l(\overline{X}(i)+(p-1)\bar{x}_i+1)}\right)}$$
$$\cdot\sum_{\substack{\text{all the different combinations}\\ \text{of } \{\mathrm{x}_1,\mathrm{x}_2,\ldots,\mathrm{x}_\mathrm{p}\} \text{ satisfying}\\ \mathrm{x}_1+\cdots+\mathrm{x}_\mathrm{p}=\bar{x}_i-1,\ 0\le \mathrm{x}_\mathrm{i}\le \bar{x}_i-1}} n_x{}^*(\mathrm{x}_1,\cdots,\mathrm{x}_\mathrm{p})$$
$$\cdot\prod_{i=1}^{p}\phi''_{(p-1)x_i+1}\left(c_{p,0}(\cdot)^{x_i};\omega_{l(\overline{X}'(i)+1)}\cdots\omega_{l(\overline{X}'(i)+(p-1)x_i+1)}\right) \tag{12.22b}$$

The recursive terminal of (12.22b) is $\bar{x}_i=1$. Substituting (12.20b) into (12.20a) and substituting (12.22a,b) into (12.21a), the lemma can be obtained. This completes the proof. □

For convenience, define an operator "*" for $\mathrm{sgn}_c(.)$ satisfying

$$\mathrm{sgn}_c(\upsilon_1)*\mathrm{sgn}_c(\upsilon_2)=\left[\mathrm{sgn}_r(\mathrm{Re}(\upsilon_1\upsilon_2))\quad \mathrm{sgn}_r(\mathrm{Im}(\upsilon_1\upsilon_2))\right]$$

for any $\upsilon_1,\upsilon_2\in\mathbb{C}$. It is obvious $\mathrm{sgn}_c(\upsilon_1)*\mathrm{sgn}_c(\upsilon_2)=\mathrm{sgn}_c(\upsilon_1\upsilon_2)$.

The following lemma is straightforward.

Lemma 12.3 For $\upsilon_1, \upsilon_2, \nu \in \mathbb{C}$, suppose $\mathrm{sgn}_c(\upsilon_1) = -\,\mathrm{sgn}_c(\upsilon_2)$. If $\mathrm{Re}(\nu)\mathrm{Im}(\nu) = 0$, then $\mathrm{sgn}_c(\upsilon_1\nu) = -\,\mathrm{sgn}_c(\upsilon_2\nu)$. If $\mathrm{Re}(\nu)\mathrm{Im}(\nu) = 0$ and $\nu \neq 0$, then $\mathrm{sgn}_c(\upsilon_1/\nu) = -\,\mathrm{sgn}_c(\upsilon_2/\nu)$. □

Proposition 12.2 The output spectrum in (12.19a) is an alternating series with respect to any specific parameter $c_{p,0}(\mathrm{r}_1,\mathrm{r}_2,\ldots,\mathrm{r}_p)$ satisfying $c_{p,0}(.)>0$ and $p = 2\bar{r}+1$ for $\bar{r} = 1, 2, 3, \ldots$

(a1) if and only if

$$\mathrm{sgn}_c\left(\sum_{\omega_{k_1}+\cdots+\omega_{k_{(p-1)n+1}}=\Omega} (-1)^{n-1}\varphi_{(p-1)n+1}\left(c_{p,0}(\cdot)^n;\omega_{l(1)}\cdots\omega_{l((p-1)n+1)}\right)\right) = const, \quad \text{i.e.,}$$

$$\mathrm{sgn}_c\left(\frac{H_1(j\Omega)}{L_{(p-1)n+1}(j\Omega)}\sum_{\omega_{k_1}+\cdots+\omega_{k_{(p-1)n+1}}=\Omega}\ \sum_{\substack{\text{all the different combinations}\\ \text{of }\{\bar{\mathrm{x}}_1,\bar{\mathrm{x}}_2,\ldots,\bar{\mathrm{x}}_p\}\text{ satisfying}\\ \bar{\mathrm{x}}_1+\cdots+\bar{\mathrm{x}}_p=n-1,\ 0\le\bar{\mathrm{x}}_i\le n-1}} \left[\prod_{i=1}^{p}\varphi'_{(p-1)\bar{\mathrm{x}}_i+1}\left(c_{p,0}(\cdot)^{\bar{\mathrm{x}}_i};\omega_{l(\bar{X}(i)+1)}\cdots\omega_{l(\bar{X}(i)+(p-1)\bar{\mathrm{x}}_i+1)}\right)\cdot\frac{n_x{}^*(\bar{\mathrm{x}}_1,\cdots,\bar{\mathrm{x}}_p)}{n_r{}^*(\mathrm{r}_1,\cdots,\mathrm{r}_p)}\cdot\sum_{\substack{\text{all the different}\\ \text{permutations of}\\ \{\mathrm{r}_1,\cdots,\mathrm{r}_p\}}}\prod_{i=1}^{p}\left(j\omega_{l(\bar{X}(i)+1)}+\cdots+j\omega_{l(\bar{X}(i)+(p-1)\bar{\mathrm{x}}_i+1)}\right)^{r_i}\right]\right) = const \tag{12.23}$$

(a2) if $\mathrm{r}_1=\mathrm{r}_2=\ldots=\mathrm{r}_p=\mathrm{r}$ in $c_{p,0}(.)$, $\mathrm{Re}\left(\frac{H_1(j\Omega)}{L_{(p-1)n+1}(j\Omega)}\right)\mathrm{Im}\left(\frac{H_1(j\Omega)}{L_{(p-1)n+1}(j\Omega)}\right) = 0$, and

$$\mathrm{sgn}_c\left(\sum_{\omega_{k_1}+\cdots+\omega_{k_{(p-1)n+1}}=\Omega}\ \sum_{\substack{\text{all the different combinations}\\ \text{of } \{\overline{x}_1,\overline{x}_2,\ldots,\overline{x}_p\} \text{ satisfying}\\ \overline{x}_1+\cdots+\overline{x}_p=n-1,\ 0\le\overline{x}_i\le n-1}} \left[n_x{}^*(\overline{x}_1,\cdots,\overline{x}_p)\cdot\prod_{i=1}^{p}{}^{''}\varphi_{(p-1)\overline{x}_i+1}\left(c_{p,0}(\cdot)^{\overline{x}_i};\omega_{l(\overline{X}(i)+1)}\cdots\omega_{l(\overline{X}(i)+(p-1)\overline{x}_i+1)}\right)\right]\right)=const \tag{12.24}$$

where *const* is a two-dimensional constant vector whose elements are +1, 0 or −1.

Proof

(a1) From Lemma 12.1, any asymmetric $\varphi_{(p-1)n+1}\left(c_{p,0}(\cdot)^n;\omega_{k_1},\cdots,\omega_{k_{(p-1)n+1}}\right)$ is sufficient for the computation of $\widetilde{F}_{(p-1)n+1}(j\Omega)$. It can be obtained that

$$\mathrm{sgn}_c\left(\widetilde{F}_{(p-1)n+1}(j\Omega)\right)=\mathrm{sgn}_c\left(-j\left(\frac{F_d}{2}\right)^{(p-1)n+1}\right)$$
$$*\,\mathrm{sgn}\left(\sum_{\omega_{k_1}+\cdots+\omega_{k_{(p-1)n+1}}=\Omega}\varphi_{(p-1)n+1}\left(c_{p,0}(\cdot)^n;\omega_{k_1},\cdots,\omega_{k_{(p-1)n+1}}\right)\right)$$

From Lemma 12.3, $\mathrm{sgn}_c\left(-j\left(\frac{F_d}{2}\right)^{(p-1)n+1}\right)$ has no effect on the alternating nature of the sequence $\widetilde{F}_{(p-1)n+1}(j\Omega)$ for n=1,2,3,…. Hence, (12.19a) is an alternating series with respect to $c_{p,0}(.)$ if and only if the sequence $\sum_{\omega_{k_1}+\cdots+\omega_{k_{(p-1)n+1}}=\Omega}\varphi_{(p-1)n+1}\left(c_{p,0}(\cdot)^n;\omega_{k_1},\cdots,\omega_{k_{(p-1)n+1}}\right)$ for n=1,2,3,… is alternating. This is equivalent to

$$\mathrm{sgn}_c\left(\sum_{\omega_{k_1}+\cdots+\omega_{k_{(p-1)n+1}}=\Omega}^{(-1)n-1}\varphi_{(p-1)n+1}\left(c_{p,0}(\cdot)^n;\omega_{l(1)}\cdots\omega_{l((p-1)n+1)}\right)\right)=const$$

In the equation above, replacing $\varphi_{(p-1)n+1}\left(c_{p,0}(\cdot)^n;\omega_{k_1},\cdots,\omega_{k_{(p-1)n+1}}\right)$ by using the result in Lemma 12.2 and noting (p−1)n+1 is an odd integer, it can be obtained that

$$\mathrm{sgn}_c\left(\begin{array}{l}\displaystyle\sum_{\omega_{k_1}+\cdots+\omega_{k_{(p-1)n+1}}=\Omega}\frac{\prod_{i=1}^{(p-1)n+1}H_1(j\omega_{l(i)})}{L_{(p-1)n+1}(j\omega_{l(1)}+\cdots+j\omega_{l((p-1)n+1)})}\\ \cdot\displaystyle\sum_{\substack{\text{all the different combinations}\\ \text{of }\{\bar{x}_1,\bar{x}_2,\ldots,\bar{x}_p\}\text{ satisfying}\\ \bar{x}_1+\cdots+\bar{x}_p=n-1,\ 0\le\bar{x}_i\le n-1}}\left[\prod_{i=1}^{p}\phi'_{(p-1)\bar{x}_i+1}\left(c_{p,0}(\cdot)^{\bar{x}_i};\omega_{l(\bar{X}(i)+1)}\cdots\omega_{l(\bar{X}(i)+(p-1)\bar{x}_i+1)}\right)\right.\\ \left.\cdot\dfrac{n_x{}^*(\bar{x}_1,\cdots,\bar{x}_p)}{n_r{}^*(r_1,\cdots,r_p)}\cdot\displaystyle\sum_{\substack{\text{all the different}\\ \text{permutations of}\\ \{r_1,\cdots,r_p\}}}\prod_{i=1}^{p}\left(j\omega_{l(\bar{X}(i)+1)}+\cdots+j\omega_{l(\bar{X}(i)+(p-1)\bar{x}_i+1)}\right)^{r_i}\right]\end{array}\right)$$

$$=\mathrm{sgn}_c\left(\begin{array}{l}\dfrac{H_1(j\Omega)\prod_{i=1}^{(p-1)n/2}|H_1(j\Omega)|^2}{L_{(p-1)n+1}(j\Omega)}\cdot\\ \displaystyle\sum_{\omega_{k_1}+\cdots+\omega_{k_{(p-1)n+1}}=\Omega}\ \sum_{\substack{\text{all the different combinations}\\ \text{of }\{\bar{x}_1,\bar{x}_2,\ldots,\bar{x}_p\}\text{ satisfying}\\ \bar{x}_1+\cdots+\bar{x}_p=n-1,\ 0\le\bar{x}_i\le n-1}}\left[\prod_{i=1}^{p}\phi'_{(p-1)\bar{x}_i+1}\left(c_{p,0}(\cdot)^{\bar{x}_i};\omega_{l(\bar{X}(i)+1)}\cdots\omega_{l(\bar{X}(i)+(p-1)\bar{x}_i+1)}\right)\right.\\ \left.\cdot\dfrac{n_x{}^*(\bar{x}_1,\cdots,\bar{x}_p)}{n_r{}^*(r_1,\cdots,r_p)}\cdot\displaystyle\sum_{\substack{\text{all the different}\\ \text{permutations of}\\ \{r_1,\cdots,r_p\}}}\prod_{i=1}^{p}\left(j\omega_{l(\bar{X}(i)+1)}+\cdots+j\omega_{l(\bar{X}(i)+(p-1)\bar{x}_i+1)}\right)^{r_i}\right]\end{array}\right)=const$$

Note that $\prod_{i=1}^{(p-1)n/2}|H_1(j\Omega)|^2$ has no effect on the equality above from using Lemma 12.3, then the equation above is equivalent to (12.23).

(a2) If additionally, $r_1=r_2=\ldots=r_p=r$ in $c_{p,0}(.)$, then using the result in Lemma 12.2, (12.23) can be written as

$$\mathrm{sgn}_c\left(\begin{array}{l}\dfrac{(j\Omega)^r H_1(j\Omega)}{L_{(p-1)n+1}(j\Omega)}\displaystyle\sum_{\omega_{k_1}+\cdots+\omega_{k_{(p-1)n+1}}=\Omega}\ \sum_{\substack{\text{all the different combinations}\\ \text{of }\{\bar{x}_1,\bar{x}_2,\ldots,\bar{x}_p\}\text{ satisfying}\\ \bar{x}_1+\cdots+\bar{x}_p=n-1,\ 0\le\bar{x}_i\le n-1}}\left[n_x{}^*(\bar{x}_1,\cdots,\bar{x}_p)\right.\\ \left.\cdot\displaystyle\prod_{i=1}^{p}\varphi''_{(p-1)\bar{x}_i+1}\left(c_{p,0}(\cdot)^{\bar{x}_i};\omega_{l(\bar{X}(i)+1)}\cdots\omega_{l(\bar{X}(i)+(p-1)\bar{x}_i+1)}\right)\right]\end{array}\right)=const$$

From Lemma 12.3, $(j\Omega)^r$ has no effect on this equation. Then the equation above is equivalent to

$$\mathrm{sgn}_c\left(\begin{array}{l}\dfrac{H_1(j\Omega)}{L_{(p-1)n+1}(j\Omega)}\displaystyle\sum_{\omega_{k_1}+\cdots+\omega_{k_{(p-1)n+1}}=\Omega}\ \sum_{\substack{\text{all the different combinations}\\ \text{of }\{\overline{x}_1,\overline{x}_2,\ldots,\overline{x}_p\}\text{ satisfying}\\ \overline{x}_1+\cdots+\overline{x}_p=n-1,\ 0\le\overline{x}_i\le n-1}}\left[n_x{}^*(\overline{x}_1,\cdots,\overline{x}_p)\right.\\ \left.\cdot\displaystyle\prod_{i=1}^{p}\varphi''_{(p-1)\overline{x}_i+1}\left(c_{p,0}(\cdot)^{\overline{x}_i};\omega_{l(\overline{X}(i)+1)}\cdots\omega_{l(\overline{X}(i)+(p-1)\overline{x}_i+1)}\right)\right]\end{array}\right)=const$$

If $\mathrm{Re}\left(\frac{H_1(j\Omega)}{L_{(p-1)n+1}(j\Omega)}\right)\mathrm{Im}\left(\frac{H_1(j\Omega)}{L_{(p-1)n+1}(j\Omega)}\right)=0$, then $\frac{H_1(j\Omega)}{L_{(p-1)n+1}(j\Omega)}$ has no effect, either. This gives (12.24). The proof is completed. □

Proposition 12.2 provides a sufficient and necessary condition for the output spectrum series (12.19a) to be an alternating series with respect to a specific nonlinear parameter $c_{p,0}(r_1,r_2,\ldots,r_p)$ satisfying $c_{p,0}(.)>0$ and $p=2\overline{r}+1$ for $\overline{r}=1,2,3,\ldots$. Similar results can also be established for any other nonlinear parameters. Regarding nonlinear parameter $c_{p,0}(r_1,r_2,\ldots,r_p)$ satisfying $c_{p,0}(.)>0$ and $p=2\overline{r}$ for $\overline{r}=1,2,3,\ldots$, it can be obtained from (12.19a) that

$$Y(j\Omega)=\widetilde{F}_1(\Omega)+c_{p,0}(\cdot)^2\widetilde{F}_{2(p-1)+1}(\Omega)+\cdots+c_{p,0}(\cdot)^{2n}\widetilde{F}_{2(p-1)n+1}(\Omega)+\cdots$$

$\widetilde{F}_{2(p-1)n+1}(\Omega)$ for n=1,2,3,... should be alternating so that $Y(j\Omega)$ is alternating. This yields

$$\mathrm{sgn}_c\left(\sum_{\omega_{k_1}+\cdots+\omega_{k_{2(p-1)n+1}}=\Omega}\varphi_{2(p-1)n+1}\left(c_{p,0}(\cdot)^{2n};\omega_{l(1)}\cdots\omega_{l(2(p-1)n+1)}\right)\right)$$
$$=-\mathrm{sgn}_c\left(\sum_{\omega_{k_1}+\cdots+\omega_{k_{2(p-1)(n+1)+1}}=\Omega}\varphi_{2(p-1)(n+1)+1}\left(c_{p,0}(\cdot)^{2(n+1)};\omega_{l(1)}\cdots\omega_{l(2(p-1)(n+1)+1)}\right)\right)$$

Clearly, this is completely different from the conditions in Proposition 12.2. It may be more difficult for the output spectrum to be alternating with respect to $c_{p,0}(.)>0$ with $p=2\overline{r}$ than $c_{p,0}(.)>0$ with $p=2\overline{r}+1$.

Note that (12.19a) is based on the assumption that there is only nonlinear parameter $c_{p,0}(.)$ and all the other nonlinear parameters are zero. If the effects from the other nonlinear parameters are considered, (12.19a) can be written as

$$Y(j\Omega)=\widetilde{F}'_1(\Omega)+c_{p,0}(\cdot)\widetilde{F}'_p(\Omega)+\cdots+c_{p,0}(\cdot)^n\widetilde{F}'_{(p-1)n+1}(\Omega)+\cdots\tag{12.25a}$$

where

$$\widetilde{F}'_{(p-1)n+1}(\Omega) = \widetilde{F}_{(p-1)n+1}(\Omega) + \delta_{(p-1)n+1}\left(\Omega; C_{p',q'} \backslash c_{p,0}(.)\right) \tag{12.25b}$$

$C_{p',q'}$ includes all the nonlinear parameters in the system. Based on the parametric characteristic analysis in Chap. 5 and the new mapping function $\varphi_n(CE(H_n(\cdot)); \omega_1, \cdots, \omega_n)$ defined in Chap. 11, (12.25b) can be determined consequently. For example, suppose p is an odd integer larger than 1, then $\widetilde{F}_{(p-1)n+1}(j\Omega)$ is given in (12.19c), and $\delta_{(p-1)n+1}\left(\Omega; C_{p',q'} \backslash c_{p,0}(.)\right)$ can be computed as

$$\delta_{(p-1)n+1}\left(\Omega; C_{p',q'} \backslash c_{p,0}(.)\right) = \sum_{\substack{\text{all the monomails consisting of the parameters in } C_{p',q'} \backslash c_{p,0}(\cdot) \\ \text{satisfying np} + \sum \left(\mathrm{p}_i' + \mathrm{q}_i'\right) \text{ is odd and less than N}}} \left[-j\left(\tfrac{F_d}{2}\right)^{n\left(c_{p,0}{}^n s(\cdot)\right)} \cdot \sum_{\omega_{k_1} + \cdots + \omega_{k_{n\left(c_{p,0}{}^n s(\cdot)\right)}} = \Omega} \varphi_{n\left(c_{p,0}{}^n s(\cdot)\right)}\left(c_{p,0}{}^n s\left(C_{p',q'} \backslash c_{p,0}(\cdot)\right); \omega_{k_1} \cdots \omega_{k_{n\left(c_{p,0}{}^n s(\cdot)\right)}}\right)\right]$$

where $s\left(C_{p',q'} \backslash c_{p,0}(\cdot)\right)$ denotes a monomial consisting of some parameters in $C_{p',q'} \backslash c_{p,0}(\cdot)$.

It is obvious that if (12.19a) is an alternating series, then (12.25a) can still be alternating under a proper design of the other nonlinear parameters (For example, these parameters are sufficiently small). Moreover, from the discussions above, it can be seen that whether the system output spectrum is an alternating series or not with respect to a specific nonlinear parameter is greatly dependent on the system linear parameters.

Example 12.3 To demonstrate the theoretical results above, consider again the model (12.13) in Example 12.2. Let $u(t) = F_d \sin(\Omega t)$ $(F_d > 0)$. The output spectrum at frequency Ω is given in (12.16) and (12.17). From Lemma 12.2, it can be derived for this case that

$$\varphi_{2n+1}\left(c_{3,0}(\cdot)^n; \omega_{l(1)} \cdots \omega_{l(2n+1)}\right) = \frac{(-1)^{n-1} \prod_{i=1}^{2n+1} \left[\left(j\omega_{l(i)}\right)^r H_1\left(j\omega_{l(i)}\right)\right]}{L_{2n+1}\left(j\omega_{l(1)} + \cdots + j\omega_{l(2n+1)}\right)} \cdot \sum_{\substack{\text{all the different combinations} \\ \text{of } \{\bar{\mathrm{x}}_1, \bar{\mathrm{x}}_2, \bar{\mathrm{x}}_3\} \text{ satisfying} \\ \bar{\mathrm{x}}_1 + \bar{\mathrm{x}}_2 + \bar{\mathrm{x}}_3 = n-1,\ 0 \le \bar{\mathrm{x}}_i \le \mathrm{n} - 1}} n_x{}^*(\bar{\mathrm{x}}_1, \bar{\mathrm{x}}_2, \bar{\mathrm{x}}_3) \cdot \prod_{i=1}^{3} \varphi''_{2\bar{x}_i+1}\left(c_{3,0}(\cdot)^{\bar{x}_i}; \omega_{l(\overline{X}(i)+1)} \cdots \omega_{l(\overline{X}(i)+2\bar{x}_i+1)}\right) \tag{12.26a}$$

where, if $\bar{x}_i = 0$, $\varphi''_{(p-1)\bar{x}_i+1}\left(c_{p,0}(\cdot)^{\bar{x}_i}; \omega_{l(\overline{X}(i)+1)} \cdots \omega_{l(\overline{X}(i)+(p-1)\bar{x}_i+1)}\right) = 1$, otherwise,

$$\varphi''_{2\bar{x}_i+1}\left(c_{3,0}(\cdot)^{\bar{x}_i};\omega_{l(\overline{X}(i)+1)}\cdots\omega_{l(\overline{X}(i)+2\bar{x}_i+1)}\right)$$
$$=\frac{\left(j\omega_{l(\overline{X}(i)+1)}+\cdots+j\omega_{l(\overline{X}(i)+2\bar{x}_i+1)}\right)^r}{-L_{2\bar{x}_i+1}\left(j\omega_{l(\overline{X}(i)+1)}+\cdots+j\omega_{l(\overline{X}(i)+2\bar{x}_i+1)}\right)}\cdot$$

$$\sum_{\substack{\text{all the different combinations}\\ \text{of }\{x_1,x_2,x_3\}\text{ satisfying}\\ x_1+x_2+x_3=\bar{x}_i-1,\ 0\le x_j\le\bar{x}_i-1}} n_x{}^*(x_1,x_2,x_3)\cdot\prod_{j=1}^{3}\varphi''_{2x_j+1}\left(c_{3,0}(\cdot)^{x_j};\omega_{l(\overline{X}'(j)+1)}\cdots\omega_{l(\overline{X}'(j)+2x_j+1)}\right) \quad (12.26\text{b})$$

Note that the terminal condition for (12.26a,b) is

$$\left.\varphi''_{2\bar{x}_i+1}\left(c_{3,0}(\cdot)^{\bar{x}_i};\omega_{l(\overline{X}(i)+1)}\cdots\omega_{l(\overline{X}(i)+2\bar{x}_i+1)}\right)\right|_{\bar{x}_i=1}$$
$$=\varphi''_3\left(c_{3,0}(\cdot);\omega_{l(1)}\cdots\omega_{l(3)}\right)=\frac{\left(j\omega_{l(1)}+\cdots+j\omega_{l(3)}\right)^r}{-L_3\left(j\omega_{l(1)}+\cdots+j\omega_{l(3)}\right)} \quad (12.26\text{c})$$

Therefore, from (12.26a–c) it can be shown that $\varphi_{2n+1}(c_{3,0}(\cdot)^n;\omega_1\cdots\omega_{2n+1})$ can be written as

$$\varphi_{2n+1}\left(c_{3,0}(\cdot)^n;\omega_1\cdots\omega_{2n+1}\right)=\frac{(-1)^{n-1}\prod_{i=1}^{2n+1}j\omega_i H_1(j\omega_i)}{L_{2n+1}(j\omega_1+\cdots+j\omega_{2n+1})}$$
$$\cdot\sum_{\substack{\text{all the combination }(x_1,x_2,\ldots,x_{n-1})\\ \text{satisfying } x_i\in\{2j+1|1\le j\le n-1\}\\ x_1\ge x_2\ge..\ge x_{n-1},\text{ and}\\ ''=''\text{ happens only if } x_i+x_{i+1}\le 2n-2}} r_X(x_1,x_2,\ldots,x_{n-1})\prod_{i=1}^{n-1}\frac{j\omega_{l(1)}+\cdots+j\omega_{l(x_i)}}{-L_{x_i}\left(j\omega_{l(1)}+\cdots+j\omega_{l(x_i)}\right)} \quad (12.27)$$

where $r_X(x_1,x_2,\ldots,x_{n-1})$ is a positive integer which can be explicitly determined by (12.26a,b) and represents the number of all the involved combinations which have the same $\prod_{i=1}^{n-1}\frac{j\omega_{l(1)}+\cdots+j\omega_{l(x_i)}}{-L_{x_i}\left(j\omega_{l(1)}+\cdots+j\omega_{l(x_i)}\right)}$. Therefore, according to Proposition 12.2, it can be seen from (12.27) that the output spectrum (12.16) is an alternating series only if the following two conditions hold:

(b1)

$$\mathrm{Re}\left(\frac{H_1(j\Omega)}{L_{2n+1}(j\Omega)}\right)\mathrm{Im}\left(\frac{H_1(j\Omega)}{L_{2n+1}(j\Omega)}\right)=0$$

(b2)

$$\operatorname{sgn}_c\left(\sum_{\omega_{k_1}+\cdots+\omega_{k_{2n+1}}=\Omega}\ \sum_{\substack{\text{all the combination }(x_1,x_2,\ldots,x_{n-1})\\ \text{satisfying } x_i\in\{2j+1|1\le j\le n-1\}\\ x_1\ge x_2\ge..\ge x_{n-1},\text{ and}\\ ''='' \text{ happens only if } x_i+x_{i+1}\le 2n-2}} {}_rX(x_1,x_2,\ldots,x_{n-1})\prod_{i=1}^{n-1}\frac{j\omega_{l(1)}+\cdots+j\omega_{l(x_i)}}{-L_{x_i}\left(j\omega_{l(1)}+\cdots+j\omega_{l(x_i)}\right)}\right)=const$$

Suppose $\Omega=\sqrt{\frac{k_0}{m}}$ which is a natural resonance frequency of model (12.13). It can be derived that

$$L_{2n+1}(j\Omega)=-\sum_{k_1=0}^{K}c_{1,0}(r_1)(j\Omega)^{r_1}=-\left(m(j\Omega)^2+B(j\Omega)+k_0\right)=-jB\Omega$$

$$H_1(j\Omega)=\frac{-1}{L_1(j\Omega)}=\frac{1}{jB\Omega}$$

It is obvious that condition (b1) is satisfied if $\Omega=\sqrt{\frac{k_0}{m}}$. Considering condition (b2), it can be derived that

$$\frac{j\omega_{l(1)}+\cdots+j\omega_{l(x_i)}}{-L_{x_i}\left(j\omega_{l(1)}+\cdots+j\omega_{l(x_i)}\right)}=\frac{j\varepsilon(x_i)\Omega}{-L_{x_i}(j\varepsilon(x_i)\Omega)} \tag{12.28a}$$

where $\varepsilon(x_i)\in\{\pm(2j+1)|0\le j\le\lceil n+1\rceil\}$, and $\lceil n+1\rceil$ denotes the odd integer not larger than n+1. Especially, when $\varepsilon(x_i)=\pm 1$, it yields that

$$\frac{j\omega_{l(1)}+\cdots+j\omega_{l(x_i)}}{-L_{x_i}\left(j\omega_{l(1)}+\cdots+j\omega_{l(x_i)}\right)}=\frac{\pm j\Omega}{-L_{x_i}(\pm j\Omega)}=\frac{\pm j\Omega}{\pm jB\Omega}=\frac{1}{B} \tag{12.28b}$$

when $|\varepsilon(x_i)|>1$,

$$\begin{aligned}\frac{j\omega_{l(1)}+\cdots+j\omega_{l(x_i)}}{-L_{x_i}\left(j\omega_{l(1)}+\cdots+j\omega_{l(x_i)}\right)}&=\frac{j\varepsilon(x_i)\Omega}{-L_{x_i}(j\varepsilon(x_i)\Omega)}=\frac{j\varepsilon(x_i)\Omega}{m(j\varepsilon(x_i)\Omega)^2+B(j\varepsilon(x_i)\Omega)+k_0}\\&=\frac{j\varepsilon(x_i)\Omega}{\left(1-\varepsilon(x_i)^2\right)k_0+j\varepsilon(x_i)\Omega B}=\frac{1}{B+j\left(\varepsilon(x_i)-\frac{1}{\varepsilon(x_i)}\right)\sqrt{k_0m}}\end{aligned} \tag{12.28c}$$

If $B<<\sqrt{k_0m}$, then it gives

$$\frac{j\omega_{l(1)} + \cdots + j\omega_{l(x_i)}}{-L_{x_i}\left(j\omega_{l(1)} + \cdots + j\omega_{l(x_i)}\right)} \approx \frac{1}{j\left(\varepsilon(x_i) - \frac{1}{\varepsilon(x_i)}\right)\sqrt{k_0 m}} \tag{12.28d}$$

Note that in all the combinations involved in the summation operator in (12.27) or condition (b2), *i.e.*,

$$\sum_{\omega_{k_1}+\cdots+\omega_{k_{2n+1}}=\Omega} \sum_{\substack{\text{all the combination } (x_1, x_2, \ldots, x_{n-1}) \\ \text{satisfying } x_i \in \{2j+1 \mid 1 \le j \le n-1\} \\ x_1 \ge x_2 \ge .. \ge x_{n-1}, \text{ and} \\ "=" \text{ happens only if } x_i + x_{i+1} \le 2n-2}} (\cdot)$$

There always exists a combination such that

$$\prod_{i=1}^{n-1} \frac{j\omega_{l(1)} + \cdots + j\omega_{l(x_i)}}{-L_{x_i}\left(j\omega_{l(1)} + \cdots + j\omega_{l(x_i)}\right)} = \frac{1}{B^{n-1}} \tag{12.29}$$

Note that (12.28b) holds both for $\varepsilon(x_i) = \pm 1$, thus there is no combination such that

$$\prod_{i=1}^{n-1} \frac{j\omega_{l(1)} + \cdots + j\omega_{l(x_i)}}{-L_{x_i}\left(j\omega_{l(1)} + \cdots + j\omega_{l(x_i)}\right)} = -\frac{1}{B^{n-1}}$$

Noting that $B << \sqrt{k_0 m}$, these show that

$$\max_{\substack{\text{all the involved} \\ \text{combinations}}} \left(\left| \prod_{i=1}^{n-1} \frac{j\omega_{l(1)} + \cdots + j\omega_{l(x_i)}}{-L_{x_i}\left(j\omega_{l(1)} + \cdots + j\omega_{l(x_i)}\right)} \right| \right) = \frac{1}{B^{n-1}}$$

which happens in the combination where (12.29) holds.

Because there are n+1 frequency variables to be $+\Omega$ and n frequency variables to be $-\Omega$ such that $\omega_1 + \cdots + \omega_{2n+1} = \Omega$ in (12.16) and (12.17), there are more combinations where $\varepsilon(x_i) > 0$ that is $\left(\varepsilon(x_i) - \frac{1}{\varepsilon(x_i)}\right)\sqrt{k_0 m} > 0$ in (12.28c,d). Thus there are more combinations where $\operatorname{Im}\left(\frac{j\omega_{l(1)}+\cdots+j\omega_{l(x_i)}}{-L_{x_i}\left(j\omega_{l(1)}+\cdots+j\omega_{l(x_i)}\right)}\right)$ is negative. Using (12.28b) and (12.28d), it can be shown under the condition that $B << \sqrt{k_0 m}$,

$$\max_{\substack{\text{all the involved}\\ \text{combinations}}}\left(\left|\mathrm{Im}\left(\prod_{i=1}^{n-1}\frac{j\omega_{l(1)}+\cdots+j\omega_{l(x_i)}}{-L_{x_i}\left(j\omega_{l(1)}+\cdots+j\omega_{l(x_i)}\right)}\right)\right|\right)$$

$$\approx \frac{1}{B^{n-2}\left(\varepsilon(x_i)-\frac{1}{\varepsilon(x_i)}\right)\sqrt{k_0 m}}\Bigg|_{\varepsilon(x_i)=3}=\frac{1}{2.7B^{n-2}\sqrt{k_0 m}}$$

This happens in the combinations where the argument of $\prod_{i=1}^{n-1}\frac{j\omega_{l(1)}+\cdots+j\omega_{l(x_i)}}{-L_{x_i}\left(j\omega_{l(1)}+\cdots+j\omega_{l(x_i)}\right)}$ is either $-90°$ or $+90°$. Note that there are more cases in which the arguments are $-90°$. If the argument is $-180°$, the absolute value of the corresponding imaginary part will be not more than

$$\max_{\substack{\text{the combination}\\ \text{whose argument is}\\ -180^0}}\left(\left|\mathrm{Im}\left(\prod_{i=1}^{n-1}\frac{j\omega_{l(1)}+\cdots+j\omega_{l(x_i)}}{-L_{x_i}\left(j\omega_{l(1)}+\cdots+j\omega_{l(x_i)}\right)}\right)\right|\right)$$

$$\approx \frac{1}{B^{n-4}\left(\varepsilon(x_i)-\frac{1}{\varepsilon(x_i)}\right)^3\sqrt{k_0 m}^3}\Bigg|_{\varepsilon(x_i)=3}=\frac{1}{2.7^3B^{n-4}\sqrt{k_0 m}^3}$$

which is much less than $\frac{1}{2.7B^{n-2}\sqrt{k_0 m}}$.

Therefore, if B is sufficiently smaller than $\sqrt{k_0 m}$, the following two inequalities can hold for n>1

$$\mathrm{Re}\left(\sum_{\substack{\text{all the combination }(x_1,x_2,\ldots,x_{n-1})\\ \text{satisfying } x_i\in\{2j+1|1\le j\le n-1\}\\ x_1\ge x_2\ge ..\ge x_{n-1},\text{ and}\\ ''=''\text{ happens only if } x_i+x_{i+1}\le 2n-2}} rX(x_1,x_2,\ldots,x_{n-1})\prod_{i=1}^{n-1}\frac{j\omega_{l(1)}+\cdots+j\omega_{l(x_i)}}{-L_{x_i}\left(j\omega_{l(1)}+\cdots+j\omega_{l(x_i)}\right)}\right)>0$$

$$\mathrm{Im}\left(\sum_{\substack{\text{all the combination }(x_1,x_2,\ldots,x_{n-1})\\ \text{satisfying } x_i\in\{2j+1|1\le j\le n-1\}\\ x_1\ge x_2\ge ..\ge x_{n-1},\text{ and}\\ ''=''\text{ happens only if } x_i+x_{i+1}\le 2n-2}} {}_rX(x_1,x_2,\ldots,x_{n-1})\prod_{i=1}^{n-1}\frac{j\omega_{l(1)}+\cdots+j\omega_{l(x_i)}}{-L_{x_i}(j\omega_{l(1)}+\cdots+j\omega_{l(x_i)})}\right)<0$$

That is, condition (b2) holds for $n>1$ under $B<<\sqrt{k_0m}$ and $\Omega=\sqrt{\frac{k_0}{m}}$. Hence, (12.16) is an alternating series if the following two conditions hold:

(c1) B is sufficiently smaller than $\sqrt{k_0m}$,

(c2) The input frequency is $\Omega=\sqrt{\frac{k_0}{m}}$.

Note that in example 12.1, $\Omega=\sqrt{\frac{k_0}{m}}\approx 8.165, B=296\ll\sqrt{k_0m}=1{,}959.592$. These are consistent with the theoretical results. Therefore the conclusions are verified.

12.5 Conclusions

Nonlinear influence on system output spectrum is investigated in this Chapter from a novel perspective—alternating series. For the Volterra class of system nonlinearities, it is shown for the first time that system output spectrum can be expanded into an alternating series with respect to (nonlinear) model parameters under certain conditions and this alternating series has some interesting and favourable properties for engineering practices. Although there may be several existing methods such as perturbation analysis that could achieve similar objectives for some cases in practice, this study proposes a novel and alternative viewpoint on the nonlinear effect (i.e., alternating series) and on the analysis of nonlinear effect (i.e., the GFRFs-based) in the frequency domain. As some important properties of a linear system (e.g. stability) are determined by the positions of the poles of its transfer function, the concept of alternating series could be a crucial characteristic of nonlinear behaviours in the frequency domain. Some fundamental results are therefore developed for characterizing and understanding of nonlinear effects from this novel viewpoint. Further study will be focused on more detailed design and analysis methods based on these results for practical systems.

Chapter 13
Magnitude Bound Characteristics of Nonlinear Frequency Response Functions

13.1 Introduction

In many cases, the magnitude of a frequency response function such as GFRFs can reveal important information about the system, and consequently takes a great role in the analysis of the convergence or stability of the system and the truncation error of the corresponding Volterra series. It can be used to evaluate the significant orders of nonlinearities or the significant nonlinear terms for the magnitude bound, indicate the stability of a system and provide a basis for analysis of system output frequency response. Several efforts to derive the magnitudes of the GFRFs and output frequency response have been attempted. A very simple algorithm to evaluate the magnitude bounds of the GFRFs was provided in Zhang and Billings (1996). Billings and Lang (1996) proposed a more detailed recursive algorithm to compute the gain bounds of the GFRFs and output frequency response. Notice that in these results, the relationship between the magnitude of the system frequency response functions and the system time domain model parameters is not revealed explicitly.

New bound characteristics of both the generalized frequency response functions (GFRFs) and output frequency response for the NARX model are presented in this chapter. It is shown that the magnitudes of the GFRFs and the system output spectrum can all be bounded by a polynomial function of the magnitude bound of the first order GFRF, and the coefficients of the polynomial are functions of the NARX model parameters. These new bound characteristics of the NARX model provide an important insight into the relationship between the model parameters and the magnitudes of the system frequency response functions, reveal the effect of the model parameters on the stability of the NARX model to a certain extent, and provide a useful technique for the magnitude based analysis of nonlinear systems in the frequency domain. Based on these results, truncation error and the highest order associated with Volterra series expression of nonlinear systems can be studied. Sufficient conditions for the BIBO stability of the NARX model can also be

X. Jing, Z. Lang, *Frequency Domain Analysis and Design of Nonlinear Systems based on Volterra Series Expansion*, Understanding Complex Systems,
DOI 10.1007/978-3-319-12391-2_13

established. A numerical example is given to demonstrate the effectiveness of the theoretical results. An important application of these results will be discussed in the next chapter to address an important convergence issue of Volterra series expansion.

The bound characteristics of this chapter can be further developed with less conservatism, which can be referred to Jing et al. (2008b, 2009b).

13.2 The Frequency Response Functions of Nonlinear Systems and the NARX Model

For convenience, the technical background of this study is simply given in this section. The details can be referred to Chaps. 2 and 3. Nonlinear systems with stable zero equilibrium point can be approximated in the neighbourhood of the equilibrium by the Volterra series

$$y(t) = \sum_{n=1}^{N} \int_{-\infty}^{\infty} \cdots \int_{-\infty}^{\infty} h_n(\tau_1, \cdots, \tau_n) \prod_{i=1}^{n} u(t - \tau_i) d\tau_i \tag{13.1}$$

where $h_n(\tau_1,\cdots,\tau_n)$ is called the nth order Volterra kernel, which is a real valued function of $\tau_1,\cdots,\tau_n$, N is the maximum order of the system nonlinearity, which may need to be large enough to guarantee required accuracy of approximation. The output frequency response of the system can be described as

$$Y(j\omega) = \sum_{n=1}^{N} \frac{1}{\sqrt{n}(2\pi)^{n-1}} \int_{\omega_1+\cdots+\omega_n=\omega} H_n(j\omega_1, \cdots, j\omega_n) \prod_{i=1}^{n} U(j\omega_i) d\sigma_{\omega n} \tag{13.2}$$

where $\sigma_{\omega n}$ denotes a small unite in the n dimensional hyperplane $\omega_1+\cdots+\omega_n=\omega$, and

$$H_n(j\omega_1,\cdots,j\omega_n) = \int_{-\infty}^{\infty} \cdots \int_{-\infty}^{\infty} h_n(\tau_1, \cdots, \tau_n) \exp\left(-j(\omega_1\tau_1 + \cdots + \omega_n\tau_n)\right) d\tau_1 \cdots d\tau_n \tag{13.3}$$

is the nth order GFRF of system (13.1). When the system is subjected to a multi-tone input described by

$$u(t) = \sum_{i=1}^{K} |F_i| \cos\left(\omega_i t + \angle F_i\right) \tag{13.4}$$

the system output spectrum can be written as

$$Y(j\omega) = \sum_{n=1}^{N} \frac{1}{2^n} \sum_{\omega_{k_1}+\cdots+\omega_{k_n}=\omega} H_n(j\omega_{k_1},\cdots,j\omega_{k_n})F(\omega_{k_1})\cdots F(\omega_{k_n}) \tag{13.5}$$

where,

$$F(\omega) = \begin{cases} |F_i|e^{j\angle F_i} & \text{if } \omega \in \{\omega_k, k=\pm 1,\cdots,\pm K\} \\ 0 & \text{else} \end{cases} \tag{13.6}$$

The NARX model of nonlinear systems is given by

$$y(t) = \sum_{m=1}^{M} y_m(t) \tag{13.7a}$$

$$y_m(t) = \sum_{p=0}^{m} \sum_{k_1,k_{p+q}=1}^{K} c_{p,q}(k_1,\cdots,k_{p+q}) \prod_{i=1}^{p} y(t-k_i) \prod_{i=p+1}^{p+q} u(t-k_i) \tag{13.7b}$$

where $y_m(t)$ is the mth-order output of the system, and $p+q=m$, $k_i=1,\ldots, K$, $\sum_{k_1,k_{p+q}=1}^{K}(\cdot) = \sum_{k_1=1}^{K}(\cdot)\cdots\sum_{k_{p+q}=1}^{K}(\cdot)$. A recursive algorithm can be used to compute as follows:

$$\begin{aligned} &L_n(\omega)\cdot H_n(j\omega_1,\cdots,j\omega_n) \\ &= \sum_{k_1,k_n=1}^{K} c_{0,n}(k_1,\cdots,k_n)\exp(-j(\omega_1 k_1+\cdots+\omega_n k_n)) \\ &+\sum_{q=1}^{n-1}\sum_{p=1}^{n-q}\sum_{k_1,k_{p+q}=1}^{K} c_{p,q}(k_1,\cdots,k_{p+q})\exp(-j(\omega_{n-q+1}k_{p+1}+\cdots+\omega_n k_{p+q}))H_{n-q,p}(j\omega_1,\cdots,j\omega_{n-q}) \\ &+\sum_{p=2}^{n}\sum_{k_1,k_p=1}^{K} c_{p,0}(k_1,\cdots,k_p)H_{n,p}(j\omega_1,\cdots,j\omega_n) \end{aligned} \tag{13.8}$$

$$H_{n,p}(\cdot) = \sum_{i=1}^{n-p+1} H_i(j\omega_1,\cdots,j\omega_i)H_{n-i,p-1}(j\omega_{i+1},\cdots,j\omega_n)\exp(-j(\omega_1+\cdots+\omega_i)k_p) \tag{13.9}$$

$$H_{n,1}(j\omega_1,\cdots,j\omega_n) = H_n(j\omega_1,\cdots,j\omega_n)\exp(-j(\omega_1+\cdots+\omega_n)k_1) \tag{13.10}$$

where $L_n(\omega) = 1 - \sum_{k_1=1}^{K} c_{1,0}(k_1)\exp(-j\omega k_1)$ and $\omega=\omega_1+\cdots+\omega_n$. Moreover, $H_{n,p}(j\omega_1,\cdots,j\omega_n)$ in (13.9) can also be written as

$$H_{n,p}(j\omega_1,\cdots,j\omega_n) = \sum_{\substack{r_1\cdots r_p = 1 \\ \sum r_i = n}}^{n-p+1} \prod_{i=1}^{p} H_{r_i}\left(j\omega_{r_{X+1}},\cdots,j\omega_{r_{X+r_i}}\right)$$

$$\times\exp\left(-j\left(\omega_{r_{X+1}} + \cdots + j\omega_{r_{X+r_i}}\right)k_i\right), \text{ where } X = \sum_{x=1}^{i-1} r_x \quad (13.11)$$

Based on (13.8)–(13.11), the GFRFs of the NARX model (13.7a,b) of any order can be obtained. The objective of this chapter is to investigate the bound characteristics of the GFRFs and the output spectrum of nonlinear systems described by the NARX model to provide an important insight into the effects of the model parameters on these system frequency response functions. Note that the bounded-input bounded-output (BIBO) stability can be guaranteed by the frequency domain property of bounded-input and bounded-output spectrum. The bound characteristics of the NARX model are also significant for the system BIBO stability. Sufficient bounded stability criteria of the NARX model can be derived from the bound characteristics of system output spectrum.

13.3 Bound Characteristics of NARX Model in the Frequency Domain

In this section, some notations and useful operators are introduced first. Then bound characteristics of the GFRFs of the NARX model are derived using these notations and operators. Finally, the bound characteristics of system output spectrum are developed.

13.3.1 Notations and Operators

Let $\underline{L} = \inf_{\omega \in I_\omega}\{|L_n(\omega)|\}$, where I_ω is the non-negative frequency region of the outputspectrum of a NARX model. In what follows, let

$$C(p,q)=\begin{cases}\sum\limits_{k_1,k_{p+q}=1}^{K}\left|c_{p,q}(k_1,\cdots,k_{p+q})\right|, & 1\le q\le n-1,1\le p\le n-q\\ \sum\limits_{k_1,k_n=1}^{K}\left|c_{0,n}(k_1,\cdots,k_n)\right|, & q=n,p=0\\ \sum\limits_{k_1,k_p=1}^{K}\left|c_{p,0}(k_1,\cdots,k_p)\right|, & q=0,2\le p\le n\\ 0, & else\end{cases} \tag{13.12}$$

Obviously, $C(p,q)$ is a nonnegative function of the coefficients $c_{pq}(.)$ defined on all $0\le p,q\le n$. Moreover, let

$$\begin{cases}\overline{H}_{n,p}=\sup\limits_{\omega_1\cdots\omega_n\in R_\omega}\left(\left|H_{n,p}(\cdot)\right|\right),\ H_{0,0}(\cdot)=1\\ H_{n,0}(\cdot)=0\ \text{for}\ n>0\\ H_{n,p}(\cdot)=0\ \text{for}\ n<p\\ \overline{H}_n=\sup\limits_{\omega_1\cdots\omega_n\in R_\omega}\left(\left|H_n(\cdot)\right|\right)\end{cases} \tag{13.13}$$

where R_ω is the input frequency range of a NARX model.

In order to develop the bound characteristics of the GFRFs of the NARX model, define two operators as follows. Consider two polynomials of degree n and m respectively,

$$f_a=a_0+a_1h+\cdots+a_nh^n=a\cdot\hbar_n^T,\ \text{and}\ f_b=b_0+b_1h+\cdots+b_nh^m=b\cdot\hbar_m^T$$

where the coefficients $a_0, a_1, \ldots, a_n$; $b_0, b_1, \ldots, b_m$ are all real numbers, h stands for a real or complex valued function, $a=[a_0, a_1, \ldots, a_n]$, $b=[b_0, b_1, \ldots, b_m]$, and $\hbar_i=[1, h, \ldots, h^i]$.

Define a multiplication operator "$\otimes$" as

$$a\otimes b=c,$$

where c is an $n+m+1$-dimension vector,

$$c(k)=\sum_{\substack{i+j=k\\0\le i\le n,\,0\le j\le n}} a_ib_j\ \text{for}\ 0\le k\le m+n.$$

Denote $(a\otimes b)(k)=\sum\limits_{\substack{i+j=k\\0\le i\le n,\,0\le j\le m}} a_ib_j.$

From this operator it follows that, for example, $f_a \cdot f_b = a \otimes b \cdot \hbar_{n+m}^T$. Similarly, define an addition operator "$\oplus$" as

$$a \oplus b = c,$$

where c is an x-dimension vector,

$x = \max\{m,n\}$, $c(k) = a(k) + b(k)$ for $0 \le k \le x$.

If $k > n$ or m, then $a(k) = 0$ or $b(k) = 0$, accordingly.

From the operator "$\oplus$" it follows that, for example, $f_a + f_b = a \oplus b \cdot \hbar_{\max(n,m)}^T$. Moreover, let $\underset{(*)}{\otimes}(\cdot)$ and $\underset{(*)}{\oplus}(\cdot)$ denote the multiplication and addition in terms of the operator "$\otimes$" and "$\oplus$" for the series (.) satisfying (*), respectively.

Note that the operators "$\otimes$" and "$\oplus$" are different from those defined in Chap. 4. Here they are used for bound computation with a special physical meaning.

13.3.2 Bound Characteristics of the GFRFs

The bound characteristics of the GFRFs are derived in this section. A fundamental result is given in Lemma 13.1, which shows that the magnitude bound of the nth order GFRF can be recursively determined from the magnitude bounds of the lower order GFRFs. Then based on Lemma 13.1, Theorem 13.1 is established which describes the magnitude bound of the GFRFs as a polynomial function of the magnitude bound of the first order GFRF $H_1(j\omega)$.

Lemma 13.1

$$\overline{H}_n \le \frac{1}{L} \sum_{m=2}^{n} \sum_{\substack{p+q=m \\ 0 \le p,q \le m}} C(p,q) \overline{H}_{n-q,p}$$

$$\overline{H}_{n-q,p} \le \sup \sum_{\substack{r_1 \cdots r_p = 1 \\ \sum r_i = n-q}}^{n-q-p+1} \prod_{i=1}^{p} \left| H_{r_i}\left(j\omega_{r_{X+1}}, \cdots, j\omega_{r_{X+r_i}}\right) \right| = \sum_{\substack{r_1 \cdots r_p = 1 \\ \sum r_i = n-q}}^{n-q-p+1} \prod_{i=1}^{p} \overline{H}_{r_i}$$

where, $n > 1$, $X = \sum_{x=1}^{i-1} r_x$, $\sum_{\substack{p+q=m \\ 0 \le p,q \le m}} (\cdot)$ or $\sum_{\substack{p,q=0 \\ p+q=m}}^{m} (\cdot)$ denotes the sum of the corresponding terms with respect to all the combinations of (p,q) satisfying $p+q=m$ and $0 \le p,q \le m$. □

Note that $0 \le p,q \le m$ denotes that $0 \le p \le m$ and $0 \le q \le m$, and $r_1 \cdots r_p = 1$ means that $r_1 = 1, \cdots, r_p = 1$.

Proof of Lemma 1 From (13.8), (13.12), (13.13), and noting $\underline{L}$ is the lower bound of$L_n(\omega)$, it follows

$$|H_n(j\omega_1,\cdots,j\omega_n)| \le \frac{1}{\underline{L}}\sum_{k_1,k_n=1}^{K}|c_{0,n}(k_1,\cdots,k_n)||H_{0,0}(j\omega_1,\cdots,j\omega_n)|$$

$$+\frac{1}{\underline{L}}\sum_{q=1}^{n-1}\sum_{p=1}^{n-q}\sum_{k_1,k_n=1}^{K}\left|c_{p,q}\left(k_1,\cdots,k_{p+q}\right)\right|\left|H_{n-q,p}\left(j\omega_1,\cdots,j\omega_{n-q}\right)\right|$$

$$+\frac{1}{\underline{L}}\sum_{p=2}^{n}\sum_{k_1,k_p=1}^{K}\left|c_{p,0}\left(k_1,\cdots,k_p\right)\right|\left|H_{n,p}(j\omega_1,\cdots,j\omega_n)\right|$$

$$\le \frac{1}{\underline{L}}C(0,n)\overline{H}_{0,0}+\frac{1}{\underline{L}}\sum_{q=1}^{n-1}\sum_{p=1}^{n-q}C(p,q)\overline{H}_{n-q,p}+\frac{1}{\underline{L}}\sum_{p=2}^{n}C(p,0)\overline{H}_{n,p}$$

$$=\frac{1}{\underline{L}}\sum_{q=0}^{n}\sum_{p=0}^{n-q}C(p,q)\overline{H}_{n-q,p} \tag{13.14}$$

It can be easily seen that $\sum_{q=0}^{n}\sum_{p=0}^{n-q}C(p,q)\overline{H}_{n-q,p}$ includes all the permutations of (p,q)satisfying $p+q=m$, $0\le p,q\le m$, and m=2,. . .,n. Hence, it follows $\sum_{q=0}^{n}\sum_{p=0}^{n-q}C(p,q)\overline{H}_{n-q,p}=\sum_{m=2}^{n}\sum_{\substack{p+q=m\\0\le p,q\le m}}C(p,q)\overline{H}_{n-q,p}$

From (13.11), it can be derived that

$$\overline{H_{np}}=\sup\left|H_{np}(j\omega_1,\cdots,j\omega_n)\right|=\sup\left|\sum_{\substack{r_1\cdots r_p=1\\\sum r_i=n}}^{n-p+1}\prod_{i=1}^{p}\left|H_{r_i}\left(j\omega_{r_{X+1}},\cdots,j\omega_{r_{X+r_i}}\right)\exp\left(-j\left(\omega_{r_{X+1}}+\cdots+\omega_{r_{X+r_i}}\right)k_i\right)\right|\right|$$

$$\le \sup\sum_{\substack{r_1\cdots r_p=1\\\sum r_i=n}}^{n-p+1}\prod_{i=1}^{p}\left|H_{r_i}\left(j\omega_{r_{X+1}},\cdots,j\omega_{r_{X+r_i}}\right)\right|=\sum_{\substack{r_1\cdots r_p=1\\\sum r_i=n}}^{n-p+1}\prod_{i=1}^{p}\overline{H}_{r_i}$$

Therefore $\overline{H}_{n-q,p}\le\sup\sum_{\substack{r_1\cdots r_p=1\\\sum r_i=n-q}}^{n-q-p+1}\prod_{i=1}^{p}\left|H_{r_i}\left(j\omega_{r_{X+1}},\cdots,j\omega_{r_{X+r_i}}\right)\right|=\sum_{\substack{r_1\cdots r_p=1\\\sum r_i=n-q}}^{n-q-p+1}\prod_{i=1}^{p}\overline{H}_{r_i}.$

This completes the proof. □

Although Lemma 13.1 shows essentially the same result as those obtained in Zhang and Billings (1996) and Billings and Lang (1996), Lemma 13.1 provides a general expression for the magnitude bound of the nth-order GFRF in terms of the model parameters and $\overline{H}_1 \cdots \overline{H}_{n-1}$, and compared with the result in Billings and Lang (1996), it is much simpler in form and derived in a more systematic approach. Based on Lemma 13.1 and by using the new operators defined in Sect. 13.3.1, a more comprehensive result about the bound of the GFRFs of the NARX model can be obtained.

Theorem 13.1 Consider the *n*th-order GFRF for the NARX model (13.7a,b). There exists a series of scalar positive real numbers $b_{n0}, b_{n1}, \cdots, b_{nn}$, such that

$$|H_n(j\omega_1, \cdots, j\omega_n)| \leq b_{n0} + b_{n1}\overline{H}_1 + b_{n2}\overline{H}_1^{\,2} + \cdots + b_{nn}\overline{H}_1^{\,n} \tag{13.15a}$$

where the coefficients $b_{n0}, b_{n1}, \ldots, b_{nn}$, can be recursively determined as follows (denote $b_n = [\, b_{n0} \quad b_{n1} \quad \cdots \quad b_{nn} \,]$):

$$b_{nk} = \frac{1}{\underline{L}} C(k, n-k) + \frac{1}{\underline{L}} \left(\bigoplus_{m=2}^{n-1} \bigoplus_{\substack{p+q=m \\ 0 \leq p,q \leq m}} \left(C(p,q) \cdot \bigoplus_{\substack{\sum r_i = n-q \\ 1 \leq r_1 \cdots r_p \leq n-m+1}} \left(\bigotimes_{i=1}^{p} b_{r_i} \right) \right) \right)(k)$$

$$\text{for } 0 \leq k \leq n \tag{13.15b}$$

$$b_2 = [b_{20}, b_{21}, b_{22}] = \left[\frac{1}{\underline{L}} C(0,2), \frac{1}{\underline{L}} C(1,1), \frac{1}{\underline{L}} C(2,0) \right] \tag{13.15c}$$

$$b_1 = [b_{10}, b_{11}] = [0, 1] \tag{13.15d}$$

Moreover, $\bigotimes_{i=1}^{p} b_{r_i} = 0$ if $p<1$, and $\bigoplus_{m=2}^{n} (\cdot) = 0$ if $n<2$.

Proof Use the induction method. For the second and third order GFRFs, it is easy to obtain from Lemma 13.1 that

$$\begin{aligned}
|H_2(j\omega_1, j\omega_2)| &\leq \frac{1}{\underline{L}} \sum_{m=2}^{2} \sum_{\substack{p+q=m \\ 0 \leq p,q \leq m}} C(p,q) \overline{H}_{2-q,p} \\
&= \frac{1}{\underline{L}} \left(C(0,2) + C(1,1)\overline{H}_{1,1} + C(2,0)\overline{H}_{2,2} \right) \\
&= \frac{1}{\underline{L}} \left(C(0,2) + C(1,1)\overline{H}_1 + C(2,0)\overline{H}_1^{\,2} \right) = b_2 \cdot h_2^T
\end{aligned}$$

$$
\begin{aligned}
|H_3(j\omega_1, j\omega_2, j\omega_3)| &\le \frac{1}{L}\sum_{m=2}^{3}\sum_{\substack{p+q=m\\0\le p,q\le m}} C(p,q)\overline{H}_{3-q,p}\\
&= \frac{1}{L}\left(\begin{array}{l}\left(C(0,3)+\frac{1}{L}C(1,1)C(0,2)\right)+\left(C(1,2)+\frac{1}{L}C(1,1)^2+\frac{2}{L}C(2,0)C(0,2)\right)\overline{H}_1+\\ \left(C(2,1)+\frac{1}{L}C(1,1)C(2,0)+\frac{2}{L}C(2,0)C(1,1)\right)\overline{H}_1{}^2+\left(C(3,0)+\frac{2}{L}C(2,0)^2\right)\overline{H}_1{}^3\end{array}\right)\\
&= b_3\cdot h_3^T
\end{aligned}
$$

Hence, the theorem holds for n=2 and 3. Consider the nth order GFRF under the assumption that the theorem holds for all the GFRFs of orders less than n. From Lemma 13.1,

$$
|H_n(j\omega_1,\cdots,j\omega_n)| \le \frac{1}{L}\sum_{m=2}^{n}\sum_{\substack{p+q=m\\0\le p,q\le m}} C(p,q)\sup\left(\sum_{\substack{1\le r_1\cdots r_p\le n-m+1\\ \sum r_i=n-q}}\prod_{i=1}^{p}|H_{r_i}(.)|\right) \tag{13.16}
$$

Note $1\le n-m+1\le n-1$ and $0\le\sum r_i=n-q\le n$, each $|H_{r_i}(.)|$ is bounded by a polynomial of the form of (13.15a) with degree $r_i(\le n-1)$, and $\prod_{i=1}^{p}|H_{r_i}(.)|$ is therefore bounded by a polynomial of the form (13.15a) with degree $n-q(\le n)$. It follows from inequality (13.16) that $|H_n(j\omega_1,\cdots,j\omega_n)|$ must be bounded by a polynomial of the form (13.15a) with degree n.

The explicit expression for the coefficients in (13.15a) is derived as follows. It follows from (13.16) that

$$|H_n(j\omega_1,\cdots,j\omega_n)| \le \frac{1}{\underline{L}} \sum_{\substack{p+q=n \\ 0\le p,q\le n}} C(p,q)\left(\sup \sum_{\substack{1\le r_1\cdots r_p\le 1 \\ \sum r_i = n-q}} \prod_{i=1}^{p} |H_{r_i}(.)| \right)$$

$$+\frac{1}{\underline{L}} \sum_{m=2}^{n-1} \sum_{\substack{p+q=m \\ 0\le p,q\le m}} C(p,q)\left(\sup \sum_{\substack{1\le r_1\cdots r_p\le n-m+1 \\ \sum r_i = n-q}} \prod_{i=1}^{p} |H_{r_i}(.)| \right)$$

$$=\frac{1}{\underline{L}}\left(C(0,n)+C(1,n-1)\overline{H}_1+\cdots+C(n,0)\overline{H}_1{}^n\right)$$

$$+\frac{1}{\underline{L}} \sup \sum_{m=2}^{n-1} \sum_{\substack{p+q=m \\ 0\le p,q\le m}} C(p,q)\left(\sum_{\substack{1\le r_1\cdots r_p\le n-m+1 \\ \sum r_i = n-q}} \prod_{i=1}^{p} |H_{r_i}(.)| \right) \tag{13.17}$$

Because

$$\left|H_{r_i}\left(j\omega_{r_{X+1}},\cdots,j\omega_{r_{X+r_i}}\right)\right| \le b_{r_i,0}+b_{r_i,1}\overline{H}_1+\cdots+b_{r_i,r_i}\overline{H}_1{}^{r_i} = b_{r_i}\cdot h_{r_i}^T \quad \text{for } 1\le r_i\le n-m+1$$

where $b_{r_i} = \begin{bmatrix} b_{r_i,0} & b_{r_i,1} & \cdots & b_{r_i,r_i} \end{bmatrix}$ and $h_{r_i} = \begin{bmatrix} 1 & \overline{H}_1 & \cdots & \overline{H}_1{}^{r_i} \end{bmatrix}$, it can be derived by using the operators "$\otimes$" and "$\oplus$" that

$$\sum_{\substack{1\le r_1\cdots r_p\le n-m+1 \\ \sum r_i = n-q}} \prod_{i=1}^{p} \left|H_{r_i}\left(j\omega_{r_{X+1}},\cdots,j\omega_{r_{X+r_i}}\right)\right|$$

$$= \sum_{\substack{1\le r_1\cdots r_p\le n-m+1 \\ \sum r_i = n-q}} \bigotimes_{i=1}^{p} b_{r_i}\cdot h_{n-q} = \left(\bigoplus_{\substack{\sum r_i = n-q \\ 1\le r_1\cdots r_p\le n-m+1}} \left(\bigotimes_{i=1}^{p} b_{r_i} \right) \right)\cdot h_{n-q}$$

Therefore,

$$\sum_{m=2}^{n-1} \sum_{\substack{p+q=m \\ 0 \le p,q \le m}} C(p,q) \left(\sum_{\substack{1 \le r_1 \cdots r_p \le n-m+1 \\ \sum r_i = n-q}} \prod_{i=1}^{p} \left| H_{r_i}\left(j\omega_{r_{X+1}}, \cdots, j\omega_{r_{X+r_i}}\right) \right| \right)$$

$$= \left(\bigoplus_{m=2}^{n-1} \bigoplus_{\substack{p+q=m \\ 0 \le p,q \le m}} \left(C(p,q) \cdot \bigoplus_{\substack{\sum r_i = n-q \\ 1 \le r_1 \cdots r_p \le n-m+1}} \left(\bigotimes_{i=1}^{p} b_{r_i} \right) \right) \right) \cdot h_n$$

and (13.17) can be written as

$$|H_n(j\omega_1, \cdots, j\omega_n)| \le \frac{1}{\underline{L}} \left(C(0,n) + C(1,n-1)\overline{H}_1 + \cdots + C(n,0)\overline{H}_1{}^n \right)$$

$$+ \frac{1}{\underline{L}} \left(\bigoplus_{m=2}^{n-1} \bigoplus_{\substack{p+q=m \\ 0 \le p,q \le m}} \left(C(p,q) \cdot \bigoplus_{\substack{\sum r_i = n-q \\ 1 \le r_1 \cdots r_p \le n-m+1}} \left(\bigotimes_{i=1}^{p} b_{r_i} \right) \right) \right) \cdot h_n$$

This proves (13.15b). Equation (13.15c) follows from the first two steps of the recursive computation. The proof of Theorem 13.1 is thus completed. □

Theorem 13.1 throws that the magnitude of the nth-order GFRF can be bounded by a polynomial function of the magnitude bound of the first order GFRF $H_1(j\omega_1)$ of degree n, and the coefficients of the polynomial are the functions of the model parameters. This reveals an explicit relationship between the NARX model parameters and the magnitude bound of the nth-order GFRF, and is therefore important for the system analysis. From Theorem 13.1, the magnitude bounds of any order GFRFs for the NARX model can readily be computed from the model parameters and the first order GFRF.

13.3.3 Bound Characteristics of the Output Spectrum

Based on Theorem 13.1, a bound function in polynomial form can be derived for the system output spectrum in terms of the magnitude bound of $H_1(j\omega_1)$, and a sufficient condition for the convergence of the bound function can be obtained in terms of the system model parameters which can guarantee the BIBO stability of the NARX model. The results for the boundedness of the output spectrum of the NARX model (13.7a,b) when subjected to a general input are given in the following theorem.

Theorem 13.2 Assume the input of the NARX model (13.7a,b) is a general input with spectrum $U(j\omega)$ defined by $U(j\omega)=\begin{cases} U(j\omega) & \omega\in R_\omega \\ 0 & \text{otherwise}\end{cases}$. Then the output spectrum of the NARX model is bounded by

$$|Y(j\omega)|\leq \overset{N}{\underset{n=1}{\oplus}}\frac{1}{(2\pi)^{n-1}}\cdot b_n\cdot h_n^T\cdot |U|*\cdots*|U(j\omega)| = \left(\overset{N}{\underset{n=1}{\oplus}}\alpha_n b_n\right)\cdot h_N^T \quad (13.18\text{a})$$

and the series on the right side of (13.18a) is convergent if the model parameters satisfy

$$\lim_{\substack{N\to\infty\\ k\to\infty}}\sqrt[k]{\left(\overset{N}{\underset{n=1}{\oplus}}\alpha_n b_n\right)(k)}<\frac{1}{\overline{H}_1} \quad (13.18\text{b})$$

In (13.18a,b), $h_N=\left[1\ \overline{H}_1\ \cdots\ \overline{H}_1{}^N\right]$, $b_n=[b_{n0}\ b_{n1}\ \cdots\ b_{nn}]$, $\alpha_n=(2\pi)^{1-n}\underbrace{|U|*\cdots*|U(j\omega)|}_{n}$, and $\underbrace{|U|*\cdots*|U(j\omega)|}_{n}=\frac{1}{\sqrt{n}}\int_{\omega_1+\cdots+\omega_n=\omega}\prod_{i=1}^{n}|U(j\omega_i)|d\sigma_{\omega n}$.

Proof It can be derived from (13.2) that

$$\begin{aligned}|Y(j\omega)| &\leq \sum_{n=1}^{N}\frac{|H_n(j\omega_1^*,\cdots,j\omega_n^*)|}{\sqrt{n}(2\pi)^{n-1}}\left|\int_{\omega_1+\cdots+\omega_n=\omega}\prod_{i=1}^{n}U(j\omega_i)d\sigma_\omega\right| \\ &\leq \sum_{n=1}^{N}\frac{|H_n(j\omega_1^*,\cdots,j\omega_n^*)|}{\sqrt{n}(2\pi)^{n-1}}\int_{\omega_1+\cdots+\omega_n=\omega}\prod_{i=1}^{n}|U(j\omega_i)|d\sigma_{\omega n} \\ &= \sum_{n=1}^{N}\frac{1}{(2\pi)^{n-1}}|H_n(j\omega_1^*,\cdots,j\omega_n^*)|\,\underbrace{|U|*\cdots*|U(j\omega)|}_{n} \end{aligned} \quad (13.19)$$

where $(j\omega_1^*,\cdots,j\omega_n^*)$ is a point on the hyper-plane $\omega_1+\cdots+\omega_n=\omega$ satisfying the mean value principle. According to Theorem 13.1,

$$|H_n(.)|\leq b_n\cdot h_n^T = b_{n0}+b_{n1}\overline{H}_1+b_{n2}\overline{H}_1{}^2+\cdots+b_{nn}\overline{H}_1{}^n$$

Thus using the operator "$\oplus$", inequality (13.19) yields

$$|Y(j\omega)|\leq \overset{N}{\underset{n=1}{\oplus}}\frac{1}{(2\pi)^{n-1}}\cdot b_n\cdot h_n^T\cdot |U|*\cdots*|U(j\omega)| = \left(\overset{N}{\underset{n=1}{\oplus}}\alpha_n b_n\right)\cdot h_N^T$$

which can be rewritten as

$$|Y(j\omega)| \le \overline{Y} = \left(\overset{N}{\underset{n=1}{\oplus}} \alpha_n b_n\right)(0) + \left(\overset{N}{\underset{n=1}{\oplus}} \alpha_n b_n\right)(1)\overline{H}_1 + \left(\overset{N}{\underset{n=1}{\oplus}} \alpha_n b_n\right)(2)\overline{H}_1^2$$
$$+ \cdots + \left(\overset{N}{\underset{n=1}{\oplus}} \alpha_n b_n\right)(k)\overline{H}_1^k + \cdots \tag{13.20}$$

The bound of the output spectrum is in general an infinite series as given by (13.20). The convergence of the series guarantees the stability of the NARX model. According to Cauchy's criterion (Weisstein 1999) for convergence, a sufficient condition for the convergence of the series in (13.20) is $\lim_{\substack{N\to\infty \\ k\to\infty}} \sqrt[k]{\left(\overset{N}{\underset{n=1}{\oplus}} \alpha_n b_n\right)(k)\overline{H}_1^k} = \overline{H}_1 \lim_{k\to\infty} \sqrt[k]{\left(\overset{\infty}{\underset{n=1}{\oplus}} \alpha_n b_n\right)(k)} < 1$. This completes the proof. □

Note that in Theorem 13.2, $b_n = \begin{bmatrix} b_{n0} & b_{n1} & \cdots & b_{nn} \end{bmatrix}$ can be determined according to Theorem 13.1, and $\underbrace{|U| * \cdots * |U(j\omega)|}_{n} = \frac{1}{\sqrt{n}} \int_{\omega_1+\cdots+\omega_n=\omega} \prod_{i=1}^{n} |U(j\omega_i)| d\sigma_{\omega n}$ can be calculated by an algorithm given in Billings and Lang (1996). Similarly, the following result can be obtained for the output spectrum of the NARX model (13.7a,b) when the input is a multi-tone signal.

Theorem 13.3 Assume the input of the NARX model (13.7a,b) is the multi-tone signal (13.4). Then the output spectrum of the NARX model is bounded by

$$|Y(j\omega)| \le \overset{N}{\underset{n=1}{\oplus}} \left(2^{-n} \cdot b_n \cdot h_n^T \cdot \sum_{\omega_{k_1}+\cdots+\omega_{k_n}=\omega} |F(\omega_{k_1}) \cdots F(\omega_{k_n})| \right)$$
$$= \left(\overset{N}{\underset{n=1}{\oplus}} \beta_n b_n\right) \cdot h_N^T \tag{13.21a}$$

and the series on the right side of (13.21a) is convergent if the system model parameters satisfy

$$\lim_{\substack{N\to\infty \\ k\to\infty}} \sqrt[k]{\left(\overset{N}{\underset{n=1}{\oplus}} \beta_n b_n\right)(k)} < \frac{1}{\overline{H}_1} \tag{13.21b}$$

In (13.21a,b) $h_N = \begin{bmatrix} 1 & \overline{H}_1 & \cdots & \overline{H}_1^N \end{bmatrix}$, $b_n = \begin{bmatrix} b_{n0} & b_{n1} & \cdots & b_{nn} \end{bmatrix}$ which can bedetermined according to Theorem 13.1, $\beta_n = 2^{-n} \sum_{\omega_{k_1}+\cdots+\omega_{k_n}=\omega} |F(\omega_{k_1}) \cdots F(\omega_{k_n})|$.

Proof From (13.5), it follows that

$$|Y(j\omega)| \le \sum_{n=1}^{N} \frac{1}{2^n} \sum_{\omega_{k_1}+\cdots+\omega_{k_n}=\omega} |H_n(j\omega_{k_1},\cdots,j\omega_{k_n})||F(\omega_{k_1})\cdots F(\omega_{k_n})|$$
$$\le \sum_{n=1}^{N} \left(\overline{H}_n \cdot 2^{-n} \sum_{\omega_{k_1}+\cdots+\omega_{k_n}=\omega} |F(\omega_{k_1})\cdots F(\omega_{k_n})| \right)$$

According to Theorem 13.1, and following a similar process as the proof of Theorem 13.2, the conclusion of the theorem can be reached. □

In order to illustrate the results above, consider a specific but frequently encountered case of the NARX model (13.7a,b). When there are only pure output nonlinearities in (13.7a,b), the NARX model is

$$y(t) = \sum_{m=p=1}^{M} \left(\sum_{k_1,k_p=1}^{K} c_{p,0}(k_1,\ldots,k_p) \prod_{i=1}^{p} y(t-k_i) + \delta(m-1) \sum_{k_1=1}^{K} c_{0,1}(k_1) u(t-k_1) \right) \tag{13.22}$$

where $\delta(m) = \begin{cases} 1, & m=0 \\ 0, & \text{else} \end{cases}$. For many engineering systems, this model can be regarded as a general linear/nonlinear state feedback system, and consequently hassignificance in the analysis and synthesis of feedback control systems in practical applications (see Chap. 10). When the input is only a sinusoidal signal $u(t)=F_d\sin(\omega_0 t)$ $(F_d>0)$, then $F(\omega_{k_l}) = -jk_l F_d$ for $k_l = \pm 1$, $\omega_{k_l} = k_l\omega_0$, and $l=1,\cdots,n$ in (13.5). In this case, the following result can be achieved.

Corollary 13.1 Assume the nonlinear system described by NARX model (13.22) is subjected to the input signal $u(t)=F_d\sin(\omega_0 t)$ $(F_d>0)$. The nth-order GFRF for this nonlinear system is bounded by

$$|H_n(j\omega_1,\cdots,j\omega_n)| \le b_{nn}{\overline{H}_1}^n \tag{13.23a}$$

and the output spectrum of the NARX model is bounded by

$$|Y(j\omega)| \le \sum_{n=0}^{\lfloor N-1/2 \rfloor} C_{2n+1}^{n} \left(\frac{F_d}{2}\right)^{2n+1} b_{2n+1,2n+1}{\overline{H}_1}^{2n+1} \tag{13.23b}$$

which is convergent if the system model parameters satisfy

$$\lim_{n\to\infty} \sqrt[2n+1]{C_{2n+1}^{n} b_{2n+1,2n+1}} < \frac{2}{F_d \overline{H}_1} \tag{13.23c}$$

where $b_{nn} = \frac{1}{\underline{L}}C(n,0) + \frac{1}{\underline{L}}\sum_{m=2}^{n-1}\left(C(m,0)\sum_{\substack{\sum r_i = n \\ 1\le r_1\cdots r_p \le n-m+1}}\prod_{i=1}^{m} b_{r_i r_i}\right)$, $\lfloor\cdot\rfloor$ is to take the integer part of (.).

Proof According to (13.15b) in Theorem 13.1,

$$b_{nk} = \frac{1}{\underline{L}}\left(\mathop{\oplus}_{m=2}^{n-1} C(m,0) \mathop{\oplus}_{\substack{\sum r_i = n \\ 1\le r_1\cdots r_p \le n-m+1}} \left(\mathop{\otimes}_{i=1}^{m} b_{r_i}\right)\right)(k) \ for \ 0 \le k < n \tag{13.24a}$$

$$b_{nn} = \frac{1}{\underline{L}}C(n,0) + \frac{1}{\underline{L}}\left(\mathop{\oplus}_{m=2}^{n-1} C(m,0) \mathop{\oplus}_{\substack{\sum r_i = n \\ 1\le r_1\cdots r_p \le n-m+1}} \left(\mathop{\otimes}_{i=1}^{m} b_{r_i}\right)\right)(n) \tag{13.24b}$$

Note b_1=[0,1] and $b_2 = \left[0, 0, \frac{1}{L_2(\omega)}C(2,0)\right]$. It is easy to show that b_{nk}=0 for $0 \le k < n$ in (13.24a). Hence (13.24b) can be written as

$$b_{nn} = \frac{1}{\underline{L}}C(n,0) + \frac{1}{\underline{L}}\sum_{m=2}^{n-1}\left(C(m,0)\sum_{\substack{\sum r_i = n \\ 1\le r_1\cdots r_p \le n-m+1}}\prod_{i=1}^{m} b_{r_i r_i}\right) \tag{13.24c}$$

Hence, from Theorem 13.1 $|H_n(j\omega_1,\cdots,j\omega_n)| \le b_{nn}\overline{H}_1{}^n$. From (13.21a), it followsthat

$$|Y(j\omega)| \le \left(\mathop{\oplus}_{n=1}^{N} \beta_n b_n\right)\cdot h_N^T = \sum_{n=1}^{N} \beta_n b_{nn} \overline{H}_1{}^n \tag{13.25}$$

Note that, when the input is a single tone function,

$$\beta_n = 2^{-n} \sum_{\omega_{k_1}+\cdots+\omega_{k_n}=\omega} |F(\omega_{k_1})\cdots F(\omega_{k_n})|$$
$$= \begin{cases} \left(\frac{F_d}{2}\right)^n \sum\limits_{\omega_{k_1}+\cdots+\omega_{k_n}=\omega} 1, & \omega \in \left\{ \omega_{k_1} + \cdots + \omega_{k_n} \middle| \begin{array}{l} \omega_{k_l} = k_l\omega_0, k_l = \pm 1, \\ 1 \le l \le n \end{array} \right\} \\ 0 & \text{else} \end{cases} \tag{13.26a}$$

Consider the frequency of interest is $\omega = \omega_0$. It is easy to verify that

$$\sum_{\omega_{k_1}+\cdots+\omega_{k_n}=\omega_0} 1 = \begin{cases} C_n^{n-1/2} & n = 2k+1, k = 0, 1, 2 \ldots \\ 0 & \text{else} \end{cases} \tag{13.26b}$$

where, $C_n^m = \frac{n\cdot(n-1)\cdot\cdots\cdot(n-m+1)}{m\cdot(m-1)\cdot\cdots\cdot 2} = \frac{n!}{m!(n-m)!}$. Note that β_n is zero if n is an even number, it is derived from (13.24c) and (13.25) that

$$|Y(j\omega)| \le \sum_{n=0}^{\lfloor N-1/2 \rfloor} C_{2n+1}^n \left(\frac{F_d}{2}\right)^{2n+1} b_{2n+1,2n+1} \overline{H}_1^{\,2n+1}$$

From Cauchy's criterion, if (13.23c) holds, the bound of $|Y(j\omega)|$ is convergent. This completes the proof. □

Corollary 13.1 gives a very clear and simple expression for the boundedness of the frequency response of the NARX model (13.22) in terms of the model parameters and the bound of the first order GFRF. The effect of the system model parameters on the boundedness of the system output spectrum and consequently the BIBO stability of the NARX model can be analysed through checking the inequality (13.23c). This simple analytical bound expression for the output frequency response function also provides a very useful and simple method to evaluate the truncation error associated with the Volterra series expression of nonlinear systems and the highest order N needed in the Volterra series' approximation. Although the check of the stability for a nonlinear system theoretically involves the computation of a limitation as given in (13.18b) or (13.21b) or (13.23c), the result obtained for a sufficiently large N and K or n should be sufficient to provide a significant indication of the system stability.

13.4 A Numerical Example

Consider a nonlinear system

$$\begin{aligned} y(t) = 0.15y(t-2) + 0.1u(t-1) - 0.05y(t-1)y(t-2) \\ - 0.02y(t-1)^2 - 0.01y(t-1)^3 \end{aligned} \tag{13.27}$$

which can be written into the form (13.22) with $c_{1,0}(2)=0.15, c_{0,1}(1)=0.1, c_{2,0}(1,2)=-0.05, c_{2,0}(1,1)=-0.02, c_{3,0}(1,1,1)=-0.01$ else $c_{p,q}(\cdot)=0$, and $K=2$, $M=3$. There are only pure output nonlinear terms in this model.

Compute the magnitude bound of the GFRFs up to fifth order for system (13.27) according to Corollary 13.1. From (13.8), it can be obtained

$$|H_1(j\omega)| = \left|\frac{0.1\exp^{-j\omega}}{1-0.15\exp^{-j2\omega}}\right| = \frac{0.1}{L}$$

where $L = \left|1-0.15\exp^{-j2\omega}\right| = \sqrt{(1-0.15\cos 2\omega)^2 + (0.15\sin 2\omega)^2}$. It is easy to have $\underline{L} = 0.7225$ and thus $\overline{H}_1 = 0.1384$. According to Corollary 13.1, only $b_{n,n}$ is needed for evaluating the magnitude bounds of the GFRFs:

For n=1 and 2, $b_{1,1}$=1, $b_{2,2} = \frac{1}{\underline{L}}C(2,0) = 0.07/\underline{L} = 0.09689$, and thus

$$|H_2(j\omega_1, j\omega_2)| \le b_2 \cdot h_2^T = \frac{0.07}{\underline{L}}\overline{H}_1^{\,2} = 0.001856$$

For n=3,

$$b_{3,3} = \frac{1}{\underline{L}}C(3,0) + \frac{1}{\underline{L}}\sum_{m=2}^{2}\left(C(m,0) \sum_{\substack{\sum r_i = 3 \\ 1 \le r_1 \cdots r_p \le 2-m+2}} \prod_{i=1}^{m} b_{r_i r_i} \right)$$

$$= \frac{0.01}{\underline{L}} + \frac{1}{\underline{L}}C(2,0) \sum_{\substack{\sum r_i = 3 \\ 1 \le r_1 \cdots r_p \le 2-2+2}} \prod_{i=1}^{2} b_{r_i r_i} = \frac{0.01}{\underline{L}} + \frac{0.07}{\underline{L}}(2b_{11}b_{22}) = 0.03261$$

thus $|H_3(j\omega_1,\cdots,j\omega_3)| \le b_3 \cdot h_3^T = 0.03261\overline{H}_1{}^3 = 0.0000864609$

For n=4,

$$b_{44} = \frac{1}{\underline{L}}C(4,0) + \frac{1}{\underline{L}}\sum_{m=2}^{3}\left(C(m,0) \sum_{\substack{\sum r_i = 4 \\ 1 \le r_1 \cdots r_p \le 4-m+1}} \prod_{i=1}^{m} b_{r_i r_i} \right)$$

$$= \frac{1}{\underline{L}}\left(C(2,0) \sum_{\substack{\sum r_i = 4 \\ 1 \le r_1 \cdots r_p \le 4-2+1}} \prod_{i=1}^{2} b_{r_i r_i} + C(3,0) \sum_{\substack{\sum r_i = 4 \\ 1 \le r_1 \cdots r_p \le 4-3+1}} \prod_{i=1}^{3} b_{r_i r_i} \right)$$

$$= \frac{1}{\underline{L}}(0.07(2b_{33} + b_{22}^2) + 0.01(3b_{22})) = 0.01125$$

thus

$$|H_4(j\omega_1,\cdots,j\omega_4)| \le b_4 \cdot h_4^T = 0.01125\overline{H}_1{}^4 = 0.0000041289$$

For n=5,

$$b_{55} = \frac{1}{L}C(5,0) + \frac{1}{L}\sum_{m=2}^{4}\left(C(m,0) \sum_{\substack{\sum r_i = 5 \\ 1 \le r_1 \cdots r_p \le 5-m+1}} \prod_{i=1}^{m} b_{r_i r_i} \right)$$

$$= \frac{1}{L}\left(C(2,0) \sum_{\substack{\sum r_i = 5 \\ 1 \le r_1 \cdots r_p \le 5-2+1}} \prod_{i=1}^{2} b_{r_i r_i} + C(3,0) \sum_{\substack{\sum r_i = 5 \\ 1 \le r_1 \cdots r_p \le 5-3+1}} \prod_{i=1}^{3} b_{r_i r_i} + C(4,0) \sum_{\substack{\sum r_i = 5 \\ 1 \le r_1 \cdots r_p \le 5-4+1}} \prod_{i=1}^{4} b_{r_i r_i} \right)$$

$$= \frac{1}{L}\left(0.07(2b_{22}b_{33} + 2b_{44}) + 0.01 \cdot 3\left(b_{33} + b_{22}^2\right)\right) = 0.004537$$

thus

$$|H_5(j\omega_1, \cdots, j\omega_5)| \le b_5 \cdot h_5^T = 0.004537\overline{H}_1^{\,5} \le 0.00000023036$$

Carrying on with the above recursive calculation process, the magnitude bound of the GFRFs of any order can be obtained according to Corollary 13.1. It should be noted from the above computation that, with the order n going larger, b_{nn} is becoming smaller, and so is the magnitude bound of the nth order GFRF. These information can be used to determine the truncation error of the Volterra series expression of system (13.27) and to determine the largest order N in the Volterra series approximation (Billings and Lang 1997).

To demonstrate the bound characteristics of the system output spectrum of the NARX model, consider system (13.27) is subjected to input $u(t) = 10\sin(\omega_0 t)$ $(F_d > 0)$. Then, according to Corollary 13.1,

$$|Y(j\omega_0)| \le \sum_{n=0}^{\lfloor N-1/2 \rfloor} C_{2n+1}^{n} \left(\frac{F_d}{2}\right)^{2n+1} b_{2n+1,2n+1}\overline{H}_1^{\,2n+1}$$

$$= \frac{F_d}{2}\overline{H}_1 + \frac{3F_d^3}{8}0.03262\overline{H}_1^{\,3} + \frac{5F_d^5}{16}0.004537\overline{H}_1^{\,5} + \frac{35F_d^7}{128}0.00086719\overline{H}_1^{\,5} + \cdots$$

$$+ C_{2n+1}^{n}\left(\frac{F_d}{2}\right)^{2n+1} b_{2n+1,2n+1}\overline{H}_1^{\,2n+1}$$

To check the convergence of this series in the bound expression, the condition

$$\lim_{n\to\infty} \sqrt[2n+1]{C_{2n+1}^{n} b_{2n+1,2n+1}} < \frac{2}{F_d \overline{H}_1}$$

should be analysed. Note that if

$$\lim_{n\to\infty} b_{2n+1,2n+1} < \left(\frac{2}{F_d\overline{H}_1}\right)^{2n+1} \frac{1}{C_{2n+1}^n} = \frac{1.4451^{2n+1}}{C_{2n+1}^n}$$

then the convergent condition must hold. Let $b(n)=b_{2n+1,2n+1}$ and $bb(n) = \frac{1.4451^{2n+1}}{C_{2n+1}^n}$, which can be easily computed for any n by a computer program. Obviously, if $b(n)<bb(n)$, then the bound series is convergent. The result is shown in Fig. 13.1, which indicates the convergence of the bound series where $b(n)=b_{2n+1,2n+1}$ is computed up to the 41st order. Figure 13.1 indicates a very quick convergent rate of the bound series in this specific case.

Moreover, it shall be noted that through symbolic manipulations, an analytical expression for the bound expressions of both the GFRFs and the output spectrum of system (13.27) can be obtained in terms of model parameters $c_{p,q}(\cdot)$. Thus the magnitude of the GFRFs and output spectrum can be optimized and analysed with respect to considered model parameters. This issue will be discussed in later publications.

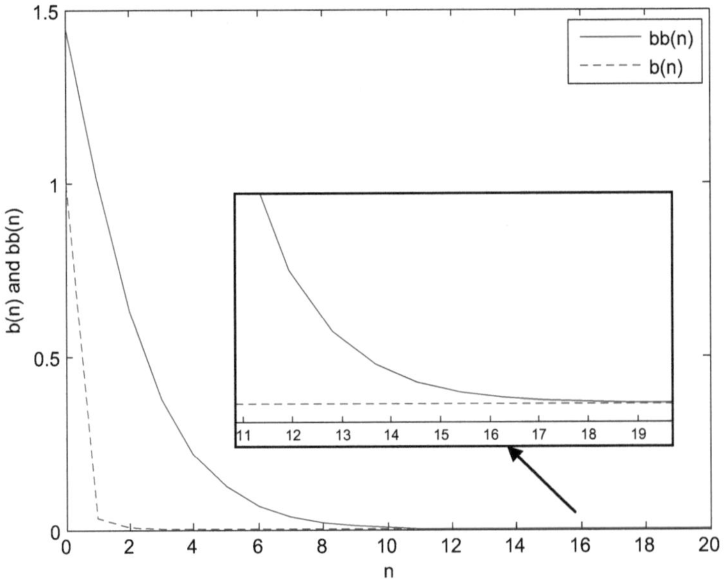

Fig. 13.1 Boundedness of the output spectrum (Jing et al. 2007a)

13.5 Magnitude Bound Characteristics of the SIDO NARX System

To apply the parametric boundedness approach above, this section provides an evaluation of the magnitude bound of $Y(j\omega)$ for the SIDO NARX system in (2.28), which is significant in many cases where only the magnitude of $Y(j\omega)$ is needed to obtain some information of a system without computing the complicated analytical functions in (2.30–2.33) in multi-dimensional complex space.

It can be derived from (6.54) that

$$\begin{aligned}|Y(j\omega)| &= \left|\sum_{n=1}^{N}\frac{1}{\sqrt{n}(2\pi)^{n-1}}\int_{\omega_1+\cdots+\omega_n=\omega} H_n^y(j\omega_1,\cdots,j\omega_n)\prod_{i=1}^{n}U(j\omega_i)d\sigma_\omega\right| \\ &= \left|\sum_{n=1}^{N}\frac{H_n^y(j\omega_1^*,\cdots,j\omega_n^*)}{\sqrt{n}(2\pi)^{n-1}}\int_{\omega_1+\cdots+\omega_n=\omega}\prod_{i=1}^{n}U(j\omega_i)d\sigma_\omega\right| \\ &\le \sum_{n=1}^{N}\frac{|H_n^y(j\omega_1^*,\cdots,j\omega_n^*)|}{\sqrt{n}(2\pi)^{n-1}}\int_{\omega_1+\cdots+\omega_n=\omega}\prod_{i=1}^{n}|U(j\omega_i)|d\sigma_\omega \\ &= \sum_{n=1}^{N}\frac{1}{(2\pi)^{n-1}}|H_n^y(j\omega_1^*,\cdots,j\omega_n^*)|\underbrace{|U|*\cdots*|U(j\omega)|}_{n}\end{aligned} \tag{13.28a}$$

Denote $Y_n(j\omega)=\frac{1}{\sqrt{n}(2\pi)^{n-1}}\int_{\omega_1+\cdots+\omega_n=\omega} H_n^y(j\omega_1,\cdots,j\omega_n)\prod_{i=1}^{n}U(j\omega_i)d\sigma_\omega$ representing the nth-order output frequency response. Then

$$|Y_n(j\omega)| \le \frac{1}{(2\pi)^{n-1}}|H_n^y(j\omega_1^*,\cdots,j\omega_n^*)|\underbrace{|U|*\cdots*|U(j\omega)|}_{n} \tag{13.28b}$$

Note that $\underbrace{|U|*\cdots*|U(j\omega)|}_{n}$ can be computed by an algorithm in Billings and Lang (1996). Thus from (13.28a,b), it can be seen that $|H_n^y(j\omega_1,\cdots,j\omega_n)|$ should be evaluated first in order to obtain the magnitude bound for $Y(j\omega)$. For this purpose, the following notations are introduced.

$$\widetilde{C}(p,q)=\begin{cases}\sum_{k_1,k_{p+q}=0}^{K}\left|\widetilde{c}_{p,q}(k_1,\cdots,k_{p+q})\right|, & 1\le q\le n-1, 1\le p\le n-q\\ \sum_{k_1,k_n=0}^{K}\left|\widetilde{c}_{0,n}(k_1,\cdots,k_n)\right|, & q=n, p=0\\ \sum_{k_1,k_p=0}^{K}\left|\widetilde{c}_{p,0}(k_1,\cdots,k_p)\right|, & q=0, 1\le p\le n\\ 0, & else\end{cases} \tag{13.29}$$

$\overline{C}(p,q)$ has the similar definition as (13.29), except $\overline{C}(1,0)=0$. Let

$$\underline{L}=\inf_{\omega=\omega_1+\cdots+\omega_n}\{|L_n(\omega)|\} \tag{13.30}$$

Moreover, let

$$\begin{cases}\overline{H}_{n,p}=\sup_{\omega_1\cdots\omega_n\in R_\omega}\left(\left|H_{n,p}(\cdot)\right|\right), \; H_{0,0}(\cdot)=1\\ H_{n,0}(\cdot)=0 \text{ for } n>0\\ H_{n,p}(\cdot)=0 \text{ for } n<p\\ \overline{H}_n^x=\sup_{\omega_1\cdots\omega_n\in R_\omega}\left(\left|H_n^x(\cdot)\right|\right)\end{cases} \tag{13.31}$$

where R_ω is the input frequency range. Furthermore, two operations "•" and "∘" are needed in the evaluation of magnitude bound, which are "⊗" and "⊕" defined above respectively.

Proposition 13.1 Considering system (2.28), for $\omega_1+\cdots+\omega_i\neq 0$ (i=1,2,...,n), the magnitude of $H_n^y(j\omega_1,\cdots,j\omega_n)$ for system (2.28) is bounded by

$$\left|H_n^y(j\omega_1,\cdots,j\omega_n)\right|\le\widetilde{C}(0,n)+\left(\mathop{\circ}_{q=0}^{n-1}\mathop{\circ}_{p=0}^{n-q}\widetilde{C}(p,q)\cdot\mathop{\circ}_{\substack{\sum r_i=n-q\\ 1\le r_1\cdots r_p\le n-p-q+1}}\left(\mathop{\bullet}_{i=1}^{p}b_{r_i}\right)\right)\cdot h_n \tag{13.32}$$

where

$$h_n=\left[1\;\;(\overline{H}_1^x)^1\;\;\cdots\;\;(\overline{H}_1^x)^n\right] \text{ and } b_{r_i}=\left[b_{r_i,0}\;\;b_{r_i,1}\;\;\cdots\;\;b_{r_i,r_i}\right] \tag{13.33}$$

b_{nk} for $0\le k\le n$ can be recursively computed as follows,

$$b_{nk} = \frac{1}{\underline{L}}\overline{C}(k, n-k) + \frac{1}{\underline{L}}\left(\mathop{\circ}_{m=2}^{n} \mathop{\circ}_{\substack{p+q=m \\ 0\le p,q\le m}} \left(\overline{C}(p,q) \cdot \mathop{\circ}_{\substack{\sum r_i = n-q \\ 1\le r_1 \cdots r_p \le n-m+1}} \left(\mathop{\bullet}_{i=1}^{p} b_{r_i} \right) \right) \right)(k) \quad (13.34)$$

$$b_2 = [b_{20}, b_{21}, b_{22}] = \left[\frac{1}{\underline{L}}\overline{C}(0,2), \frac{1}{\underline{L}}\overline{C}(1,1), \frac{1}{\underline{L}}\overline{C}(2,0) \right] \quad (13.35)$$

$$b_1 = [b_{10}, b_{11}] = [0, 1] \quad (13.36)$$

Moreover, $\mathop{\bullet}_{i=1}^{p} b_{r_i} = 0$ if $p<1$, and $\mathop{\circ}_{m=2}^{n}(\cdot) = 0$ if $n<2$.

Proof See the proof in Sect. 13.7. □

The bound in (13.32) provides another explicit analytical expression for the relationship between system GFRFs and model parameters. The magnitude bound of the nth-order GFRF can directly be described by an n-degree polynomial function of $\overline{H}_1$. Different order of the GFRFs has a different degree polynomial of $\overline{H}_1$, and has no crossing effect with each other. Using (13.28a,b) and (13.32), it can be derived that

$$|Y(j\omega)| \le \sum_{n=1}^{N} \left\{ \frac{1}{(2\pi)^{n-1}} \underbrace{|U| * \cdots * |U(j\omega)|}_{n} \cdot \left(\widetilde{C}(0,n) \circ \mathop{\circ}_{q=0}^{n-1} \mathop{\circ}_{p=0}^{n-q} \widetilde{C}(p,q) \cdot \right.\right.$$

$$\left.\left. \times \mathop{\circ}_{\substack{\sum r_i = n-q \\ 1\le r_1 \cdots r_p \le n-p-q+1}} \left(\mathop{\bullet}_{i=1}^{p} b_{r_i} \right) \right) \cdot h_n \right\}$$

$$= \left\{ \mathop{\circ}_{n=1}^{N} \frac{1}{(2\pi)^{n-1}} \underbrace{|U| * \cdots * |U(j\omega)|}_{n} \cdot \left(\widetilde{C}(0,n) \circ \mathop{\circ}_{q=0}^{n-1} \mathop{\circ}_{p=0}^{n-q} \widetilde{C}(p,q) \cdot \right.\right.$$

$$\left.\left. \times \mathop{\circ}_{\substack{\sum r_i = n-q \\ 1\le r_1 \cdots r_p \le n-p-q+1}} \left(\mathop{\bullet}_{i=1}^{p} b_{r_i} \right) \right) \right\} \cdot h_N = \left(\mathop{\circ}_{n=1}^{N} (\alpha_n \cdot B_n) \right) \cdot h_N \quad (13.37)$$

$$|Y_n(j\omega)| \le \alpha_n \cdot B_n \cdot h_n \quad (13.38)$$

where

$$\alpha_n = \frac{1}{(2\pi)^{n-1}} \underbrace{|U| * \cdots * |U(j\omega)|}_{n} \tag{13.39}$$

$$B_n = \widetilde{C}(0,n) \circ \overset{n-1}{\underset{q=0}{\circ}} \overset{n-q}{\underset{p=0}{\circ}} \widetilde{C}(p,q) \cdot \underset{\substack{\sum r_i = n-q \\ 1 \le r_1 \cdots r_p \le n-p-q+1}}{\circ} \left(\overset{p}{\underset{i=1}{\bullet}} b_{r_i} \right) \tag{13.40}$$

Similarly, when the input of (2.28) is a multi-tone signal (3.2), then the output spectrum of system (2.28) is bounded by

$$|Y(j\omega)| \le \left(\overset{N}{\underset{n=1}{\circ}} (\beta_n \cdot B_n) \right) \cdot h_N \tag{13.41}$$

$$|Y_n(j\omega)| \le \beta_n \cdot B_n \cdot h_n \tag{13.42}$$

$$\beta_n = \frac{1}{2^n} \sum_{\omega_{k_1} + \cdots + \omega_{k_n} = \omega} F(\omega_{k_1}) \cdots F(\omega_{k_n}) \tag{13.43}$$

The magnitude of a frequency response function for a system usually reveals some important information about the system, and consequently takes a great role in the convergence or stability analysis of the system and the truncation error of the corresponding series. Therefore, the magnitude bound results developed in this section can be used to measure the significant orders of nonlinearities or to find the significant nonlinear terms, indicating the stability of a system and providing a basis for the analysis and optimization of system output frequency response.

Example 13.1 Consider the following system, i.e.,

$$\begin{aligned} &mx(t-2) + a_1 x(t-1) + a_3 x^3(t-1) + kx(t) = u(t) \\ &y(t) = a_1 x(t-1) + a_3 x^3(t-1) + kx(t) \end{aligned} \tag{13.44}$$

and let $u = A\sin(\Omega t)$. Assume that m, a_1, a_3, and k are all positive. There are only two nonlinear parameters, *i.e.*, $\overline{c}_{3,0}(111) = -a_3/k$ and $\overline{c}_{3,0}(111) = a_3$. Before the magnitude bound of the output spectrum is evaluated, the parametric characteristics of the GFRFs for $y(t)$ are checked first. In this case, the parametric characteristics for the GFRFs can be computed according to (6.72). It is noted from (6.78–6.82) that

$$CE(H_{2i}^{y}(\cdot)) = 0 \quad \text{for } i \ge 1 \tag{13.45}$$

thus

$$H_{2i}^{y}(\cdot) = 0 \text{ for } i \ge 1 \tag{13.46}$$

according to Proposition 6.5. Hence, only $|H_{2i-1}^{y}(\cdot)|$ for $i \ge 1$ are needed to be evaluated for the magnitude of $Y(j\omega)$. Since the input is a sinusoidal signal, the magnitude of $Y(j\omega)$ can be evaluated by (13.41)–(13.43), which can be written in this case as

$$|Y(j\omega)| \leq \left(\overset{\lfloor N+1/2 \rfloor}{\underset{i=1}{\circ}} (\beta_{2i-1} \cdot B_{2i-1}) \right) \cdot h_{\lfloor N+1/2 \rfloor} \tag{13.47}$$

and

$$|Y_{2i-1}(j\omega)| \leq \beta_{2i-1} \cdot B_{2i-1} \cdot h_{2i-1} \tag{13.48}$$

Note that $u = A\sin(\Omega t)$ is a single tone signal, then

$$\beta_n = 2^{-n} \sum_{\omega_{k_1}+\cdots+\omega_{k_n}=\omega} |F(\omega_{k_1}) \cdots F(\omega_{k_n})|$$
$$= \begin{cases} \left(\frac{A}{2}\right)^n \sum\limits_{\omega_{k_1}+\cdots+\omega_{k_n}=\omega} 1, & \omega \in \left\{ \omega_{k_1} + \cdots + \omega_{k_n} \middle| \begin{array}{l} \omega_{k_l} = k_l\Omega, k_l = \pm 1, \\ 1 \leq l \leq n \end{array} \right\} \\ 0 & \text{else} \end{cases} \tag{13.49}$$

From (13.34)–(13.36) it can be obtained that

$$b_{2\text{i}} = 0 \text{ for i} = 1, 2, 3, \ldots \tag{13.50}$$

and for n=2i−1, i=1,2,3,...

$$b_{nk} = 0 \;\text{ for }\; 0 \leq k < n \tag{13.51}$$

$$b_{11} = 1, b_{33} = \frac{1}{\underline{L}}\overline{C}(3,0),$$

$$b_{nn} = \frac{1}{\underline{L}}\overline{C}(3,0) \sum_{\substack{\sum r_i = n \\ 1 \leq r_1 \cdots r_3 \leq n-3+1}} \prod_{i=1}^{3} b_{r_i r_i} \;\text{ for n} > 3 \tag{13.52}$$

Therefore,

$$B_1 = \overset{0}{\underset{q=0}{\circ}} \overset{1-q}{\underset{p=0}{\circ}} \underset{\substack{\sum r_i = 1 \\ 1 \leq r_1 \cdots r_p \leq 2-p-q}}{\circ} \left(\widetilde{C}(p,q) \cdot \left(\overset{p}{\underset{i=1}{\bullet}} b_{r_i} \right) \right) = \widetilde{C}(1,0) \cdot b_1 = \left(|\widetilde{c}_{1,0}(1)| \right.$$
$$\left. + |\widetilde{c}_{1,0}(0)| \cdot b_1 = [0, a_1 + k] \right. \tag{13.53}$$

and for n=2i−1, i=2,3...

$$B_n = \mathop{\circ}_{q=0}^{n-1} \mathop{\circ}_{p=0}^{n-q} \mathop{\circ}_{\substack{\sum r_i = n-q \\ 1 \le r_1 \cdots r_p \le n-p-q+1}} \left(\tilde{C}(p,q) \cdot \left(\mathop{\bullet}_{i=1}^{p} b_{r_i} \right) \right) = \left(\tilde{C}(1,0) \cdot b_n \right) \circ \left(\tilde{C}(3,0) \cdot \mathop{\circ}_{\substack{\sum r_i = n \\ 1 \le r_1 \cdots r_3 \le n-2}} \left(\mathop{\bullet}_{i=1}^{3} b_{r_i} \right) \right) = ((a_1 + k) \cdot b_n) \circ \left(a_3 \cdot \mathop{\circ}_{\substack{\sum r_i = n \\ 1 \le r_1 \cdots r_3 \le n-2}} \left(\mathop{\bullet}_{i=1}^{3} b_{r_i} \right) \right) \quad (13.54)$$

According to (13.54) and (13.51), (13.52), B_n can be computed up to any high orders. For example,

$$B_3 = ((a_1 + k) \cdot b_3) \circ \left(a_3 \cdot \mathop{\circ}_{\substack{\sum r_i = 3 \\ 1 \le r_1 \cdots r_3 \le 3-2}} \left(\mathop{\bullet}_{i=1}^{3} b_{r_i} \right) \right) = ((a_1 + k) \cdot b_3) \circ (a_3 \cdot b_1 \bullet b_1 \bullet b_1)$$
$$= \left((a_1 + k) \cdot \left[0, 0, \frac{a_3}{k\underline{L}} \right] \right) \circ (a_3 \cdot [0, 0, 1]) = \left[0, 0, \frac{a_3(a_1 + k)}{k\underline{L}} + a_3 \right] \quad (13.55)$$

Let $B_n=[B_{n0},B_{n1},\ldots,B_{nn}]$. Hence, using (13.52) and (13.54),

$$B_{nk} = 0 \text{ for } 0 \le k < n \quad (13.56)$$

$$B_{nn} = \left((a_1 + k) \cdot \frac{1}{\underline{L}} \tilde{C}(3,0) \sum_{\substack{\sum r_i = n \\ 1 \le r_1 \cdots r_3 \le n-3+1}} \prod_{i=1}^{3} b_{r_i r_i} \right) \circ \left(a_3 \cdot \mathop{\circ}_{\substack{\sum r_i = n \\ 1 \le r_1 \cdots r_3 \le n-2}} \left(\mathop{\bullet}_{i=1}^{3} b_{r_i r_i} \right) \right)$$
$$\text{for } n = 2\mathrm{i} - 1,\ \mathrm{i} = 1, 2, 3 \ldots \quad (13.57)$$

Since only the last element in B_n is nonzero, (13.47)–(13.48) can be rewritten as

$$|Y(j\omega)| \le \mathop{\circ}_{i=1}^{\lfloor N+1/2 \rfloor} \left(\beta_{2i-1} \cdot B_{2i-1,2i-1} \cdot \left(\overline{H}_1^x \right)^{2i-1} \right) \quad (13.58)$$

and

$$|Y_{2i-1}(j\omega)| \le \beta_{2i-1} \cdot B_{2i-1,2i-1} \cdot \left(\overline{H}_1^x \right)^{2i-1} \quad (13.59)$$

Note from (13.30)–(13.31) that

$$\underline{L} = \inf\left|1 + \frac{m}{k}\exp\left(-j2(\omega_1 + \cdots + \omega_n\right) + \frac{a_1}{k}\exp\left(-j(\omega_1 + \cdots + \omega_n\right)\right| \quad (13.60)$$

$$\overline{H}_1^x = \sup\left|\frac{1}{k + m\exp(-j2\omega_1) + a_1\exp(-j\omega_1)}\right| \quad (13.61)$$

Based on (13.58)–(13.61), the magnitude bound of the output spectrum of system (13.44) can be evaluated readily. For instance,

$$|Y_1(j\Omega)| \le \beta_1 \cdot B_{1,1} \cdot \overline{H}_1^x = \frac{A(a_1 + k)}{2}\overline{H}_1^x$$

$$|Y_3(j\Omega)| \le \beta_3 \cdot B_{3,3} \cdot \left(\overline{H}_1^x\right)^3 = \frac{3A^3 a_3\left(a_1 + k + k\underline{L}\right)}{8k\underline{L}}\left(\overline{H}_1^x\right)^3$$

This process can be conducted for up to any higher orders, which can be used to evaluate some properties of the nonlinear system, such as the truncation error of Volterra series and system stability etc (Sect. 13.3). □

13.6 Conclusions

The bound characteristics of the frequency response functions of the NARX model including the GFRFs and the output spectrum are developed in this chapter. The magnitude bounds of the GFRFs and system output spectrum can all be expressed as a polynomial function of the magnitude bound of the first order frequency response function, and the coefficients of the polynomial are the functions of the system model parameters. These bound characteristics reveal an important relationship between model parameters and the boundedness of system frequency response functions, and provide a significant insight into magnitude based analysis of nonlinear systems in the frequency domain. Sufficient conditions for the BIBO stability of the NARX model can also be derived from these results. Note that the boundedness results derived in this chapter are based on the use of the triangular inequality. This may introduce conservatism to a certain extent. Further studies will focus on practical applications of the established theoretical results, and development of methods to reduce possible conservatism associated with these boundedness results. Some results can be referred to the next chapter and Jing et al. (2008b, 2009b).

13.7 Proof of Proposition 13.1

It is derived from (2.32) that

$$\begin{aligned}&\left|H_n^y(j\omega_1,\cdots,j\omega_n)\right|\le\sum_{k_1,k_n=1}^{K}\left|\tilde{c}_{0,n}(k_1,\cdots,k_n)\right|\left|H_{0,0}(j\omega_1,\cdots,j\omega_n)\right|\\&+\sum_{q=1}^{n-1}\sum_{p=1}^{n-q}\sum_{k_1,k_n=1}^{K}\left|\tilde{c}_{p,q}(k_1,\cdots,k_{p+q})\right|\left|H_{n-q,p}(j\omega_1,\cdots,j\omega_{n-q})\right|+\sum_{p=1}^{n}\sum_{k_1,k_p=1}^{K}\left|\tilde{c}_{p,0}(k_1,\cdots,k_p)\right|\left|H_{n,p}(j\omega_1,\cdots,j\omega_n)\right|\\&\le\tilde{C}(0,n)\overline{H}_{0,0}+\sum_{q=1}^{n-1}\sum_{p=1}^{n-q}\tilde{C}(p,q)\overline{H}_{n-q,p}+\sum_{p=1}^{n}\tilde{C}(p,0)\overline{H}_{n,p}=\sum_{q=0}^{n}\sum_{p=0}^{n-q}\tilde{C}(p,q)\overline{H}_{n-q,p}\end{aligned}\tag{D1}$$

From Lemma 1 and Theorem 1 in Jing et al. (2007a),

$$\overline{H}_{n-q,p}\le\sum_{\substack{r_1\cdots r_p=1\\ \sum r_i=n-q}}^{n-q-p+1}\prod_{i=1}^{p}\overline{H}_{r_i}^x\qquad\text{for } p\ne 0,q\ne n\tag{D2}$$

and

$$\overline{H}_{r_i}^x=b_{r_i,0}+b_{r_i,1}\overline{H}_1^x+\cdots+b_{r_i,r_i}\left(\overline{H}_1^x\right)^{r_i}=b_{r_i}\cdot h_{r_i}^T\tag{D3}$$

where $b_{r_i}=\begin{bmatrix}b_{r_i,0} & b_{r_i,1} & \cdots & b_{r_i,r_i}\end{bmatrix}$ which can be determined by (13.33)–(13.36), and $h_{r_i}=\begin{bmatrix}1 & \overline{H}_1^x & \cdots & \left(\overline{H}_1^x\right)^{r_i}\end{bmatrix}$. Then it can be derived from (D2), (D3) that

$$\overline{H}_{n-q,p}\le\sum_{\substack{r_1\cdots r_p=1\\ \sum r_i=n-q}}^{n-p-q+1}\left(\overset{p}{\underset{i=1}{\bullet}}b_{r_i}\right)\cdot h_{n-q}=\left(\underset{\substack{\sum r_i=n-q\\ 1\le r_1\cdots r_p\le n-p-q+1}}{\circ}\left(\overset{p}{\underset{i=1}{\bullet}}b_{r_i}\right)\right)\cdot h_{n-q}\tag{D4}$$

Substituting (D4) into (D1) yields

$$\begin{aligned}\left|H_n^y(j\omega_1,\cdots,j\omega_n)\right|&\le\tilde{C}(0,n)+\sum_{q=0}^{n-1}\sum_{p=0}^{n-q}\tilde{C}(p,q)\left(\underset{\substack{\sum r_i=n-q\\ 1\le r_1\cdots r_p\le n-p-q+1}}{\circ}\left(\overset{p}{\underset{i=1}{\bullet}}b_{r_i}\right)\right)\cdot h_{n-q}\\&=\tilde{C}(0,n)+\left(\overset{n-1}{\underset{q=0}{\circ}}\overset{n-q}{\underset{p=0}{\circ}}\tilde{C}(p,q)\cdot\underset{\substack{\sum r_i=n-q\\ 1\le r_1\cdots r_p\le n-p-q+1}}{\circ}\left(\overset{p}{\underset{i=1}{\bullet}}b_{r_i}\right)\right)\cdot h_n\end{aligned}\tag{D5}$$

This completes the proof.

Chapter 14
Parametric Convergence Bounds of Volterra-Type Nonlinear Systems

14.1 Introduction

Volterra series theory has been extensively used in many different areas, for example, behavior modeling of radio frequency amplifier, telecommunication channel modeling and channel equalization, nonlinear adaptive filter design, system identification, acoustic echo cancellation, active noise control, vibration control, and even applications in biomedical engineering (Crespo-Cadenas et al. 2010; Hermann 1990; Krall et al. 2008; Batista et al. 2010; Kuech and Kellermann 2005; Li and Jean 2001; Mileounis and Kalouptsidis 2009; Jing et al. 2012) etc. To conduct nonlinear analysis and design with Volterra series theory for a given nonlinear system described by a NARX model, a fundamental issue is to ensure that the excitation magnitude and/or model parameters should be in appropriate ranges such that the NARX system has a convergent Volterra series expansion. Several attempts in the literature have been done to derive such a convergence criterion for guiding practical applications. In Boyd and Chua (1985) and Sandberg (1983), convergence criteria for fading memory systems or nonlinear operators are theoretically given but may be too general to implement in practice. Similarly in Bullo (2002), a convergence criterion for analytic systems in L^p-spaces is established. For a specific nonlinear system, such as Duffing oscillators, convergence criteria in the frequency domain are discussed in Tomlinson et al. (1996), Peng and Lang (2007) and Li and Billings (2011). But all those results are either very conservative or obviously overestimated. Recently, computation of the convergence bound for a class of input-analytic nonlinear systems is presented in Helie and Laroche (2011). However, the result is also conservative, and the systems considered are only a special case of the NARX system.

In this chapter, based on the bound characteristics of frequency response functions, evaluation of the convergence bound in the frequency domain for Volterra series expansion of nonlinear systems described by NARX models is studied. This

X. Jing, Z. Lang, *Frequency Domain Analysis and Design of Nonlinear Systems based on Volterra Series Expansion*, Understanding Complex Systems,
DOI 10.1007/978-3-319-12391-2_14

provides new convergence criteria under which the nonlinear system of interest has a convergent Volterra series expansion, and the new criteria are expressed explicitly in terms of the input magnitude, model parameters, and frequency variable. The new convergence criteria are firstly developed for harmonic inputs, which are frequency-dependent, and then extended to multi-tone and general input cases, which are frequency-independent. Based on the theoretical analysis, a general procedure for calculating the convergence bound is provided. The results provide a fundamental basis for nonlinear signal processing using the Volterra series theory. More discussions can also be referred to Xiao et al. (2013a, b).

14.2 The NARX Model and Its Volterra Series Expansion Problem

Consider the NARX model in (2.10), and suppose that the NARX system is of zero initial conditions. The input-output relationship of the NARX model can be approximated by the Volterra series with a maximum truncation order N as in (2.1). The multi-variate Fourier transform of the nth order Volterra kernel is defined as the nth order GFRF. The GFRFs for the NARX model can be computed with a recursive algorithm as given in Chap. 2. With the GFRFs, nonlinear output spectrum can therefore be evaluated (See Chaps. 2 and 3). This represents a natural and formal frequency domain solution of nonlinear systems given a specific excitation input.

It is known that, whether the input-output relationship of the system in (2.10) has a convergent Volterra series expansion is greatly dependent on the model parameters, input magnitude, and excitation frequency (i.e., characteristic parameters). Although several results have been developed to evaluate the convergence bound in the literature, most of the existing criteria focus more on the evaluation of the input bound under which a convergent series exists and usually have more or less drawbacks as mentioned before. This study aims at developing new convergent criteria, which are expressed explicitly in terms of all the characteristic parameters, for a more general nonlinear system described by the NARX model above using a frequency domain boundedness approach.

Technically, the nth order GFRF is a function of the characteristic parameters; thus the boundedness of the GFRFs and nonlinear output spectrum, given in Lemma 14.1, Lemma 14.2, and Lemma 14.3, provides a parametric insight into the output response of the system (2.10) under any given input signal (based on the results in Chap. 13); The bound of the output response is expressed into a simple infinite series form in terms of the nth order GFRF and input U, which greatly facilitates the investigation of the convergence of underlying Volterra series expansion. Based on these boundedness results, the frequency dependent convergence criteria are derived in Proposition 14.1 and Corollary 14.1 for harmonic inputs, which are then extended to the cases for any input signals in Proposition 14.3.

14.3 The Convergence Criteria

Consider the NARX system subjected to a harmonic input

$$u(t) = |A|\cos(\omega t + \angle A) = \frac{A}{2}e^{j\omega t} + \frac{A^*}{2}e^{-j\omega t} \tag{14.1}$$

The output is generally given by (3.3), i.e.,

$$Y(j\varpi) = \sum_{n=1}^{+\infty} \frac{1}{2^n} \sum_{\omega_1+\cdots+\omega_n=\varpi} H_n(j\omega_1,\cdots,\omega_n) \prod_{i=1}^{n} A(\omega_i) \tag{14.2}$$

where $\omega_i \in \{\omega, -\omega\}$, $A(\omega_i) \in \{A, A^*\}$, $|A|$ is the magnitude of $A(\omega_i)$ which is denoted by U in what follows.

The following definitions are needed.

$$\underline{L}(\omega) = \inf_{\varpi \in W_\infty} \{\|L_n(j\omega_1,\cdots,j\omega_n)\|\} \tag{14.3}$$

where $W_\infty = \bigcup_{k=1}^{\infty} W_k = \bigcup_{k=1}^{\infty} \{\varpi | \varpi = \omega_1 + \cdots + \omega_k, \omega_i \in \{\omega, -\omega\}\}$ represents the output frequency range when the NARX model is excited by (14.1) (Chap. 3), and the operator $\|\cdot\|$ means the absolute value for scalars and Euclidian norm $\|\cdot\|_2$ for vectors.

$$C(p,q) = \sum_{(k_1,\cdots,k_m)} \|c_{p,q}(k_1,\cdots,k_m)\| \tag{14.4}$$

From (14.4), $C(p,q)$ is a nonnegative function of the parameters $c_{p,q}(\cdot)$ which are the coefficients of the NARX model in (2.10). Moreover, let

$$\overline{H}_1(j\omega_1) = \|H_1(j\omega_1)\| \tag{14.5}$$

$\mathbb{N}$ denotes the nonnegative integer set, and $\mathbb{N}^+$ denotes positive integer set.

A. Boundedness of the GFRF and Nonlinear Output Spectrum

Lemma 14.1 For the upper bound of the *n*th order GFRF, it can be obtained as

$$\sup\{\|H_n(j\omega_1,\cdots,j\omega_n)\| | \forall \omega_1,\cdots,\omega_n \in \{\omega, -\omega\}\} \leq \overline{H}_n(j\omega_1,\cdots,j\omega_n)$$

$$\overline{H}_n(j\omega_1,\cdots,j\omega_n) = \frac{1}{\underline{L}(\omega)}\left[C(0,n) + \sum_{m=2}^{n}\sum_{p=1}^{m} C(p,q) \sum_{r_1,\cdots,r_p=1, \sum r_i = n-q}^{n-m+1} \prod_{i=1}^{p} \overline{H}_{r_i}(\omega_{X+1},\cdots,\omega_{X+r_i})\right] \tag{14.6}$$

The proof can be referred to Chap. 13.

Lemma 14.2 The upper bound of the output spectrum at $\overline{\omega} = k\omega, k \in \mathbb{N}$ is given by

$$\left|Y(j\Omega)\right| \leq \overline{Y}_{\Omega=k\omega}(U) = \sum_{n=1}^{\infty} \frac{C_{k+2(n-1)}^{n-1}}{2^{k+2(n-1)-1}} \overline{H}_{k+2(n-1)}\left(j\omega_1,\cdots,j\omega_{k+2(n-1)}\right) U^{k+2(n-1)} \quad k \in \mathbb{N}^+ \quad (14.7a)$$

$$\left|Y(j\Omega)\right| \leq \overline{Y}_{\Omega=k\omega}(U) = \sum_{n=1}^{\infty} \frac{C_{2n}^{n}}{2^{2n}} \overline{H}_{2n}(j\omega_1,\cdots,j\omega_{2n}) U^{2n} \quad k=0 \quad (14.7b)$$

where $C_{k+2(n-1)}^{n-1}$ means the number of $n-1$ combinations in a given $k+2(n-1)$ elements.

Proof: Following (14.1) and (14.2), for $k \in \mathbb{N}^+$,

$$\begin{aligned} Y_{\overline{\omega}=k\omega}(U) &\leq \sum_{n=1}^{+\infty} \frac{2}{2^{k+2(n-1)}} \sum_{\overline{\omega}=k\omega} \overline{H}_{k+2(n-1)}(j\omega_1,\cdots,\omega_n) \prod_{i=1}^{k+2(n-1)} \|A(\omega_i)\| \\ &\leq \sum_{n=1}^{\infty} \frac{C_{k+2(n-1)}^{n-1}}{2^{k+2(n-1)-1}} \overline{H}_{k+2(n-1)}\left(j\omega_1,\cdots,j\omega_{k+2(n-1)}\right) U^{k+2(n-1)} . \end{aligned}$$

For $k = 0$, (14.7b) is straightforward. □

See also Chap. 13 for a more general case. Here the output response bound is only for a single tone input (14.1).

Lemma 14.3 The upper bound of the output magnitude which involves all the frequencies in the output frequency range W_∞ can be given by,

$$\begin{aligned} \overline{Y}(U)_\omega &= \sum_{k=0}^{\infty} \overline{Y}_{\overline{\omega}=k\omega}(U) = \sum_{\overline{\omega}\in W_\infty} \overline{Y}_{\overline{\omega}}(U) = \sum_{n=1}^{\infty} \left\| \sum_{\overline{\omega}\in W_n} \overline{Y}_n(j\overline{\omega}) \right\| \\ &= \sum_{n=1}^{+\infty} \overline{H}_n(j\omega_1,\cdots,j\omega_n) U^n \end{aligned} \quad (14.8)$$

Proof

$$\begin{aligned}\left\|\sum_{\overline{\omega}\in W_\infty} Y(j\overline{\omega})\right\| &\leq \sum_{n=1}^{\infty}\left\|\sum_{\overline{\omega}\in W_n} Y_n(j\overline{\omega})\right\| \leq \sum_{n=1}^{\infty}\frac{1}{2^n}\sum_{\overline{\omega}\in W_n}\sum_{\omega_1+\cdots+\omega_n=\overline{\omega}}\|H_n(j\omega_1,\cdots,\omega_n)\|\prod_{i=1}^{n}\|A(\omega_i)\| \\ &\leq \sum_{k=0}^{\infty}\Big(\sum_{i=1}^{k+1}\frac{C_{2k+1}^{k+i}}{2^{2k}}\Big)\overline{H}_{2k+1}(j\omega_1,\cdots,\omega_{2k+1})U^{2k+1} \\ &+\sum_{k=1}^{\infty}\left(\sum_{i=1}^{k}\frac{C_{2k}^{k+i}}{2^{2k-1}}+\frac{C_{2k}^{k}}{2^{2k}}\right)\overline{H}_{2k}(j\omega_1,\cdots,\omega_{2k})U^{2k} \\ &=\sum_{k=0}^{\infty}\overline{H}_{2k+1}(j\omega_1,\cdots,\omega_{2k+1})U^{2k+1}+\sum_{k=1}^{\infty}\overline{H}_{2k}(j\omega_1,\cdots,\omega_{2k})U^{2k} \\ &=\sum_{n=1}^{+\infty}\overline{H}_n(j\omega_1,\cdots,j\omega_n)U^n\end{aligned}$$

This completes the proof. □

Remark 14.1 Equation (14.7a,b) is the upper bound of the output spectrum of nonlinear system (2.10) subjected to a harmonic excitation, which can be seen as a power series with nonnegative coefficients; and (14.8) is the sum of the power series presented in (14.7a,b). If (14.8) is convergent, then $\forall k \in \mathbb{N}$, (14.7a,b) is convergent, that is, the upper bound of the output spectrum is convergent. The bound expression of the output response in (14.8) takes a simple infinite series form. Obviously, the convergence of this series indicates the convergent of the Volterra series expansion.

B. Frequency Dependent Convergence Criteria

With the bound results of the nonlinear output spectrum obtained above, the bound of the output spectrum in (14.8) and the input magnitude U are shown to satisfy an equation $\overline{Y}(U)_\omega = U\Phi\big(\overline{Y}(U)_\omega, U\big)$, which gives a closed-form expression for the bound of the output frequency response, where $\Phi(*,*)$ is a function to be defined, then the Analytic Inversion Lemma (Flajolet and Sedgewick 2009) is used to develop the frequency dependent convergence criterion.

Definition 14.1 For the case that $q \in \mathbb{N}$, where q is the nonlinear degree in terms of system input in the NARX model (2.10), a formal function $\Phi(x,U)$ is defined as,

$$\Phi(x,U) = \frac{\overline{H}_1(j\omega) + \frac{1}{\underline{L}(\omega)}\sum\limits_{q=m=2}^{+\infty} C(0,m)U^{m-1}}{1 - \frac{1}{\underline{L}(\omega)}\sum\limits_{m=2}^{+\infty}\sum\limits_{p=1}^{m} C(p,q)x^{p-1}U^q} \tag{14.9}$$

where U is the input amplitude in (14.1) and x is the upper bound of the output magnitude in (14.8).

For the NARX model with only pure input nonlinearity, the whole input part which includes both the linear and nonlinear terms in terms of input can be equivalent to a new input, and then the new NARX model can be seen as a linear system. Therefore, this case is not focused on in this study. The following result presents a convergence criterion for the NARX model with output nonlinear degree larger than or equal to 1.

Proposition 14.1 Except the case that the NARX model has only nonlinearity with index $p=1$ or together with pure input nonlinearity, the convergence bound for the Volterra series expansion of the NARX model can be obtained by solving the following equations to find U,

$$\begin{cases} x(\omega,U) = U\Phi(x(\omega,U),U) \\ \Phi(x(\omega,U),U) = x(\omega,U)\dfrac{\partial \Phi(x,U)}{\partial x} \end{cases} \tag{14.10}$$

For the case that the NARX model has only nonlinearity with index $p=1$ or together with pure input nonlinearity, the bound of the infinite series in (14.8) is given by,

$$x = \frac{\overline{H}_1(j\omega)U + \frac{1}{\underline{L}(\omega)}\sum_{m=2}^{+\infty} C(0,m)U^m}{1 - \frac{1}{\underline{L}(\omega)}\sum_{q=1}^{+\infty} C(1,q)U^q} \tag{14.11a}$$

Then the convergence bound can be obtained by solving,

$$\frac{1}{\underline{L}(\omega)}\sum_{q=1}^{+\infty} C(1,q)U^q < 1 \tag{14.11b}$$

Proof An outline of the poof is given here. Firstly, it is shown that the bound of (14.8) and the input magnitude U satisfy the equation $x(\omega,U) = U\Phi(x(\omega,U),U)$ by deriving a closed form expression for the output spectrum bound (see (A1) in the proof). Then it is shown that the divergence condition of the bound of output spectrum $x(\omega,U)$ (i.e., the divergence condition of an infinite power series) is equivalent to the closest point to the expanded centre where the infinite power series becomes singular. Finally, according to the Analytic Inversion Lemma (Flajolet and Sedgewick 2009), the singular condition of the bound of output spectrum $x(\omega,U)$ can be obtained, that is, $\frac{dU}{dx} = 0$, which further leads to the conclusion. See the details in Sect. 14.6A. □

Definition 14.2 When the index q in the NARX model takes only 0 or 1, the formal function $\Phi(x)$ can also be defined as

$$\Phi(x) = \frac{\overline{H}_1(j\omega) + \frac{1}{\underline{L}(\omega)}\sum_{m=2}^{+\infty} C(m-1,1)x^{m-1}}{1 - \frac{1}{\underline{L}(\omega)}\sum_{m=2}^{+\infty} C(m,0)x^{m-1}} \tag{14.12}$$

Corollary 14.1 When $\Phi(x)$ is given in (14.12), except the case that the NARX model has only nonlinear terms like $\sum_{(k_1,k_2)} c_{1,1}(k_1,k_2)y(t-k_1)u(t-k_2)$, (14.10) still holds for the convergence bound. For the case that the NARX model only possesses nonlinearity with $p=1$, (14.11a) and (14.11b) hold for the bound of output spectrum and the convergence bound, respectively.

Proof See Sect. 14.6B. □

To compute the convergence bound, the following procedure can be used.

Algorithm 1

Step 1. Calculate $\underline{L}(\omega)$ according to (14.3), calculate the bound $\overline{H}_1(\omega)$ of the first order GFRF, and calculate $C(p,q)$ from (14.4).

Step 2. For the corresponding cases, solve (14.10) or (14.11b) to obtain the convergence bound respectively.

Moreover, using the results in Proposition 14.1 and Corollary 14.1, a bound for the truncation error of the Volterra series expansion can be assessed in the frequency domain as follows.

Proposition 14.2 Denote

$$\overline{Y}(U)_{\omega}^{N} = \sum_{n=N+1}^{\infty} \overline{Y}_n(U)_{\omega} = \sum_{n=N+1}^{+\infty} \overline{H}_n(j\omega_1,\cdots,j\omega_n)U^n \tag{14.13}$$

where N is the maximum truncation order. The truncation error bound can be obtained as

$$\left\| \sum_{n=N+1}^{\infty} Y_n(U)_{\omega} \right\| \le \overline{Y}(U)_{\omega}^{N} \tag{14.14}$$

When $\tau(\omega,\rho(\omega))$ exists, the following result holds

$$\overline{Y}(U)_{\omega}^{N} \le \tau(\omega,\rho(\omega))\left(\frac{U}{\rho(\omega)}\right)^{N+1} \Big/ \left[1 - \frac{U}{\rho(\omega)}\right] \tag{14.15}$$

Proof According to the Cauchy estimates (Stewart and Tall 1983),

$$\overline{H}_n(j\omega_1,\cdots,j\omega_n) = \frac{1}{n!}\frac{\mathrm{d}^n\overline{Y}(U)_\omega}{\mathrm{d}U^n}\bigg|_{U=0} \le \frac{\tau(\omega,\rho(\omega))}{[\rho(\omega)]^n},$$

then

$$\begin{aligned}\overline{Y}(U)_\omega^N &= \sum_{n=N+1}^{+\infty}\overline{H}_n(j\omega_1,\cdots,j\omega_n)U^n \\ &\le \sum_{n=N+1}^{+\infty}\frac{\tau(\omega,\rho(\omega))}{[\rho(\omega)]^n}U^n = \tau(\omega,\rho(\omega))\left(\frac{U}{\rho(\omega)}\right)^{N+1}\Big/\left[1-\frac{U}{\rho(\omega)}\right]\end{aligned}$$

This completes the proof. □

Remark 14.2 When the solution of the NARX model (2.10) has unique steady state, which is related to the fading memory property (Boyd and Chua 1985), the solution of the NARX model can be approximated by a convergent Volterra series. Then, the proposed criterion can give a very good estimation of the true bound under which the solution of the NARX model can be well approximated by a convergent Volterra series. When the solution of the NARX model has more than one unique steady state with the given input amplitude, the proposed criteria in Proposition 14.1 and Corollary 14.1 may lead to over estimation of the true convergence bound. However, for a given specific NARX model, the Harmonic Balance method can be used to check whether the solution of the model possesses a unique steady state or not. A further study will focus on this problem.

C. Frequency Independent Convergence Criteria

As discussed in Remark 14.2, to overcome the over estimation problem of the frequency dependent convergence criteria above, frequency independent ones can be derived. Comparing with the frequency dependent results, the frequency independent ones are more conservative. Simultaneously, the results will be generalized to more general cases for multi-tone or any input signals. Denote $L_n = \inf\{\|L_n(j\omega_1,\cdots,j\omega_n)\| \,|\forall\omega_i \in \sigma_\omega\}$, $\underline{L} = \inf\{L_n | n \in \mathrm{N}^+\}$, and $\overline{H}_n = \sup\{\|H_n(j\omega_1,\cdots,j\omega_n)\| \,|\forall\omega_i \in \sigma_\omega\}$, where σ_ω represents the whole nonnegative frequency range.

Proposition 14.3

(1) When the NARX model is subjected to a multi-tone input given by

$$u(t) = \sum_{i=1}^{N_u}|A_i|\cos(\omega_i t + \angle A_i) \tag{14.16}$$

where N_u is the number of input frequency, the frequency independent convergence bound can be obtained by solving,

$$\begin{cases} \tau = \rho\Phi(\tau, \rho) \\ \Phi(\tau, \rho) = \tau \frac{\partial \Phi(x,U)}{\partial x}\Big|_{x=\tau, U=\rho} \end{cases} \tag{14.17}$$

where $U = \sup\{|A_i| \,|\, i = 1, 2, \cdots, N_u\}$.

(2) When the NARX model is subjected to a general input given by

$$\begin{aligned} u(t) &= \frac{1}{2\pi}\int_0^{\infty} 2|U(j\omega)| \cos[\omega t + \angle U(j\omega)] d\omega \\ &= \frac{1}{2\pi}\int_{-\infty}^{\infty} U(j\omega)e^{j\omega t} d\omega = \frac{1}{2\pi}\int_{\omega \in \sigma_\omega} U(j\omega)e^{j\omega t} d\omega \end{aligned} \tag{14.18}$$

where $U(j\omega)$ represents the input spectrum, the frequency independent convergence bound of the Volterra series expansion of the NARX model with the general input (14.18) can be obtained by solving,

$$\begin{cases} \tau = \rho\Phi(\tau, \rho) \\ \Phi(\tau, \rho) = \tau \frac{\partial \Phi(x,U)}{\partial x}\Big|_{x=\tau, U=\rho} \end{cases}$$

where $U = \sup\{|U(j\omega)| \,|\, \omega \in \sigma_\omega\}$.

Proof See Sect. 14.6C. □

Remark 14.3 Let N_u in (14.16) equals to 1, then the frequency independent convergence criterion for the NARX model with single tone input can also be obtained via (14.17). It should be noted that the results in Proposition 14.3 will become the result in Helie and Laroche (2011) when the nonlinear system is restricted to be input-affine, i.e., the maximum nonlinearity degree for the input not larger than 1 and the difference order of the input limited to 1 in the NARX model. It will be shown that our results are more general, and less conservative (even for the latter case).

14.4 Examples

To demonstrate the theoretical results, four NARX models are discussed here with different input and output nonlinearities. The systems considered in these examples can be given in the following general form with zero initial conditions,

$$y(k) = c_{1,0}(1)y(k-1) + c_{1,0}(2)y(k-2) + c_{3,0}(1,1,1)y^3(k-1) + c_{2,1}(1,1,1)y^2(k-1)u(k-1) \\ + c_{1,2}(1,1,1)y(k-1)u^2(k-1) + c_{0,3}(1,1,1)u^3(k-1) + c_{0,1}(1)u(k-1) \tag{14.19}$$

Equation (14.19) can be obtained by using the backward difference method to discretize the nonlinear differential equation

$$m\ddot{y}(t) + c\dot{y}(t) + k_1 y(t) + k_{30}y^3(t) + k_{21}y^2(t)u(t) + k_{12}y(t)u^2(t) + k_{03}u^3(t) = u(t)$$

with zero initial condition, where

$m = 1, c = 0.01\omega_0, k_1 = \omega_0^2, k_{30} = 0.01\omega_0^6, k_{21} = 1.2306 * 10^7, k_{12} = 615.289,$ $k_{03} = 0.9229, \omega_0 = 20\pi, u(t) = U\cos(\omega t)$. In (14.19), set $T_s = 1/2000s$, then $u(k)$ $= U\cos(\Omega k) = U\cos(\omega T_s k)$ and $c_{1,0}(1) = 2 - \frac{cT_s}{m} - \frac{kT_s^2}{m} = 1.9987,\ c_{1,0}(2) = \frac{cT_s}{m}$ $-1 = -0.9997, c_{0,1}(1) = \frac{T_s^2}{m} = 2.5 * 10^{-7} c_{3,0}(1,1,1) = -\frac{k_{30}T_s^2}{m} = -153.8223,\ c_{2,1}$ $(1,1,1) = -\frac{k_{21}T_s^2}{m} = -3.0764, c_{1,2}(1,1,1) = -\frac{k_{12}T_s^2}{m} = -1.5382 * 10^{-4}, c_{0,3}(1,1,1)$ $= -\frac{k_{03}T_s^2}{m} = -2.3073 * 10^{-7}$.

Firstly, the NARX model in (14.19) with only pure output nonlinearity is discussed in case A, which is obtained by setting $c_{2,1}(1,1,1) = c_{1,2}(1,1,1) = c_{0,3}$ $(1,1,1) = 0$, corresponding to the discretized model of the continuous time Duffing equation. The comparison between the proposed criteria for the mentioned NARX model and other existing criteria focusing on the Duffing equation is presented. Then the NARX model with pure output nonlinearity and input-output cross nonlinearity with $q=1$ is given in case B, which is obtained by setting $c_{1,2}(1,1,1)$ $= c_{0,3}(1,1,1) = 0$ in (14.19). Similarly, the comparisons between the convergence bounds obtained with the proposed results in Proposition 14.1, Corollary 14.1, and Proposition 14.3, and with the other existing results are presented. Then the NARX model with input-output nonlinearity with $q=2$ is given in case C, which is obtained by setting $c_{2,1}(1,1,1) = c_{0,3}(1,1,1) = 0$ in (14.19). In this case, the proposed criteria in this study such as Proposition 14.1 and Proposition 14.3 can effectively provide a convergence bound, but no existing results are available. Finally, the NARX model with pure output nonlinearity and pure input nonlinearity is given in case D, which is obtained by setting $c_{2,1}(1,1,1) = c_{1,2}(1,1,1) = 0$ in (14.19). In this case, similarly to case C, the proposed criteria in this study such as Proposition 14.1 and Proposition 14.3 can effectively provide a convergence bound, but no existing results are available. Moreover, it can be clearly seen how the pure input nonlinearity affects the convergence bound when compared with case A.

In these examples, the output frequency response up to the Nth order is computed by using (14.1) and (14.2), and then the output frequency response is transformed into time domain, which is referred to as the synthesized output. The time domain output response obtained directly from simulation of the model with a Runge-Kutta method is referred to as the real output. The normalized root mean square error (NRMSE) is used to measure the difference between the synthesized output and the real output, which is defined as

$$NRMSE = \sqrt{\frac{\sum \left(y_{synthesized}(k) - y_{real}(k)\right)^2}{\sum (y_{real}(k))^2}} \tag{14.20}$$

where $y_{synthesized}(k)$ is the synthesized output, and $y_{real}(k)$ is the real output. By comparisons between the synthesized output and the real output, the validation of the proposed criteria is shown. In the discussions below, the true convergence bound of input magnitudes is obtained by numerical simulations.

A. A NARX Model with Pure Output Nonlinearity

The NARX model in (14.19) with only pure output nonlinearity is studied here. The NARX model is given by

$$y(k) = c_{1,0}(1)y(k-1) + c_{1,0}(2)y(k-2) + c_{3,0}(1,1,1)y^3(k-1) + c_{0,1}(1)u(k-1) \tag{14.21}$$

Firstly, calculate the upper bound of the linear order GFRF, that is, $\overline{H}_1(\omega) = H_1(\mathrm{j}\omega)$, and the lower bound of $L_n(\mathrm{j}\omega)$, that is, $\underline{L}(\omega)$. In this case, $\underline{L}(\omega) = \inf\{\|L(\omega)\|, \|L(3\omega)\|, \|L(5\omega)\|, \cdots\}$ (Chaps. 3, 5 and 6). Although the output frequencies happen at all odd multiples of the input frequencies (Chaps. 3, 5 and 6), the first several orders would take dominant roles and thus $\underline{L}(\omega)$ could be evaluated simply by $\underline{L}(\omega) = \inf\{\|L(\omega)\|, \|L(3\omega)\|, \|L(5\omega)\|, \|L(7\omega)\|\}$. In (14.21), $\Phi(x)$ in (14.12) is the same as that in (14.9). Thus

$$\Phi(x) = \overline{H}_1(\omega)/\{1 - [C(3,0)/\underline{L}(\omega)]x^2\}$$

Then solving the equation $x\frac{\mathrm{d}\Phi(x)}{\mathrm{d}x} = \Phi(x)$, the solution is $x(\omega) = \sqrt{\underline{L}(\omega)/[3C(3,0)]}$. Then according to the first equation of (14.10), the convergence bound can be obtained,

$$\rho(\omega) = 2\sqrt{\underline{L}(\omega)}/\left[3\sqrt{3C(3,0)}\overline{H}_1(\omega)\right]$$

which is presented in Fig. 14.1.

The existing criteria developed for the continuous time model of (14.21), i.e., the Duffing equation, are also presented in Fig. 14.1 for comparisons. Most existing criteria such as those mentioned before are obviously over-estimated. The proposed criteria in this study provide much more reliable and close estimation of the convergence bound especially for the frequency range below the resonant frequency (it can be verified for the Duffing equation that the system has unique steady state (relating to the fading memory property (Boyd and Chua 1985)) in this frequency range). The frequency independent convergence bound obtained by Proposition 3 is 0.0038. Although the frequency-independent bound is quite conservative at the other frequency (accurate at the resonance frequency), it is

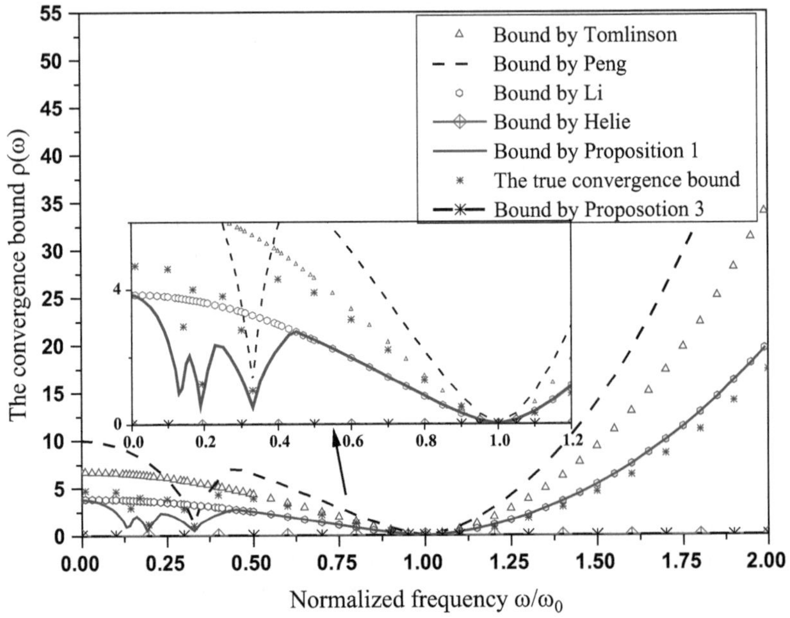

Fig. 14.1 Frequency domain convergence bound of model (14.21); in the figure, Propositions 1 and 3 refer to Propositions 14.1 and 14.3 respectively. (Xiao et al. 2013b © IEEE)

obviously less conservative than the result in Helie and Laroche (2011) (a recently developed one), while the latter provides a convergence bound as $5.3895 * 10^{-6}$.

In Fig. 14.1, at the resonant frequency the convergence bound is 0.0038 and at the other frequency for example ω=0.8, it is 0.8322. In Figs. 14.2 and 14.3, the comparisons between the synthesized output and the real output are presented. When the input amplitude is taken less than or equal to the computed convergence bound value, the synthesized output and the simulated real output has a good agreement. That is, a very small NRMSE can be seen in this case. With a little larger value of the input amplitude, the synthesized output becomes slowly divergent, with the NRMSE becoming larger and larger.

B. A NARX Model with Pure Output Nonlinearity and Input-Output Nonlinearity with q=1

In this case, the NARX model contains both pure output nonlinear term and cross nonlinear term with q=1 is given by

$$y(k) = c_{1,0}(1)y(k-1) + c_{1,0}(2)y(k-2) + c_{3,0}(1,1,1)y^3(k-1) + c_{2,1}(1,1,1)y^2(k-1)u(k-1) + c_{0,1}(1)u(k-1) \tag{14.22}$$

To compute the convergence bound, $\Phi(x, U)$ can be obtained from (14.9), that is,

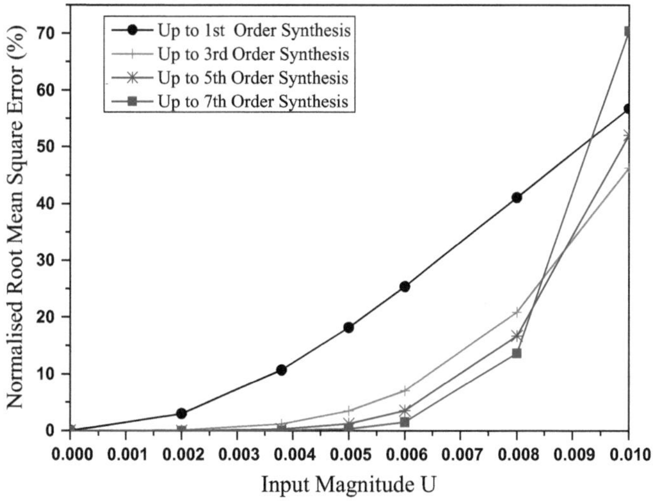

Fig. 14.2 Comparison between the synthesized output and the real output under the cosinusoidal input at $\omega=1$ (Xiao et al. 2013b)

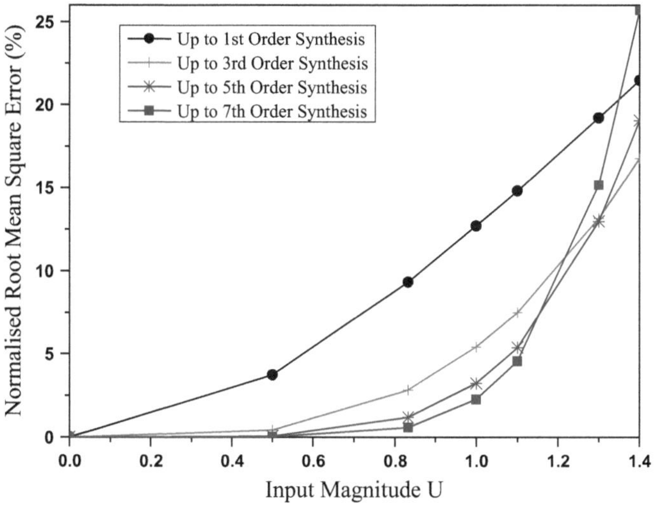

Fig. 14.3 Comparison between the synthesized output and the real output under the cosinusoidal input at $\omega=0.8$ (Xiao et al. 2013b)

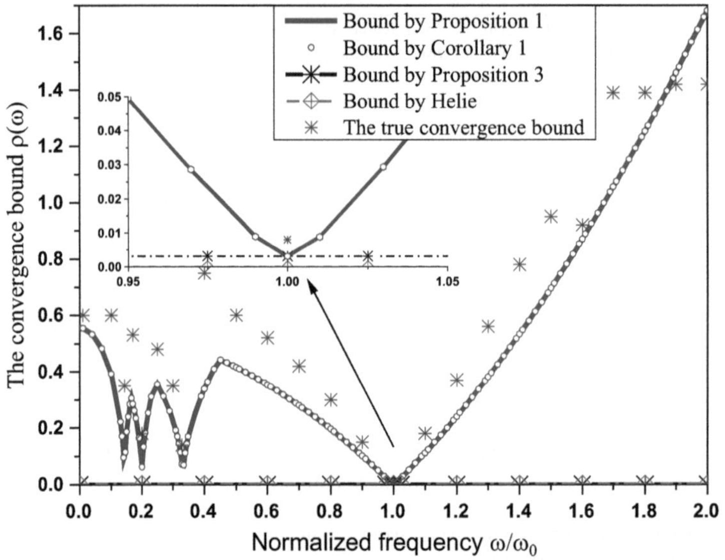

Fig. 14.4 Frequency domain convergence bound of model (14.22); in the figure, Proposition 1, Proposition 3 and Corollary 1 refer to Proposition 14.1, Proposition 14.3 and Corollary 14.1, respectively. (Xiao et al. 2013b © IEEE)

$$\Phi(x,U)=\overline{H}_1(\omega)\Big/\left[1-\frac{C(2,1)}{\underline{L}(\omega)}xU-\frac{C(3,0)}{\underline{L}(\omega)}x^2\right]$$

As a comparison, $\Phi(x)$ can also be constructed from (14.11a,b), i.e.,

$$\Phi(x)=\left[\overline{H}_1(\omega)+\frac{C(2,1)}{\underline{L}(\omega)}x^2\right]\Big/\left[1-\frac{C(3,0)}{\underline{L}(\omega)}x^2\right]$$

Solving (14.10), the convergence bounds can be obtained, which are presented in Fig. 14.4.

In Fig. 14.4, the convergence bound obtained by using Corollary 14.1 superimposes on that obtained by using Proposition 14.1, both of which provide very good estimation on the convergence bound except for the frequencies larger than 1.8, where it can be verified that the system dynamic response possesses more than one steady states. To overcome the over-estimation problem at some frequencies, the convergence bound can also be given by Proposition 14.3, which is a frequency independent bound (i.e., 0.0031). Although this frequency independent bound is conservative but much better than the bound estimated by the method in Helie and Laroche (2011), which gives the bound to be $5.3718 * 10^{-6}$. It should be noted that few existing methods could be applicable to this example in the literature. Moreover, the convergence bound obtained with Proposition 14.3 is exactly the convergence bound obtained by using Proposition 14.1 or Corollary 14.1 at the resonant frequency.

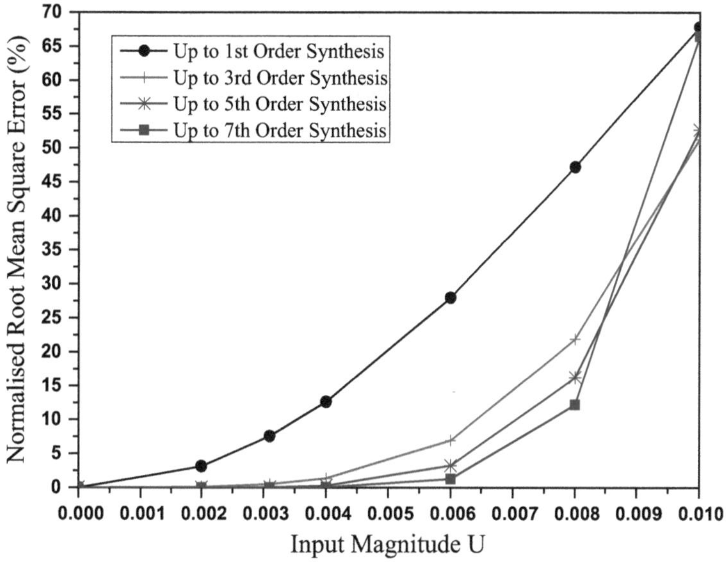

Fig. 14.5 Comparison between the synthesized output and the real output under the cosinusoidal input at $\omega=1$ (Xiao et al. 2013b)

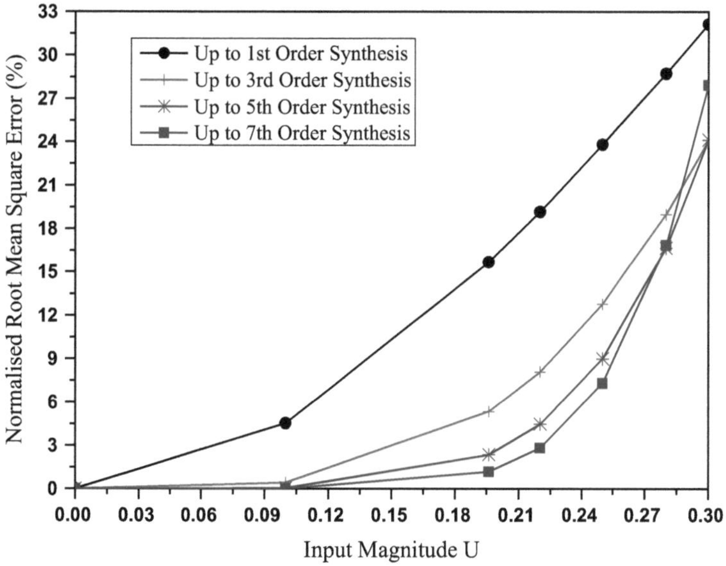

Fig. 14.6 Comparison between the synthesized output and the real output under the cosinusoidal input at $\omega=0.8$. (Xiao et al. 2013b)

In Figs. 14.5 and 14.6, when the input amplitude takes a value equal to or smaller than the computed convergence bound, that is, at $\omega=1$ in (14.1) it is given as 0.0031 and at the other frequencies for example $\omega=0.8$ it is given as 0.1960, the synthesized output has a very good match with the real output up to the seventh order. When a little larger input amplitude is used, the synthesized output becomes slowly divergent with an observed increasing NRMSE.

C. A NARX Model with Pure Output Nonlinearity and Input-Output Nonlinearity with q≥2

The NARX model in this case is given as

$$\begin{aligned} y(k) = {} & c_{1,0}(1)y(k-1) + c_{1,0}(2)y(k-2) + c_{3,0}(1,1,1)y^3(k-1) \\ & + c_{1,2}(1,1,1)y(k-1)u^2(k-1) + c_{0,1}(1)u(k-1). \end{aligned} \tag{14.23}$$

No existing results are available to compute the convergence bound for (14.23). With our results in Proposition 14.1 or Proposition 14.3, the convergence bound can be computed as follows. From (14.9), $\Phi(x,U) = \overline{H}_1(\omega)/\{1 - [C(1,2)/\underline{L}(\omega)]U^2 - [C(3,0)/\underline{L}(\omega)]x^2\}$. Solving (14.10), the convergence bound in the frequency domain can be obtained, which is presented in Fig. 14.7. With Proposition 14.3, the frequency independent convergence bound can be obtained, which is 0.0038 after calculation.

The bound results are shown in Fig. 14.7, which indicate clearly that the new convergence bound can provide a reliable estimation on the convergence condition for the Volterra series expansion at all frequency band, and the proposed one matches well the true one obtained by the numerical simulation.

In Figs. 14.8 and 14.9, when the input amplitude takes a value equal to or smaller than the computed convergence bound value, for example, it is 0.0038 for $\omega=1$,

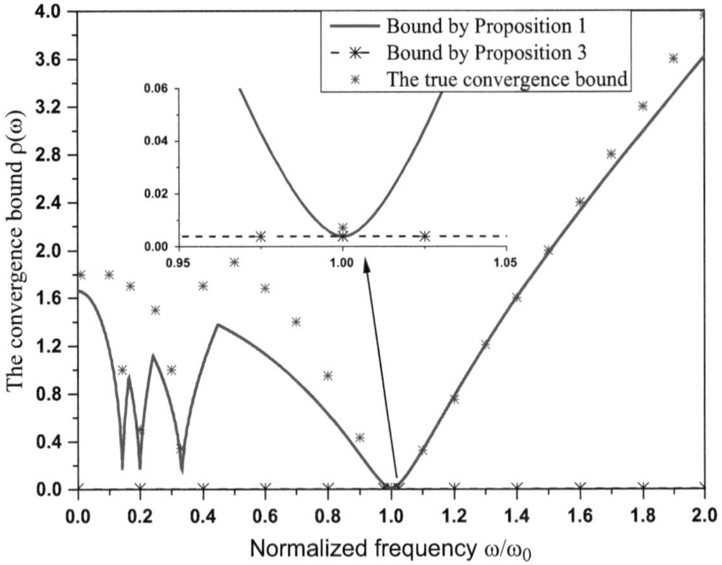

Fig. 14.7 Frequency domain convergence bound of model (14.23); in the figure, Propositions 1 and 3 refer to Propositions 14.1 and 14.3 respectively. (Xiao et al. 2013b © IEEE)

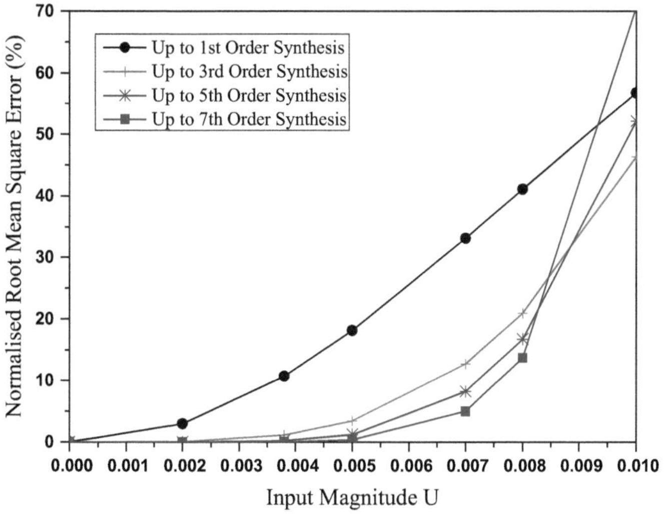

Fig. 14.8 Comparison between the synthesized output and the real output under the cosinusoidal input at $\omega = 1$ (Xiao et al. 2013b)

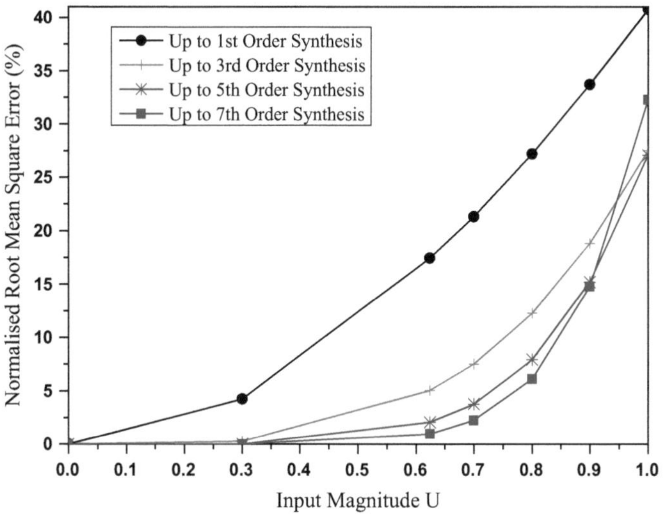

Fig. 14.9 Comparison between the synthesized output and the real output under the cosinusoidal input at $\omega = 0.8$ (Xiao et al. 2013b)

and 0.6283 for $\omega = 0.8$, the synthesized output up to the seventh order has a very small NRMSE compared with the real output; when larger input amplitude is used, the synthesized output becomes slowly divergent with an obvious increasing NRMSE.

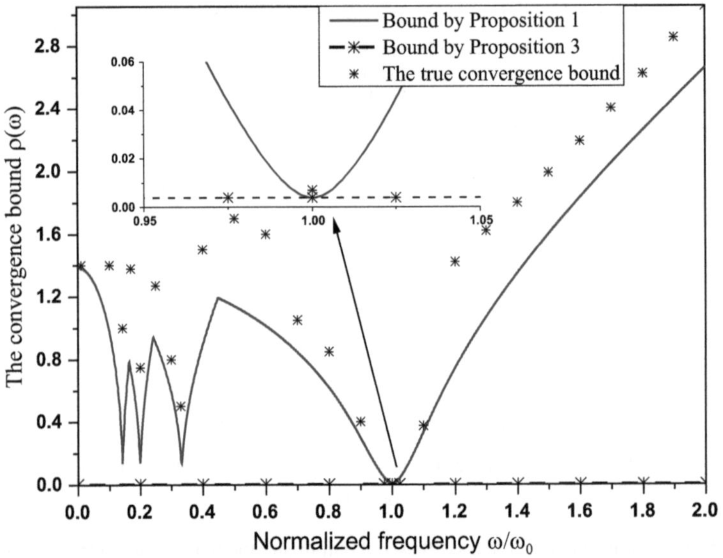

Fig. 14.10 Frequency domain convergence bound of model (14.24); in the figure, Propositions 1 and 3 refer to Propositions 14.1 and 14.3 respectively. (Xiao et al. 2013b © IEEE)

D. A NARX Model with Pure Output Nonlinearity and Pure Input Nonlinearity with q≥2

The NARX model in this example is given as

$$y(k) = c_{1,0}(1)y(k-1) + c_{1,0}(2)y(k-2) + c_{3,0}(1,1,1)y^3(k-1) + c_{0,3}(1,1,1)u^3(k-1) + c_{0,1}(1)u(k-1) \tag{14.24}$$

Similar to Example C, no existing results are available to compute the convergence bound. From (14.9), $\Phi(x,U) = \{\overline{H}_1(\omega) + [C(0,3)/\underline{L}(\omega)]U^2\}/\{1 - [C(3,0)/\underline{L}(\omega)]x^2\}$. According to (14.10) the convergence bound can be obtained, which is presented in Fig. 14.10. The frequency independent bound can be obtained according to Proposition 14.3, which is 0.0038.

From Fig. 14.10, it can be seen that the proposed criterion still has good estimation of the true convergent bound. Moreover, it shall be noted that, when the NARX model with not only pure input nonlinearity but also other nonlinearity in the output, the pure input nonlinearity can significantly affects the convergence bound of the NARX model, which can be seen clearly by comparing example D with example A. Example D is obtained from example A by introducing pure input nonlinearity with nonlinear order $q=3$, and the estimated convergence bound in example D is totally different from that in example A.

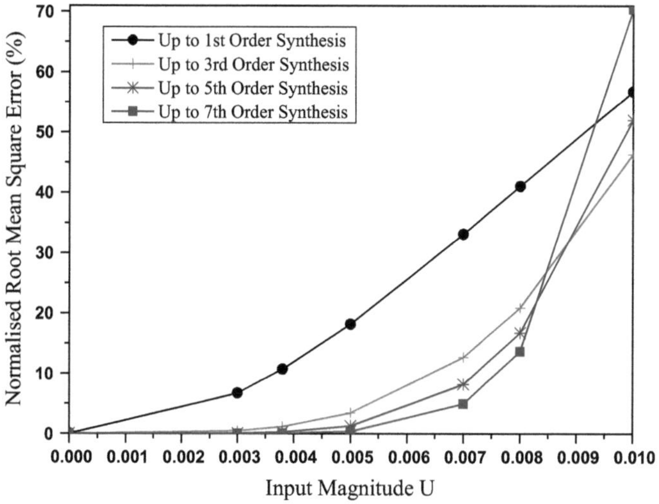

Fig. 14.11 Comparison of the synthesized output and the real output under the cosinusoidal input at $\omega=1$ (Xiao et al. 2013b)

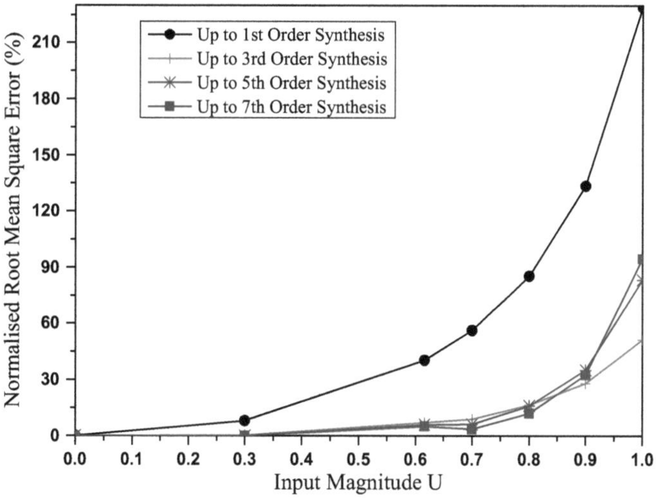

Fig. 14.12 Comparison of the synthesized output and the real output under the cosinusoidal input at $\omega=0.8$ (Xiao et al. 2013b)

In Figs. 14.11 and 14.12, when the input amplitude takes equal to or lower than the computed convergence bound, for example, 0.0038 at $\omega=1$ and 0.6162 at $\omega=0.8$, the synthesized output up to the seventh order has very good approximation to the real output; when a little larger input amplitude is used, the synthesized output becomes slowly divergent with increasing NRMSE.

14.5 Conclusions

Volterra series has been extensively used in various areas including filter design, signal processing, system identification, and control etc. The NARX model is known as a general model for nonlinear systems, which also has been frequently used in practice for system identification, signal processing, and control etc. Based on the Volterra series theory, the associated frequency domain theory and methods for the NARX model developed in the past decade can greatly facilitate the nonlinear analysis and design. Applications of these results can be found in different engineering practices such as signal processing, filter design, vibration control, fault detection, and neuronal systems. The results of this study solve an important issue related to the application of the Volterra series based theory and methods mentioned above. From a very engineering point of view, the new results attempt to systematically answer a long-existing problem, that is, under what parametric conditions a given nonlinear system (described by the NARX model) could have a convergent Volterra series expansion for a given testing input signal. Obviously, these results could provide a significant guidance for nonlinear analysis and design in nonlinear signal processing and control with the Volterra series based theory and methods. To demonstrate the results, several examples are given and discussed. It is shown that, the new convergence criteria can provide a better evaluation on the convergent region in which the targeted nonlinear system dynamics can be well approximated by a convergent Volterra series expansion in most frequency range. Given a nonlinear system, if the nonlinearity is examined to be a non-Volterra-type with the developed results, many methods can be used (for example, introduction of feedback control or decrease of input magnitude etc) to ensure the system to be a Volterra type (that is, allowing a convergent Volterra series expansion). Note that a Volterra-type nonlinearity is well-defined and mild nonlinear phenomenon, which is much easier for analysis, design and control with many developed theory and methods both in time and frequency domain, compared with other complicated nonlinear behaviors such as chaos and bifurcation etc (Nayfeh and Mook 2008; Sanders and Verhulst 1985; Buonomo and Lo Schiavo 2005).

The parametric convergence bound presented in this Chapter can also be used to indicate to what extent a given nonlinear system has a convergence Volterra series expansion in terms of any characteristic parameter. This leads to a new concept—convergence margin defined in Xiao et al. (2014). More effective parametric convergence bound can also be developed by considering the frequency and wave form of multi-tone inputs, which will be reported in Jing and Xiao (2014).

14.6 Proofs

A. Proof of Proposition 14.1

$$
\begin{aligned}
&\frac{1}{\underline{L}(\omega)}\sum_{m=2}^{+\infty}\sum_{p=1}^{m}U^{q}C(p,q)\left(\overline{Y}(U)_{\omega}\right)^{p}+\frac{1}{\underline{L}(\omega)}\sum_{q=m=2}^{+\infty}C(0,m)U^{m}\\
&=\frac{1}{\underline{L}(\omega)}\sum_{m=2}^{+\infty}\sum_{p=1}^{m}C(p,q)\left(\sum_{i=1}^{+\infty}\overline{H}_{i}(j\omega_{1},\cdots,j\omega_{i})U^{i}\right)^{p}U^{q}+\frac{1}{\underline{L}(\omega)}\sum_{q=m=2}^{+\infty}C(0,m)U^{m}\\
&=\sum_{n=2}^{+\infty}\left(\frac{1}{\underline{L}(\omega)}\sum_{m=2}^{n}\sum_{p=1}^{m}C(p,q)\sum_{r_{i}=1,\sum r_{i}=n-q}^{n-m+1}\prod_{i=1}^{p}\overline{H}_{r_{i}}(j\omega_{1},\cdots,j\omega_{r_{i}})+C(0,n)\right)U^{n}\\
&=\sum_{n=2}^{+\infty}\overline{H}_{n}(j\omega_{1},\cdots,j\omega_{n})U^{n}
\end{aligned}
$$

then,

$$
\frac{1}{\underline{L}(\omega)}\sum_{m=2}^{+\infty}\sum_{p=1}^{m}U^{q}C(p,q)\left(\overline{Y}(U)_{\omega}\right)^{p}+\frac{1}{\underline{L}(\omega)}\sum_{q=m=2}^{+\infty}C(0,m)U^{m}=\overline{Y}(U)_{\omega}-\overline{H}_{1}(j\omega)U \quad \text{(A1)}
$$

From (A1), $\overline{Y}(U)_{\omega}=U\Phi\left(\overline{Y}(U)_{\omega},U\right)$ holds, and denote $x(\omega,U)=\overline{Y}(U)_{\omega}=\sum_{n=1}^{+\infty}\overline{H}_{n}(j\omega_{1},\cdots,j\omega_{n})U^{n}$, then the equation can be rewritten as

$$
x=U\Phi(x,U) \quad \text{(A2)}
$$

Note that here (A2) gives a closed-form expression for the magnitude bound of output frequency response compared with the power series form in (14.8). From the above, $x=\overline{Y}(U)_{\omega}$ is an infinite power series of U expanded at 0. The infinite power series is analytic in the convergence region, which means that in the convergence region there does not exist any singularity. Thus the closest point to the expanded center which makes the bound of output spectrum $x=\overline{Y}(U)_{\omega}$ singular can be seen as the convergence bound of the infinite power series.

In Flajolet and Sedgewick (2009), the Analytic Inversion Lemma is stated as: An analytic function locally admits an analytic inverse near any point where its first derivative is non-zero. However, a function cannot be analytically inverted in a neighborhood of a point where its first derivative vanishes.

According to the Analytic Inversion Lemma, the singular condition of the bound of output spectrum $x=\overline{Y}(U)_{\omega}$ can be obtained as $\frac{dU}{dx}=0$.

Denote $r(\omega)$ as the convergence radius of $\Phi(x,U)$.

For the case that the NARX model does not only possess nonlinearity with index $p=1$ or together with pure input nonlinearity, $\frac{\partial\Phi(x,U)}{\partial x}$ exists and $\frac{\partial\Phi(x,U)}{\partial x}\neq 0$ if $U\neq 0$.

Because $x\frac{\partial\Phi(x,U)}{\partial x}/\Phi(x,U)$ is 0 at $x=0$, $\lim\limits_{x\to r(\omega)} x\frac{\partial\Phi(x,U)}{\partial x}/\Phi(x,U)\to\infty$ and $x\frac{\partial\Phi(x,U)}{\partial x}/\Phi(x,U)$ is an increasing function of x for $0<x<r(\omega)$, so there exists a unique solution $0<\tau(\omega,\rho(\omega))<r(\omega)$ makes $x\frac{\partial\Phi(x,U)}{\partial x}/\Phi(x,U)=1$, that is,

$$\Phi(x,U)=x\frac{\partial\Phi(x,U)}{\partial x} \tag{A3}$$

where $x=\tau(\omega,\rho(\omega)),U=\rho(\omega)$.

From (A2), $1=\frac{dU}{dx}\Phi(x,U)+U\frac{\partial\Phi(x,U)}{\partial x}+U\frac{\partial\Phi(x,U)}{\partial U}\frac{dU}{dx}$ holds. Substituting (A2) and (A3) into the equation, then $\left(\Phi(x,U)+U\frac{\partial\Phi(x,U)}{\partial U}\right)\frac{dU}{dx}=0$ holds with $x=\tau(\omega,\rho(\omega))$ and $U=\rho(\omega)$. Because $\Phi(x,U)$, $\frac{\partial\Phi(x,U)}{\partial U}$, and U are all positive, then $\left.\frac{dU}{dx}\right|_{x=\tau(\omega,\rho(\omega))}=0$ holds. Thus (A3) indicates that the singular condition of the infinite power series holds, which means that the bound of nonlinear output spectrum x is singular at $U=\rho(\omega)$.

From the analysis above, the convergence bound can be obtained via solving (14.10).

For the case that the NARX model only has nonlinear terms with index $p=1$ or together with pure input nonlinearity, from (A2), (14.11a) can be obtained directly, and then the bound results in (14.11b) is straightforward. This completes the proof. □

B. Proof of Corollary 14.1

$$\begin{aligned}
&\frac{1}{\underline{L}(\omega)}\sum_{m=2}^{+\infty}\sum_{q=0}^{1}U^qC(m-q,q)\left(\overline{Y}(U)_\omega\right)^{m-q}\\
&=\sum_{n=2}^{+\infty}\frac{1}{\underline{L}(\omega)}\sum_{m=2}^{+\infty}\sum_{q=0}^{1}C(m-q,q)\sum_{\substack{r_i=1,\\ \sum r_i=n-q}}^{n-m+1}\prod_{i=1}^{m}\overline{H}_{r_i}(j\omega_1,\cdots,j\omega_{r_i})U^n\\
&=\sum_{n=2}^{+\infty}\overline{H}_n(j\omega_1,\cdots,j\omega_n)U^n
\end{aligned}$$

then,

$$\begin{aligned}
\frac{1}{\underline{L}(\omega)}\sum_{m=2}^{+\infty}\sum_{q=0}^{1}U^qC(m-q,q)\left(\overline{Y}(U)_\omega\right)^{m-q}&=\sum_{n=2}^{+\infty}\overline{H}_n(j\omega_1,\cdots,j\omega_n)U^n\\
&=\overline{Y}(U)_\omega-\overline{H}_1(j\omega)U
\end{aligned} \tag{B1}$$

From (B1), (A2) still holds, and similar to the proof of Proposition 14.1, Corollary 14.1 holds except that when only nonlinear term $\sum\limits_{(k_1,k_2)}c_{1,1}(k_1,k_2)y(t-k_1)u(t-k_2)$

is contained in the NARX model, in this case, (14.11a,b) is the direct consequence of (A2). This completes the proof. □

C. Proof of Proposition 14.3

For the multi-tone input case, denote $\sigma_\omega = \{\omega_1, \cdots, \omega_{N_u}\}$, $\alpha_n = \sup_{\overline{\omega} \in W_n} \left\{ \frac{1}{2^n} \sum_{\omega_1 + \cdots + \omega_n = \overline{\omega}} 1 \middle| \forall \omega_i \in \sigma_\omega \cup -\sigma_\omega \right\}$, and $\alpha = \sup\{\alpha_n | n = 1, 2, \cdots\}$.

From Chap. 3, $Y(j\overline{\omega}) = \sum_{n=1}^{\infty} Y_n(j\overline{\omega}) = \sum_{n=1}^{\infty} \frac{1}{2^n} \sum_{\omega_1 + \cdots + \omega_n = \overline{\omega}} H_n(j\omega_1, \cdots, j\omega_n) \prod_{i=1}^{n} A(\omega_i)$, $\forall \omega_i \in \sigma_\omega \cup -\sigma_\omega$.

Then $\|Y_n(j\overline{\omega})\| \le \frac{1}{2^n} \sum_{\omega_1 + \cdots + \omega_n = \overline{\omega}} \|H_n(j\omega_1, \cdots, j\omega_n)\| \prod_{i=1}^{n} \|A(\omega_i)\| \le \overline{H}_n U^n \left(\frac{1}{2^n} \sum_{\omega_1 + \cdots + \omega_n = \overline{\omega}} 1 \right) \le \alpha \overline{H}_n U^n, \forall \overline{\omega} \in W_n$ and $\|Y(j\overline{\omega})\| \le \sum_{n=1}^{\infty} \overline{Y_n}(j\overline{\omega}) = \sum_{n=1}^{n} \alpha \overline{H}_n U^n, \forall \overline{\omega} \in W_\infty$. Denote

$$\overline{Y} = \alpha \sum_{n=1}^{n} \overline{H}_n U^n \tag{C1}$$

Equation (C1) is frequency independent, and it can be revised in the following brief form

$$\overline{Y_b} = \frac{\overline{Y}}{\alpha} = \sum_{n=1}^{n} \overline{H}_n U^n \tag{C2}$$

Because α is a nonnegative bounded constant, (C2) has the same convergence bound with that of (C1). By replacing $\underline{L}(\omega)$ with $\underline{L}$ and $\overline{H}_1(j\omega)$ with $\overline{H}_1$, $\overline{H}_n$ can be recursively computed with lower order bound according to (14.6). Then similar to the proof of Proposition 14.1, the result holds.

For the general input case, denote $\alpha_n = \sup_{\overline{\omega} \in W_n} \left\{ \frac{1}{\sqrt{n}(2\pi)^{n-1}} \int_{\omega_1 + \cdots + \omega_n = \overline{\omega}} 1 \, d\sigma_{\overline{\omega}} \middle| \forall \omega_i \in \sigma_\omega \cup -\sigma_\omega \right\}$, and $\alpha = \sup\{\alpha_n | n = 1, 2, \cdots\}$. $d\sigma_{\overline{\omega}}$ denotes the area of a minute element on the n-dimensional hyper plane $\overline{\omega} = \omega_1 + \cdots + \omega_n$.

From Chap. 3, it can be obtained that

$$Y(j\overline{\omega}) = \sum_{n=1}^{\infty} Y_n(j\overline{\omega}) = \sum_{n=1}^{\infty} \frac{1/\sqrt{n}}{(2\pi)^{n-1}} \int_{\omega_1+\cdots+\omega_n=\overline{\omega}} H_n(j\omega_1,\cdots,j\omega_n)\prod_{i=1}^{n} U(j\omega_i)d\sigma_{\overline{\omega}} \ ,\forall\omega_i \in \sigma_\omega \cup -\sigma_\omega.$$

Then

$$\begin{aligned} \|Y_n(j\overline{\omega})\| &\le \frac{1/\sqrt{n}}{(2\pi)^{n-1}} \int_{\omega_1+\cdots+\omega_n=\overline{\omega}} \|H_n(j\omega_1,\cdots,j\omega_n)\|\prod_{i=1}^{n} \|U(j\omega_i)\|d\sigma\overline{\omega} \\ &\le \overline{H}_n U^n \frac{1/\sqrt{n}}{(2\pi)^{n-1}} \int_{\omega_1+\cdots+\omega_n=\overline{\omega}} 1 d\sigma\overline{\omega} \le \alpha\overline{H}_n U^n, \forall\overline{\omega} \in W_n. \end{aligned}$$

and

$$\|Y(j\overline{\omega})\| \le \sum_{n=1}^{\infty} \overline{Y_n}(j\overline{\omega}) = \sum_{n=1}^{n} \alpha\overline{H}_n U^n, \forall\overline{\omega} \in W_\infty$$

Thus (C1) and (C2) still hold for the NARX model with general input. Then following the method similar to the multi-tone input case, the result holds. This completes the proof.□

Chapter 15
Summary and Overview

Frequency domain methods can often provide very intuitive insights into underlying mechanism of a dynamic system under study in a coordinate-free and equivalent manner, compared with corresponding time domain methods. Therefore, they are preferable to engineers, widely adopted in engineering practice, and also extensively studied in the literature. Due to complicated output frequency characteristics and dynamic behaviour of nonlinear systems, a systematic and effective frequency domain theory or method for the analysis and design of nonlinear systems has been a focused topic in the past several decades.

Nowadays, several methods are available in the literature for nonlinear analysis and design as discussed in Chap. 1, including traditional harmonic balance methods, describing function methods, absolute stability based theory and methods and so on. Among the progress, active research activities can be seen in development of more efficient describing functions such as the so-called higher order sinusoidal input describing function and sinusoidal input describing function (Rijlaarsdam et al. 2011; Pavlov et al. 2007), and more active methodology would be the Volterra series based approach. The obvious advantages of the Volterra series based frequency domain theory or method can be seen in that, it is a generic method and applicable to a considerably large class of nonlinearities but not limited to any specific nonlinear units or components; it is not restricted to any specific input signals but permissible to any input excitation; it can directly relate any system characteristic parameters to nonlinear output frequency response in an analytical and polynomial form; and it allows symbolic and parametric computations using computer programmes.

In this book, some new advances in the Volterra series based frequency domain theory or method developed in the past 10 years are summarized from a novel parametric characteristic approach. These results, including both theoretical investigation and practical application algorithms, can hopefully present a solid and important basis for further development of frequency domain theories and methods for nonlinear analysis and design to solve critical and challenging engineering issues in the literature and various engineering practices.

X. Jing, Z. Lang, *Frequency Domain Analysis and Design of Nonlinear Systems based on Volterra Series Expansion*, Understanding Complex Systems,
DOI 10.1007/978-3-319-12391-2_15

The main results included in this book are:

(a) A parametric characteristic analysis method is established for parameterized polynomial systems with separable property, which is to reveal what model parameters affect system frequency response functions and what the influence could be. Based on this technique, it is shown for the first time that, the analytical relationship between high order frequency response functions of Volterra systems and system time-domain model parameters, and also provides a novel method for understanding of the higher order GFRFs of Volterra systems. Refer to Chaps. 4–6.
(b) By using the parametric characteristic analysis, system output spectrum up to any orders can be explicitly expressed as a polynomial function of model parameters of interest, which can directly relate any characteristic parameters to system output frequency response such that nonlinear output spectrum can be analyzed, designed and optimized via these parameters. This provides a significant basis for nonlinear analysis and design in the frequency domain. Refer to Chaps. 7–10.
(c) A novel mapping function from the parametric characteristics of the GFRF to itself is established. This result enables the nth-order GFRF and output spectrum to be directly written as a polynomial forms in terms of the first order GFRF, model parameters and input, which is shown to be a new approach to understanding of higher order GFRFs. It is theoretically shown for the first time that system output spectrum can be expressed into an alternating series with respect to model parameters under certain conditions. The result reveals a significant nonlinear effect on system output dynamic behaviours in the frequency domain. Refer to Chaps. 11 and 12.
(d) The nonlinear effects on system output spectrum from different nonlinearities are also studied. This provides some novel insights into the nonlinear effect on system output spectrum in the frequency domain, such as the counteraction between different nonlinearities at some specific frequencies, periodicity property of output frequencies and so on. These results can facilitate the structure selection and parameter determination for system modelling, identification, filtering and controller design. Mainly refer to Chap. 3.
(e) New methods for analysis and design of nonlinear vibration control systems by employing potential nonlinear benefits are developed. It is a systematic frequency domain approach for exploiting nonlinearities to achieve a desired output frequency domain performance for vibration control or suppression. Refer to Chaps. 10 and 12.
(f) New parametric convergence bound criteria for Volterra series expansion of nonlinear systems described by the NARX model are developed, based on the parametric bound characteristics of frequency response functions of the Volterra class of nonlinear systems. The results solve an important issue related to the application of Volterra series based theory and methods, that is, under what parametric conditions a given nonlinear system (described by the NARX model) could have a convergent Volterra series expansion for a given testing input signal. Refer to Chaps. 13 and 14.

Although interesting and significant results have been achieved as discussed in previous chapters, there are still many tasks yet to be done for a full development of a systematic frequency domain method for nonlinear system analysis and design. The following topics would be reasonable to investigate based on those achievements.

- *Exploring nonlinear benefits in vibration control.* This is to develop theory and methods for analysis and design of nonlinear vibration control systems in active, semi-active or passive control by employing advantageous nonlinear benefits for much better vibration control/suppression/isolation performance. Nonlinear energy transfer or cancellation properties as shown in Chap. 3 could be used for this purpose, and similar topics can also be referred to Chaps. 7, 9, 10 and 12. The parametric characteristics approach provides a convenient tool for the corresponding nonlinear analysis and design aiming at a desired output frequency spectrum. The developed frequency domain method (see Chap 9) can provide a straightforward expression for the relationship between the nonlinear output spectrum and system characteristic parameters including those which define nonlinearities. An extension will be done such that this relationship can also be expressed as a straightforward function of model parameters which define linear dynamics of the underlying system.
- *Characterizing and understanding nonlinearity in the frequency domain.* How to characterize nonlinear dynamics and what the true feature is for a nonlinear behaviour of interest in the frequency domain are intriguing topics to study, since straightforward understanding of linear dynamics in the frequency domain is very well developed and preferable in practice. This topic is greatly related to fault/crack detection in non-destructive evaluation and structure health monitoring etc, feature or pattern recognition or detection in biological data or dynamic response signals from various disciplines, signal processing, nonlinear system identification and feedback design in control, and so on. Similar topics can be referred to Chaps. 3, 7 and 12.
- *Nonlinear system identification in the frequency domain.* With known output frequency characteristics and parametric characteristics of frequency response functions as demonstrated in Chaps. 3–12, nonlinear system identification would be much more convenient to conduct with only input-output experiment data. Obviously, this topic is far from development.
- *Parametric convergence bounds of Volterra series expansion and its applications.* As shown in Chaps. 13 and 14, a parametric boundedness approach to the frequency response functions provides a powerful tool for the analysis and evaluation of convergent bound of Volterra series expansion of a given parametric nonlinear model such as NARX or NDE. Given a nonlinear system, a parametric convergence bound can be used readily for evaluation of the nonlinear dynamics about whether it is a Volterra-type or not, and about how to design nonlinear feedback or model parameters so as to create a Volterra-type nonlinear system, and consequently the system can be easily analyzed and controlled with the established frequency domain methods.

- *Extension to describe complex nonlinear behaviours.* Besides designing a non-Volterra-type system to be Volterra-type as mentioned above, some theory and methods can also be developed technically to allow complex nonlinear behaviours such as Chaos and Bifurcation and/or those described by neural networks to be analyzed in the context of Volterra series expansion and in the frequency domain. Preliminary research studies already show that the Volterra series based approach can also be used to interpret some complex nonlinear behaviours (Boaghe and Billings 2003).

"The fear of the LORD is the beginning of wisdom..."

—Proverbs 9:10 King James Bible

References

Ahn KK, Anh HPH (2010) Inverse double NARX fuzzy modeling for system identification. IEEE/ASME Trans Mechatronics 15(1):136–148

Alleyne A, Hedrick JK (1995) Nonlinear adaptive control of active suspensions. IEEE Trans Control Syst Technol 3(1):94–101

Atherton DP (1975) Nonlinear control engineering. Van Nostrand Reinhold, London, Full edition 1975, student edition 1982

Bai E (2003) Frequency domain identification of Wiener models. Automatica 39:1521–1530

Batista ELO, Tobias OJ, Seara R (2010) A sparse-interpolated scheme for implementing adaptive Volterra filters. IEEE Trans Signal Process 58:2022–2035

Baumgartner S, Rugh W (1975) Complete identification of a class of nonlinear systems from steady state frequency response. IEEE Trans Circuits Syst 22:753–759

Bedrosian E, Rice SO (1971) The output properties of Volterra systems (nonlinear systems with memory) driven by harmonic and Gaussian inputs. Proc IEEE 59:1688

Bendat JS (1990) Nonlinear system analysis and identification from random data. Wiley, New York, NY

Billings SA, Lang ZQ (1996) A bound of the magnitude characteristics of nonlinear output frequency response functions. Int J Control, Part 1, 65(2):309–328 and Part 2, 65(3):365–384

Billings SA, Lang ZQ (1997) Truncation of nonlinear system expansions in the frequency domain. Int J Control 68(5):1019–1042

Billings SA, Lang ZQ (2002) Nonlinear systems in the frequency domain: energy transfer filters. Int J Control 75(14):1066–1081

Billings SA, Peyton-Jones JC (1990) Mapping nonlinear integro-differential equation into the frequency domain. Int J Control 54:863–879

Billings SA, Korenberg M, Chen S (1988) Identification of nonlinear output-affine systems using an orthogonal least-squares algorithm. Int J Syst Sci 19:1559–1568

Bloemen HHJ et al (2001) Wiener model identification and predictive control for dual composition control of a distillation column. J Process Control 11:601–620

Boaghe OM, Billings SA (2003) Subharmonic oscillation modeling and MISO Volterra series. IEEE Trans Circuits Syst I Fund Theor Appl 50(7):877–884

Boutabba N, Hassine L, Loussaief N, Kouki F, Bouchriha H (2003) Volterra series analysis of the photocurrent in an Al/6T/ITO photovoltaic device. Org Electron 4:1–8

Boyd S, Chua L (1985) Fading memory and the problem of approximating nonlinear operators with Volterra series. IEEE Trans Circuits Syst CAS-32(11):1150–1160

X. Jing, Z. Lang, *Frequency Domain Analysis and Design of Nonlinear Systems based on Volterra Series Expansion*, Understanding Complex Systems,
DOI 10.1007/978-3-319-12391-2

Boyd S, Ghaoui LE, Feron E, Balakrishnan V (1994) Linear matrix inequalities in system and control theory. The Society for Industrial and Applied Mathematics, Philadelphia, PA

Brilliant MB (1958) Theory of the analysis of non-linear systems. Technical Report 345, MIT, Research Laboratory of Electronics, Cambridge, MA, 3 March 1958

Bromwich TJ (1991) An introduction to the theory of infinite series, American mathematical society. AMS Chelsea Publishing, Providence, RI

Bullo F (2002) Series expansions for analytic systems linear in control. Automatica 38:1425–1432

Buonomo A, Lo Schiavo A (2005) Perturbation analysis of nonlinear distortion in analog integrated circuits. IEEE Trans Circuits Syst I Reg Papers 52:1620–1631

Bussgang JJ, Ehrman L, Graham JW (1974) Analysis of nonlinear systems with multiple inputs. Proc IEEE 62(8):1088–1119

Case D, Taheri B, Richer E (2012) Design and characterization of a small-scale magnetorheological damper for tremor suppression. IEEE/ASME Trans Mechatronics 18:96–103. doi:10.1109/TMECH.2011.2151204

Chantranuwathanal S, Peng H (1999) Adaptive robust control for active suspensions. Proceedings of the American Control Conference, San Diego, CA, June, pp 1702–1706

Chatterjee A (2010) Structural damage assessment in a cantilever beam with a breathing crack using higher order frequency response functions. J Sound Vib 329:3325–3334

Chen S, Billings SA (1989) Representation of non-linear systems: the NARMAX model. Int J Control 49:1012–1032

Chen S, Billings SA, Luo W (1989) Orthogonal least squares methods and their application to non-linear system identification. Int J Control 50:1873–1896

Chen Y, Jing X, Cheng L (2013) Frequency domain analysis and design of nonlinear vehicle suspension systems, Chapter 12. In: Liu H, Gao H, Li P (eds) Handbook of vehicle suspension control systems. IET control book series. ISBN: 978-1-84919-633-8

Chua LO, Ng CY (1979) Frequency domain analysis of nonlinear systems: general theory. IEEE J Electron Circ Syst 3(4):165–185

Crama P, Schoukens J (2001) Initial estimates of Wiener and Hammerstein systems using multisine excitation. IEEE Trans Instrum Meas 50:1791–1795

Crespo-Cadenas C, Reina-Tosina J, Madero-Ayora MJ, Munoz-Cruzado J (2010) A new approach to pruning Volterra models for power amplifiers. IEEE Trans Signal Process 58:2113–2120

Daley S, Hatonen J, Owens DH (2006) Active vibration isolation in a 'smart spring' mount using a repetitive control approach. Control Eng Pract 14:991–997

Dixit RK, Buckner GD (2011) Sliding mode observation and control for semiactive vehicle suspensions. Veh Syst Dyn 43(2):83–105

Doyle FJ III, Pearson RK, Ogunnaike BA (2002) Identification and control using Volterra models. Springer, Berlin

Elizalde H, Imregun M (2006) An explicit frequency response function formulation for multi-degree-of-freedom non-linear systems. Mech Syst Signal Process 20:1867–1882

Engelberg S (2002) Limitations of the describing function for limit cycle prediction. IEEE Trans Autom Control 47(11):1887–1890

Fard RD, Karrari M, Malik OP (2005) Synchronous generator model identification for control application using Volterra series. IEEE Trans Energy Convers 20(4):852–858

Flajolet P, Sedgewick R (2009) Analytic combinatorics. Cambridge University Press, Cambridge, U.K.

Frank WA (1996) Sampling requirements for Volterra system identification. IEEE Signal Process Lett 3(9):266–268

French S (1976) Practical nonlinear system analysis by Wiener Kernel estimation in the frequency domain. Biol Cybern 24:111–119

Friston KJ, Mechelli A, Turner R, Price CJ (2000) Nonlinear responses in fMRI: the Balloon model, Volterra kernels, and other hemodynamics. Neuroimage 12:466–477

Gelb A, Vander Velde WE (1968) Multiple-input describing functions and nonlinear system design. McGraw-Hill, New York, NY

George DA (1959) Continuous nonlinear systems. Technical Report 355, MIT Research Laboratory of Electronics, Cambridge, MA, July 24
Gilmore RJ, Steer MB (1991) Nonlinear circuit analysis using the method of harmonic balance—a review of the art. Part I. Introductory concepts. Int J Microw Millimet Wave Comput Aided Eng 1:22–27
Glass JW, Franchek MA (1999) Frequency based nonlinear controller design of regulating systems subjected to time domain constraints. Proceedings of the American control conference, San Diego, CA, June 1999, pp 2082–2086
Graham D, McRuer D (1961) Analysis of nonlinear control systems. Wiley, New York, NY
Helie T, Laroche B (2011) Computation of convergence bounds for Volterra series of linear-analytic single-input systems. IEEE Trans Autom Control 56:2062–2072
Hermann R (1990) Volterra modeling of digital magnetic saturation recording channels. IEEE Trans Magn 26:2125–2127
Housner GW, Bergman LA, Cuaghey TK, Chassiakos AG, Claus RO, Masri SF et al (1997) Structural control: past, present, and future. ASCE J Eng Mech 123(9):897–971
Hrovat D (1997) Survey of advanced suspension developments and related optimal control applications. Automatica 33(10):1781–1817
Hu T, Teel AR, Lin Z (2005) Lyapunov characterization of forced oscillations. Automatica 41:1723–1735
Huang A, Tanskanen JMA, Hartimo IO (1998) Design of optimum power estimator based on Wiener model applied to mobile transmitter power control. Proc IEEE Int Symp Circuits Syst 5:249–252
Hunter IW, Korenberg MJ (1986) The identification of nonlinear biological systems: Wiener and Hammerstein cascade models. Biol Cybern 55:135–144
Ibrahim RA (2008) Recent advances in nonlinear passive vibration isolators. J Sound Vib 314:371–452
Isidori A (1999) Nonlinear control systems II. Springer, London
Jing XJ (2011) Frequency domain analysis and identification of block-oriented nonlinear systems. J Sound Vib 330(22):5427–5442
Jing XJ (2012) Truncation order and its effect in a class of nonlinear systems. Automatica 48 (11):2978–2985. doi:10.1016/j.automatica.2012.08.004
Jing XJ (2014) Nonlinear characteristic output spectrum for nonlinear analysis and design. IEEE/ASME Trans Mechatronics 19:171–183
Jing XJ, Lang ZQ (2009a) On the generalized frequency response functions of Volterra systems. Trans ASME J Dyn Syst Meas Control 131(6), 061002 (8 pages)
Jing XJ, Lang ZQ (2009b) Frequency domain analysis of a dimensionless cubic nonlinear damping system subject to harmonic input. Nonlinear Dyn 58:469–485
Jing XJ, Xiao ZL (2014) An input-dependent convergence bound of Volterra series expansion of nonlinear systems. Submitted
Jing XJ, Lang ZQ, Billings SA, Tomlinson GR (2006) The parametric characteristic of frequency response functions for nonlinear systems. Int J Control 79(12):1552–1564
Jing XJ, Lang ZQ, Billings SA (2007a) New bound characteristics of NARX model in the frequency domain. Int J Control 80(1):140–149
Jing XJ, Lang ZQ, Billings SA (2007b) Some typoes in 'New Bound Characteristics of NARX Model in the Frequency Domain'. Int J Control 80(3):492–494
Jing XJ, Lang ZQ, Billings SA (2008a) Frequency domain analysis for suppression of output vibration from periodic disturbance using nonlinearities. J Sound Vib 314:536–557
Jing XJ, Lang ZQ, Billings SA (2008b) Magnitude bounds of generalized frequency response functions for nonlinear Volterra systems described by NARX model. Automatica 44:838–845
Jing XJ, Lang ZQ, Billings SA (2008c) Frequency domain analysis for nonlinear Volterra systems with a general nonlinear output function. Int J Control 81(2):235–251
Jing XJ, Lang ZQ, Billings SA (2008d) Output frequency response function based analysis for nonlinear Volterra systems. Mech Syst Signal Process 22:102–120

Jing XJ, Lang ZQ, Billings SA (2008e) Mapping from parametric characteristics to generalized frequency response functions of nonlinear systems. Int J Control 81(7):1071–1088

Jing XJ, Lang ZQ, Billings SA (2009a) Parametric characteristic analysis for generalized frequency response functions of nonlinear systems. Circuits Syst Signal Process 28:699–733

Jing XJ, Lang ZQ, Billings SA (2009b) Frequency-dependent magnitude bounds of the generalized frequency response functions for NARX model. Eur J Control 15(1):68–83

Jing XJ, Lang ZQ, Billings SA et al (2009c) Theoretical study of the effects of nonlinear viscous damping on vibration isolation of SDOF systems. J Sound Vib 323:352–365

Jing XJ, Lang ZQ, Billings SA (2009d) Determination of the analytical parametric relationship for output spectrum of Volterra systems based on its parametric characteristics. J Math Anal Appl 351:694–706

Jing XJ, Lang ZQ, Billings SA (2010) Output frequency properties of nonlinear systems. Int J Nonlinear Mech 45(7):681–690

Jing XJ, Lang ZQ, Billings SA (2011) Nonlinear influence in the frequency domain: alternating series. Syst Control Lett 60(5):295–309

Jing XJ, Simpson D, Allen R, Newland P (2012) Understanding neuronal systems in movement control using Wiener/Volterra kernels: a dominant feature analysis. J Neurosci Methods 203 (1):220–232

Judd KL (1998) Numerical methods in economics. MIT Press, Cambridge, MA

Kalafatis A et al (1995) A new approach to the identification of pH processes based on the Wiener model. Chem Eng Sci 50(23):3693–3701

Karnopp D (1995) Active and semi-active vibration isolation. ASME J Mech Des 117:177–185

Kay SM, Nagesha V (1994) Maximum likelihood estimation of signals in autoregressive noise. IEEE Trans Signal Process 42:88–101

Kim KI, Powers EJ (1988) A digital method of modelling quadratically nonlinear systems with a general random input. IEEE Trans Acoust Speech Signal Process 36:1758–1769

Korenberg MJ (1982) Statistical identification of parallel cascades of linear and nonlinear systems. IFAC Symposium on System Identification Parameter Estimation 1:580–585

Kotsios S (1997) Finite input/output representative of a class of Volterra polynomial systems. Automatica 33:257–262

Krall C, Witrisal K, Leus G, Koeppl H (2008) Minimum mean-square error equalization for second-order Volterra systems. IEEE Trans Signal Process 56:4729–4737

Krzyzak A (1996) On nonparametric estimation of nonlinear dynamic systems by the Fourier series estimate. Signal Process 52:299–321

Kuech F, Kellermann W (2005) Partitioned block frequency-domain adaptive second-order Volterra filter. IEEE Trans Signal Process 53:564–575

Lang ZQ, Billings SA (1996) Output frequency characteristics of nonlinear systems. Int J Control 64:1049–1067

Lang ZQ, Billings SA (1997) Output frequencies of nonlinear systems. Int J Control 57(5):713–730

Lang ZQ, Billings SA (2000) Evaluation of output frequency responses of nonlinear systems under multiple inputs. IEEE Trans Circuits Syst II Analog Digital Signal Process 47(1):28–38

Lang ZQ, Billings SA (2005) Energy transfer properties of nonlinear systems in the frequency domain. Int J Control 78:345–362

Lang ZQ, Peng ZK (2008) A novel approach for nonlinearity detection in vibrating systems. J Sound Vib 314:603–615

Lang ZQ, Billings SA, Tomlinson GR, Yue R (2006) Analytical description of the effects of system nonlinearities on output frequency responses: a case study. J Sound Vib 295:584–601

Lang ZQ, Billings SA, Yue R, Li J (2007) Output frequency response functions of nonlinear Volterra systems. Automatica 43:805–816

Lee RCH, Smith MC (2000) Nonlinear control for robust rejection of periodic disturbances. Syst Control Lett 39:97–107

Leonov GA, Ponomarenko DV, Smirnova VB (1996) Frequency-domain methods for nonlinear analysis, theory and applications. World Scientific, Singapore

Levi EC (1959) Complex curve fitting. IRE Trans Autom Control 4:37–43

Li LM, Billings SA (2011) Analysis of nonlinear oscillators using Volterra series in the frequency domain. J Sound Vib 330:337–355

Li T, Jean J (2001) Adaptive Volterra filters for active control of nonlinear noise processes. IEEE Trans Signal Process 49:1667–1676

Li G, Wen C, Zheng WX, Chen Y (2011) Identification of a class of nonlinear autoregressive models with exogenous inputs based on kernel machines. IEEE Trans Signal Process 59:2146–2159

Liu CC, Jing XJ, Daley S, Li FM (2015) Recent advances in micro-vibration isolation. Mech Syst Signal Process 56–57:55–80

Ljung L (1999) System identification: theory for the user, 2nd edn. Prentice Hall, Upper Saddle River, NJ

Logemann H, Townley S (1997) Low gain control of uncertain regular linear systems. SIAM J Control Optim 35:78–116

McWhorter LT, Scharf LL (1995) Nonlinear maximum likelihood estimation of autoregressive time series. IEEE Trans Signal Process 43:2909–2919

Mees AI (1981) Dynamics of feedback systems. Wiley, New York, NY

Mileounis G, Kalouptsidis N (2009) Blind identification of second order Volterra systems with complex random inputs using higher order cumulants. IEEE Trans Signal Process 57:4129–4135

Nam SW, Powers EJ (1994) Application of higher-order spectral analysis to cubically nonlinear-system identification. IEEE Trans Signal Process 42(7):1746–1765

Nayfeh AH, Mook DT (2014) Nonlinear oscillations. Wiley-VCH Verlag GmbH & Co. KGaA

Nuij PWJM, Bosgra OH, Steinbuch M (2006) Higher-order sinusoidal input describing functions for the analysis of non-linear systems with harmonic responses. Mech Syst Signal Process 20:1883–1904

Ogota K (1996) Modern control engineering, 3rd edn. Prentice-Hall, Upper Saddle River, NJ

Orlowski P (2007) Frequency domain analysis of uncertain time-varying discrete-time systems. Circuits Syst Signal Process 26(3):293–310

Pavlov A, Van De Wouw N, Nijmeijer H (2007) Frequency response functions for nonlinear convergent systems. IEEE Trans Autom Control 52(6):1159–1165

Peng Z, Lang ZQ (2007) On the convergence of the Volterra-series representation of the Duffing's oscillators subjected to harmonic excitations. J Sound Vib 305:322–332

Peyton Jones JC (2003) Automatic computation of polyharmonic balance equations for non-linear differential systems. Int J Control 76(4):355–365

Peyton Jones JC (2007) Simplified computation of the Volterra frequency response functions of nonlinear systems. Mech Syst Signal Process 21(3):1452–1468

Peyton Jones JC, Billings SA (1989) Recursive algorithm for computing the frequency response of a class of nonlinear difference equation models. Int J Control 50(5):1925–1940

Peyton Jones JC, Billings SA (1990) Interpretation of non-linear frequency response functions. Int J Control 52:319–346

Pintelon R, Schoukens J (2001) System identification: a frequency domain approach. IEEE Press, Piscataway, NJ

Raz GM, Van Veen BD (1998) Baseband Volterra filters for implementing carrier based non-linearities. IEEE Trans Signal Process 46(1):103–114

Rijlaarsdam D, Nuij P, Schoukens J, Steinbuch M (2011) Spectral analysis of block structured nonlinear systems and higher order sinusoidal input describing functions. Automatica 47 (12):2684–2688

Rugh WJ (1981) Nonlinear system theory: the Volterra/Wiener approach. Johns Hopkins University Press, Baltimore, MD

Sandberg IW (1982) Volterra expansions for time-varying nonlinear systems. Bell Syst Tech J 61 (2):201–225

Sandberg I (1983) On Volterra expansions for time-varying nonlinear systems. IEEE Trans Circuits Syst 30:61–67

Sanders SR (1993) On limit cycles and the describing function method in periodically switched circuits. IEEE Trans Circuits Syst I Fund Theor Appl 40(9):564–572

Sanders JA, Verhulst F (1985) Averaging methods in nonlinear dynamical systems. In: Sanders JA, Verhulst F (eds) Applied mathematical sciences, vol 59. Springer, New York. 10+ 247 pp. Price DM 84.00 (1985). ISBN 0-387-96229-8 (USA), ISBN 3-540-96229-8 (FR Germany). vol 1, 1985

Schetzen M (1980) The Volterra and wiener theories of nonlinear systems. Wiley, New York, NY

Schmidt G, Tondl A (1986) Non-linear vibrations. Cambridge University Press, Cambridge

Schoukens J, Nemeth J, Crama P, Rolain Y, Pintelon R (2003) Fast approximate identification of nonlinear systems. In: 13th IFAC symposium on system identification, Rotterdam, The Netherlands, 27–29 August 2003, pp 61–66

Shah MA, Franchek MA (1999) Frequency-based controller design for a class of nonlinear systems. Int J Robust Nonlinear Control 9(12):825–840

Sheng L, Chon KH (2003) Nonlinear autoregressive and nonlinear autoregressive moving average model parameter estimation by minimizing hypersurface distance. IEEE Trans Signal Process 51:3020–3026

Solomou M, Evans C, Rees D, Chiras N (2002) Frequency domain analysis of nonlinear systems driven by multiharmonic signals. In: Proceedings of the 19th IEEE conference on instrumentation and measurement technology, vol 1, pp 799–804

Stewart I, Tall DO (1983) Complex analysis, the hitchhiker's guide to the plane. Cambridge University Press, Cambridge

Swain AK, Billings SA (2001) Generalized frequency response function matrix for MIMO nonlinear systems. Int J Control 74(8):829–844

Taylor JH (1999) Describing functions, an article in the Electrical Engineering Encyclopedia. Wiley, New York, NY

Taylor JH, Strobel KL (1985) Nonlinear compensator synthesis via sinusoidal-input describing functions. In: Proceedings of the American control conference, Boston, MA, June 1985, pp 1242–1247

Tomlinson GR, Manson G, Lee GM (1996) A simple criterion for establishing an upper limit to the harmonic excitation level of the Duffing oscillator using the Volterra series. J Sound Vib 190:751–762

Van De Wouw N, Nijmeijer H, Van Campen DH (2002) A Volterra series approach to the approximation of stochastic nonlinear dynamics. Nonlinear Dyn 27:397–409

Van Moer W, Rolain Y, Geens A (2001) Measurement-based nonlinear modeling of spectral regrowth. IEEE Trans Instrum Meas 50(6):1711–1716

Volterra V (1959) Theory of functionals and of integral and integro-differential equations. Dover, New York, NY

Wang J, Sano A, Chen T, Huang B (2010) Identification of Hammerstein systems without explicit parameterization of nonlinearity. Int J Control 82(5):937–952

Wei HL, Billings SA (2008) Model structure selection using an integrated forward orthogonal search algorithm assisted by squared correlation and mutual information. Int J Model Ident Control 3(4):341–356

Wei H-L, Lang Z-Q, Billings SA (2007) An algorithm for determining the output frequency range of Volterra models with multiple inputs. IEEE Trans Circuits Syst II Express Brief 54(6):532–536

Weisstein EW (1999) Cauchy criterion. From MathWorld–a Wolfram web resource. http://mathworld.wolfram.com/CauchyCriterion.html

Worden K, Tomlinson GR (2001) Non-linearity in structural dynamics: detection, identification and modeling. Institute of Physics Publishing, Bristol

Wu XF, Lang ZQ, Billings SA (2007) Analysis of the output frequencies of nonlinear systems. IEEE Trans Signal Process 55(7):3239–3246

Xiao ZL, Jing XJ, Cheng L (2013a) The transmissibility of vibration isolators with cubic nonlinear damping under both force and base excitations. J Sound Vib 332(5):1335–1354

Xiao ZL, Jing XJ, Cheng L (2013b) Parameterized convergence bounds for Volterra series expansion of NARX models. IEEE Trans Signal Process 61(20):5026–5038

Xiao ZL, Jing XJ, Cheng L (2014) Estimation of parametric convergence bounds for Volterra series expansion of nonlinear systems. Mech Syst Signal Process 45(1):28–48

Yang J, Tan SX-D (2006) Nonlinear transient and distortion analysis via frequency domain Volterra series. Circuits Syst Signal Process 25(3):295–314

Young PC (1985) The instrumental variable method: a practical approach to identification and parameter estimation. In: The 7th IFAC/IFORS symposium on identification and system parameter estimation, York, UK, pp 1–14

Yuan F, Opal A (2001) Distortion analysis of periodically switched nonlinear circuits using time-varying Volterra series. IEEE Trans Circuits Syst I Fund Theory Appl 48(6):726–738

Yue R, Billings SA, Lang Z-Q (2005) An investigation into the characteristics of non-linear frequency response functions. Part 1: understanding the higher dimensional frequency spaces. Int J Control, 78(13):1031–1044; and Part 2: new analysis methods based on symbolic expansions and graphical techniques. Int J Control 78:1130–1149

Zapateiro M, Pozo F, Karimi HR, Luo N (2012) Semiactive control methodologies for suspension control with magnetorheological dampers. IEEE/ASME Trans Mechatronics 17(2):370–380

Zhang H, Billings SA (1996) Gain bounds of higher order nonlinear transfer functions. Int J Control 64(4):767–773

Zhang H, Billings SA, Zhu QM (1995) Frequency response functions for nonlinear rational models. Int J Control 61:1073–1097

Zhou L, Misawa EA (2005) Low frequency vibration suppression shape filter and high frequency vibration suppression shape filter. In: American control conference, Portland, OR, 8–10 June 2005, pp 4742–4747

Zhu Y (1999) Distillation column identification for control using Wiener model. In: American control conference, 1999. Proceedings of the 1999 (vol 5, pp 3462–3466), San Diego, CA, 02 Jun 1999–04 Jun 1999

Zhu WQ, Yang ZG, Song TT (2001) An optimal nonlinear feedback control strategy for randomly excited structural systems. Nonlinear Dyn 24:31–51

MIX
Papier aus verantwortungsvollen Quellen
Paper from responsible sources
FSC® C105338

If you have any concerns about our products,
you can contact us on
ProductSafety@springernature.com

In case Publisher is established outside the EU,
the EU authorized representative is:
Springer Nature Customer Service Center GmbH
Europaplatz 3, 69115 Heidelberg, Germany

Printed by Libri Plureos GmbH
in Hamburg, Germany